ESSAYS IN HONOUR OF JAAKKO HINTIKKA

SYNTHESE LIBRARY

STUDIES IN EPISTEMOLOGY,

LOGIC, METHODOLOGY, AND PHILOSOPHY OF SCIENCE

Guest Managing Editor:

DONALD DAVIDSON, *University of Chicago*

Editors:

ROBERT S. COHEN, *Boston University*

GABRIËL NUCHELMANS, *University of Leyden*

WESLEY C. SALMON, *University of Arizona*

VOLUME 124

PROFESSOR JAAKKO HINTIKKA

ESSAYS
IN HONOUR OF
JAAKKO HINTIKKA

On the Occasion of
His Fiftieth Birthday on January 12, 1979

Edited by

ESA SAARINEN

RISTO HILPINEN

ILKKA NIINILUOTO

MERRILL PROVENCE HINTIKKA

D. REIDEL PUBLISHING COMPANY

DORDRECHT : HOLLAND / BOSTON : U.S.A.

LONDON : ENGLAND

Library of Congress Cataloging in Publication Data

Main Entry under Title:
Essays in Honour of Jaakko Hintikka.
   (Synthese Library ; v. 124)
   Bibliography: p.
   Includes index.
   1. Philosophy–Addresses, Essays, Lectures. 2. Logic–Addresses,
Essays, Lectures. 3. Semantics (Philosophy)–Addresses, Essays,
Lectures. 4. Hintikka, Kaarlo Jaakko Juhani, 1929-   – Addresses,
Essays, Lectures. I. Saarinen, 1953-   II. Hintikka, Kaarlo Jaakko
Juhani, 1929-
B29.E797   100   -11364
ISBN-13: 978-94-009-9862-9   e-ISBN-13:978-94-009-9860-5
DOI: 10.1007/ 978-94-009-9860-5

Published by D. Reidel Publishing Company,
P.O. Box 17, Dordrecht, Holland

Sold and distributed in the U.S.A., Canada, and Mexico
by D. Reidel Publishing Company, Inc.
Lincoln Building, 160 Old Derby Street, Hingham,
Mass. 02043, U.S.A.

TABLE OF CONTENTS

V. EPISTEMOLOGY

VI. PHILOSOPHICAL AESTHETICS

VII. HISTORY OF PHILOSOPHY

PREFACE

Jaakko Hintikka was born on January 12th, 1929. He received his doctorate from the University of Helsinki under the supervision of Professor G. H. von Wright at the age of 24 in 1953. Hintikka was appointed Professor of philosophy at the University of Helsinki in 1959. Since the late 50s, he has shared his time between Finland and the U.S.A. He was appointed Professor of philosophy at Stanford University in 1964. As from 1970 Hintikka has been permanent research professor of the Academy of Finland. He has published 13 books and about 200 articles, not to mention the various editorial and organizational activities he has played an active role in.

The present collection of essays has been edited to honour Jaakko Hintikka on the occasion of his fiftieth birthday. By dedicating a *Festschrift* to Jaakko Hintikka, the contributors wish to pay homage to this remarkable man whom they see not only as a scholar of prodigious energy and insight, but as a friend, colleague and former teacher. The contributors hope the essays collected here will bring pleasure to the man they are intended to honour. All of the essays touch upon topics Hintikka has taken an direct or indirect interest in, ranging from technical problems of mathematical logic and applications of formal methods through philosophical logic, philosophy of language, epistemology and history of philosophy to philosophical aesthetics.

Helsinki, Finland, March 1978

THE EDITORS

PUBLISHER'S NOTE

We take great pride in publishing this dedication to

Professor Jaakko Hintikka

our friend, advisor and colleague. From its earliest days, this house has benefited immeasurably from his insights, contributions and direction. His roots run deep here! We thank him for all that he has done for us and extend to him our continuing best wishes on this very special occasion.

B. E. VANCE
Publisher

I

SEMANTICS OF NATURAL LANGUAGE

JULIUS M. E. MORAVCSIK

GRAMMAR AND MEANING*

In this paper I shall sketch a novel conception of the syntax and semantics of natural languages. According to this view, the semantics of a natural language cannot be presented solely in terms of the technical notions of truth and satisfaction, and yet formalizations of certain parts of natural languages within that framework can lead to theoretical insights. Thus my view will not please either the relentless formalizer, nor the unregenerate philosopher of ordinary language. The view also upholds the pretheoretic and empirical significance of syntactic evidence and syntactic mode of argumentation. Finally, my construal of natural language as a biological phenomenon will lead to a defense of the partial independence of the grammars of natural languages from semantic considerations, and it will attribute to such grammars mental reality of a sort, under idealizations. This part of the view will please neither the practitioners of Montague-grammars, nor the most influential schools of thought concerning the issue of 'psychological reality' for rules of language; i.e., it will please neither those who regard such questions beneath their (mathematical) dignity, nor those who aim at a 'real time process' model of linguistic performance.

The predictable lack of popularity of these views is due to the fact that at this stage theories of language are still in their initial state; thus proposals embodying extreme stands and slogan-like principles are more likely to command attention than views aiming at viable conceptual frameworks for scientific study. The aim of this paper will have been well served if in the long run the proposals it contains will pave the way towards an empirical theory of natural languages and language understanding.

I. CONCEPTS OF 'FORMAL LANGUAGE' IN EMPIRICAL SCIENCE

In his introduction to 'English as a Formal Language' Montague compared his 'formal' treatment of natural languages to the 'formal' treatments

3

E. Saarinen, R. Hilpinen, I. Niiniluoto, and M. Provence Hintikka (eds.), Essays in Honour of Jaakko Hintikka, 3–15. All Rights Reserved.

of generative grammar, and spoke disparagingly about the latter. From Montague's polemical remarks one might be tempted to infer that there is a single sense of 'formal' in which both Montague-grammars and transformational grammars are formal. This turns out not to be the case. Montague uses the expression 'formal language' in the sense used by his famous teacher Alfred Tarski[1]. According to this conception a formal language must meet the following conditions: (i) There must be a way of specifying all of the well-formed sentences that can be true or false. This is done by listing the syntactic primitives and the formation rules. (ii) There must be a specification of semantic primitives, definitions, and rules of inference. (iii) In order for the truth and satisfaction conditions to be specifiable, the semantic complexes must be built up compositionally, and the syntactic rules must correspond to the semantic ones. (I shall bypass the issue of 'conditions of assertability'.)

None of these conditions are constitutive of the sense in which transformational grammars represent a formal approach to language. To be sure, this approach too advocates the specification of syntactic primitives and formation rules; but the syntax thus specified applies to the whole of a natural language – the subset of sentences to which truth and falsity apply does not play a special role. Tarski's conditions on the semantic component and its relation to the syntax are motivated by the aim of specifying meaning in terms of conditions for truth and satisfaction; transformational grammar remains uncommitted on the question of what specific shape semantic representation will take.

The formal approach represented by transformational grammar involves the assignment of mathematical properties to types of grammars so as to make a hierarchical ordering of grammars, in terms of complexity, possible. In this way, grammatical theory can achieve a partial characterization of the human mind, for it can establish minimal complexity for some of the structures that a human mind can process.

These differences in the respective senses of 'formal' bear some relationship to the differences in the aims that Montague and Chomsky set for themselves. Montague's aim is to represent English, and other natural languages, as a formal language. His sole concern is to specify the truth-bearing elements of English and their parts, and give their meanings in terms of conditions of truth and satisfactions. His methodological desiderata are uniformity, generality, and elegance.

Chomsky's aim is to capture what the grammars of all natural languages have in common in such a way as to show also what – if anything – is unique about the grammars of this class of languages. Furthermore, he is interested in presenting a theory of syntax that can become a component in an empirical theory of natural language acquisition, and in a theory of how humans process language.

It has been pointed out earlier by Kasher[2] that Chomsky and Montague have different aims in constructing syntactic theories for natural languages. So far I have tried to locate explicitly the differences in the respective aims, and show that the different aims led to different conceptions of what a formal approach to natural languages is.

Recent work, such as that of Barbara H. Partee, should remind us that the difference between aims of practitioners can be separated from differences in technical frameworks. It is possible for one to share Chomsky's aims and still use the technical framework of a Montague-grammar. This involves seeking restrictions on Montague-grammar that would enable one to delineate what is unique about the grammars of natural languages, and also enable one to view a version of Montague-grammar as a component in a theory of language acquisition.

Conversely, one could try to take a transformational grammar and use its generated 'trees' as the syntactic basis for a semantic analysis in terms of truth and satisfaction. (This program would be facilitated by the assumption that transformations do not affect semantic representation.)

These reflexions show that the key issues in choices between types of grammars concern not the differences in aims or technical frameworks. Rather, they boil down to the following: (a) In what way is Montague's claim that English is a formal language subject to empirical confirmation or disconfirmation? (b) What are the likely factors that shape the semantics and syntax of natural languages, and how is their impact likely to affect the relationship between semantic and grammatical structures?

In order to answer these questions, we have to reflect on what the empirical content of a Montague-grammar is. Since the syntax is designed solely to accommodate the semantic representation, the empirical tests must come down to the confirmation and disconfirmation of semantic predictions following from a Montague-grammar. These predictions are in terms of entailment relationships, analysis of ambiguities and in terms of the assignments of denotation ranges to the relevant elements. But even if this characterization is accepted, difficulties remain. An analysis in terms of a Montague-grammar should yield the logical forms of the types of sentences that are true or false; i.e. that set of structures that depends solely on the form set by the behaviour of the logical vocabulary within these sentences. It is not clear, however, how one is to define the logical vocabulary and logical distinctions within a natural language. E.g., is there one logical form for all simple declarative sentences of the form: noun-phrase, copula, adjective phrase? One might reply negatively, pointing to the difference between 'this house is red' and 'this house is big' (in short, the attributive vs. predicative distinction); or one might answer affirmatively, and add to the system 'meaning postulates' that will place restrictions on entailments involving specific lexical entries, thus yielding only the desired inferences within the language to be analyzed. Unless we can test for the difference between logical entailment and semantic entailment, a Montague-grammar is subject to empirical test only via the joint inferential power of the logical and semantic framework.

Let us grant that the semantics of a Montague-grammar can be subjected to empirical testing in the manner sketched above. Does this enable us to give empirical content to claim that the semantics of a language like English has the structure required of a Montague-type treatment?

Possible disconfirmation is provided by some of the literature on semantics that had currency some 20 years ago, and is today shunted aside, without anyone having refuted it. E.g. Waismann[3] claimed that certain terms in English – terms to which the Montague-semantics assigns sharply delineated denotation ranges across possible worlds – are 'open-textured'; i.e., they can be understood only against a certain set of general background assumptions, such as the constancy of material objects, the relative permanence of causal relations, etc. That is to say, this is not a matter of vagueness, and cannot be dismissed by the reliance on currently fashionable doctrines about 'fuzzy sets'.

There is also Austin's claim that assessments from the point of view of veracity are non-linguistic context dependent. E.g. 'France is hexagonal' is true within the context of a geography book, and false when used as an illustration in a geometry book.

Other relevant matters include: the semantics of words like 'only', 'even', and 'because', and the possibility that many definite descriptions e.g. 'the screwdriver' in the context: 'I am missing the screwdriver'; or adverbs, like 'slowly' as in the context of 'Jumbo swam slowly' (where the interpretation depends on what class of swimming creatures we take to be the comparison class) depend on non-linguistic contexts for interpretation, and that there is no systematic formal way of representing these contexts via indices in the semantic formalism.

Thus it seems that quite apart from issues concerning the relationship between syntax and semantics, there are good reasons to suppose that English does not have a completely formally representable semantics. This observation should leave us receptive to alternatives to set-theoretic semantics. It should also help us to see that semantic representation in terms of senses need not commit us to the view that most declarative sentences of English express definite propositions – a tenet of Montague-semantics.

At the same time, partial formalizations of English in set-theoretic terms can serve useful theoretic aims. E.g. attention to the tense system or the complexity of quantificational structures could leave us with minimal requirements on semantic complexity.

II. EVIDENCE, FACT, AND SYNTACTIC ARGUMENTATION

There are signs that the nature of empirical syntactic modes of argumentation are not well understood by philosophers. Thus, e.g., in the paper cited above, Montague wrote:

Some linguists ... have proposed that syntax – that is, the analysis of the notion of a (correctly formed) sentence – be attacked first, and that only after the completion of a syntactical theory consideration be given to semantics, which would then be developed on the basis of that theory. Such a program has almost no prospect of success. There will often be many ways of syntactically generating a given set of sentences, but only a few of these will have semantic relevance; and these will be sometimes less simple, and hence less superficially appealing, than certain of the semantically uninteresting modes of generation. Thus the construction of syntax and semantics must proceed hand in hand.[4]

Montague seems to assume here that as long as we have a number of syntactic alternatives, the only relevant consideration determining priorities will be that of the requirements of the semantics. Syntax that is not purely semantically motivated appears to Montague to be arbitrary. This alleged arbitrariness, however, vanishes if we can show that there are empirical syntactic facts that will restrict the number of ways in which the set of well formed sentences of English should be generated; and even on a more abstract, conceptual level, any suggestion of arbitrariness disappears if we can show why general considerations of what a natural language is would render plausible the claim that the syntactic structure is to an extent independent of semantic considerations.

Philosophers and logicians are not alone in entertaining the mistaken view that for the analysis of a natural language semantics is more fundamental than syntax. Linguists apparently have a tendency to share the same superstition. In her critical survey of such views Bresnan writes:

This view is sometimes supported by a kind of reductionism that maintains that meaning and sound are the only 'observables' of language, hence that semantic and phonetic structures alone are empirically justifiable or 'natural', and therefore that transformations must directly convert natural semantic or 'logical' structures into surface structures. . . .[5]

In order to untangle these matters, let us draw distinctions between evidence, fact, and theory. We can ask what the nature of the evidence is that supports or disconfirms hypotheses in a given empirical science. As we move away from the 'observables', the notion of evidence becomes relative; what is established on the basis of evidence E' will itself become evidence E'' supporting higher level generalizations. Some facts exist regardless of theoretical assumptions; if we cannot start with (pre-theoretical) facts to be explained, it is not clear how a scientific enterprise can be directed towards objective goals. Theories are both based on (certain) facts, and explain facts.

Evidence in the most general sense must be specified within terms that describe what is taken – at least for the purposes of a given science – as observable. But the facts that we wish to explain need not be specified on the level of observability. Not only are theories distinct from the evidence on which they are based; the facts that we wish to explain are, typically, distinct from the evidence that we give for the postulation of such facts, and for the confirmation of theories explaining these facts. E.g. one of the facts that an adequate linguistic theory should account for is the fact of creative language use. But saying this does not commit one to maintaining that the fact of creativity is one of the observables.

It seems that one of the reasons for the popularity of ontological behaviourism was the confusion of fact with evidence. From the correct observation that behavioural evidence is one of the important factors used in confirming or disconfirming hypotheses it was falsely concluded that behaviour is what we are supposed to explain or account for. To be sure, in the long run we should be able to give an account of human behaviour, but many of the theoretically more interesting facts of our mental lives, such as the fact of intentionality, are not on the level of observability; yet these are the facts that our theories are primarily responsible for.

Turning to the attitude described by Bresnan, even if we were to admit that evidence for semantic interpretation, such as evidence as to what people can make sense of, lies closer to the observational surface, than the evidence bearing on syntactic representation, it would not follow that semantic *facts* are any more 'observable' than syntactic facts. E.g. that '*x* is a man' entails that '*x* is an animal' is no more 'observable' than the fact that '2 more small man' is not a well formed sentence of English.

Interestingly enough, from the point of view of what people regard as 'basic evidence' in philosophy we encounter a prejudice that is the opposite of the one entertained by some linguists. Quine and others have argued that from the point of view of behavioural evidence facts about meanings are inscrutable. Neither Quine nor his opponents have given us a clear conception of what they mean by 'behaviour'. E.g., in sketching the development of children's referential frameworks J. Brunner talks about a child's learning 'to focus on what someone calls attention to'. Would Quine or his opponents call this a matter of 'behaviour'? Quine is hardly in a position to object on the ground that the phenomenon described by Brunner has an intentional component, and that intentionality cannot be defined behaviourally. For the same type of intentionality affects what Quine does regard as basic evidence, namely the assent or dissent of subjects in face of presented correlations of language and reality. Assent and dissent are not strictly observable notions; only against background assumptions about intention, purpose, and their conventional manifestations can we interpret certain types of behaviour as expressing assent or dissent.

To sum up, there are no good arguments showing either that semantic evidence is more 'observational' than syntactic evidence, or the other way around. Within semantics, the claim that matters of reference are closer to verifiability than matters of meaning is also suspect. Quine has yet another argument designed to show that notions of meaning, synonymy, etc. are less clear than notions of reference, truth, etc. In *Word and Object* and elsewhere Quine often complains about the circularity of explanations of intensional notions: i.e., the fact that the notions of meaning, synonymy, intension, analyticity, and necessity can be defined only in terms of each other. Against this I maintain that the same situation holds with regards to the extensional notions; the notions that Quine favours for empirical theories of semantics. We can define truth in terms of the technical notion of satisfaction. Satisfaction, in turn, can be explained in terms of the notions of reference, denotation, and extension. But these notions can be explained only in terms of each other, or in terms of their contribution to truth or falsehood. Thus when we attempt to explain the extensional notions, we traverse a circle analogous to the one that Quine finds when he asks for an explanation of intensional notions. We have two circles, and these can be illuminated only by mutual contrast. If someone were to claim that the notions of set and satisfaction are clearer from the point of view of mathematical logic, then we can reply that from the point of view of specifying procedures for computations the notions of intension and instantiation are clearer.

So far we see syntax, extensional semantics, and intensional semantics as falling into the same category as far as empirical evidential backing is concerned. Let us now turn to *facts*, and consider the claim that there are pre-theoretical, purely syntactic facts, that we need to account for within an adequate grammar of English.

We can get at such facts by considering pairs of sequences both of which admit of semantic interpretation but only one of which is a well formed formula of English. E.g.:

      (1a) I am more angry than sad.
   *(1b) I am angrier than sad.
      (2a) John dined on a steak.
   *(2b) John dined a steak.
      (3a) The baby looked asleep.
   *(3b) The baby looked sleeping.

The reason these contrasts constitute purely syntactic facts is that it is not only by analogy with the well formed sentences that we can give semantic content to the ill formed sequences; viewed purely semantically, the members of the pairs are on par; it is no more easy to give semantic interpretation to one than it is to give to the other.

There are other facts that defy semantic analysis or justification. These include the ways in which 'want to' can contract to 'wanna', and the ways in which we form tag-questions in English (e.g. 'John is here, isn't he?' vs. * 'John is here, isn't John?'). Nor is it clear that constraints on movements – such as the complex NP constraint, according to which if an NP properly contains an S, then nothing can be moved out of that S – can be given always semantic explanations.

Thus explaining these facts is a matter of pure syntax; and arguments leading to the postulation of phrase structure constituents and transforma- tions which are based on facts such as the ones listed are purely syntactic, empirical arguments. If such argumentations lead to syntactic structures that coincide with the ones that a Montague-type grammar would posit, that is merely a case of 'pre-established harmony', and such coincidence is to be expected with about as much probability as other cases of pre-established harmony.

What has been shown so far does not deny that there will be lots of cases in which the line between syntax and semantics will be drawn within a given theory, and that with regards to certain phenomena pretheoretical intuitions will not enable us to classify things as syntactic or semantic. But these phenomena are not strewn randomly across the conceptual map. They arise either as the result of the interaction of logic and grammar, thus yielding problems of quantifier-scope and quantificational ambiguity, or they emerge at the point at which grammar intersects with conceptual categorial schemes, thus yielding category-mistakes (e.g. 'this stone is thinking of Vienna').

The existence of these fields of phenomena do not warrant our talking about a syntax-semantics continuum. Paradigm cases of continua are, e.g., colours arranged between red and yellow, or lengths arranged between two arbitrarily selected measurements. We can hardly regard something as a continuum when we have plenty of clear cases on both sides of the dividing line, and the classes of 'in-between' cases admit of independent characterization that explains their twilight status.

On the basis of these reflexions I conclude that both the notions of pre-theoretical distinctions between syntactic and semantic facts, and the notion of purely syntactic empirical argumentation are viable.

Returning now to our main theme, let me illustrate both the claim that syntactic structure might not mirror semantic structure and the claim that while syntactic analysis might be underdetermined by the available data, the same holds also of semantic analysis.

In terms of possibilities of deletions, insertions of material, etc. there are good syntactic grounds for positing the syntactic categories of PREPOSITIONAL PHRASE, and ADVERBIAL PHRASE. But it is very doubtful that either of these categories serve any semantic purpose. Prepositions like 'of', 'for', 'in', etc. have very different semantic roles: some are equivalent to verbs, some are not, some represent relations, some do not, etc. The same applies to adverbs; some of these are semantically more like certain adjectives while others are more like modals that occur syntactically in the auxilliary.

Thus we must not assume that syntactic constituent structure will coincide with the structure generated by semantic categorial analysis.

Adverbs illustrate also nicely the underdetermination of semantic analysis. One can avail oneself either of Reichenbach's analysis in terms of second-order logic, in which the semantic units correlated to adverbs become properties of properties, or one can use Davidson's analysis according to which verbs denote events, and adverbs are first-order properties. In terms of entailment and other semantic tests, the two theories are equivalent. The differences are due to disagreements over issues of intensions, ontology, etc. One can only hope that independent empirical syntactic considerations might help to cut down the number of semantic alternatives.

Syntax and semantics are, thus, mutually underdetermined. One might add on behalf of Montague-grammars that the (syntactically unmotivated) mere multiplication of syntactic categories would not create any real difficulties in a grammar. But one can also point out that there may be some very simple devices that would render the presentation of a semantics that is only indirectly linked to syntactic structure, quite plausible.

Such a device is a 'semantic transformation'. In giving an analysis of the semantics of tense and aspect, one can take the syntactic post-transformational constituent structure, and simply have an interpretative rule that takes the elements of the auxiliary and construes these as semantically playing the role of sentential operators.

Again, when determining the scope of negation in sentences with adverbials, we can leave the syntactic structure intact, and introduce an interpretive rule that will restrict the scope of negation to the adverbial.

E.g.   (a) He didn't walk slowly,

receives the semantic interpretation of

          (b) He walked not-slowly.

A grammar within which semantic and syntactic structures are only indirectly linked need not be more complex than systems in which – as Montague might put it – the syntax and the semantics go 'hand in hand'.

III. NATURAL LANGUAGE AS A BIOLOGICAL PHENOMENON

Having argued for the viability of a conception of syntax that makes it partially independent of semantic structures, let us now consider the thesis that natural language is a biological phenomenon, and see how it can be related to the partial independence of syntax.

To say that a natural language is a biological phenomenon is to say that language use and language processing are basically species-invariant phenomena that are largely determined by the biological mechanisms that are responsible for the processes in question. In this way, we construe language use as analogous to a biological process like perception. Though cultural factors influence to some extent our perceptual abilities, sense perception is by and large a species-invariant phenomenon, determined by the biological mechanisms – eye, ear, brain, etc. – that are responsible for the processing.

Just as the structure of perceptual organization does not depend merely on what we perceive, but also on the biological organism that does the perceiving, so the structure of language will not depend merely on what language can express and can be used to communicate, but also on the nature of the biological organs that play a role in language understanding. Some of the factors determining the shape of syntactic structures will be functional ones. E.g., obviously, the grammar of a natural language – unlike the grammars of formal, artificial languages – will be determined partly by the fact that natural languages are primarily spoken languages, and thus the load placed upon the perceptual mechanism must stay within certain limits. Hence the difference between the syntax of definite descriptions in natural languages and the syntax of the same type of expression in a system such as that of Russell. But going beyond this, just as functional explanations have their limits in anatomy, so they will have their limitations in syntactic theory. There is no more reason to assume that syntactic structure can be given a purely functional explanation than there is for assuming that the human bone structure can be given functional accounts down to the last small bone in the human hand or leg.

What shall we contrast with 'biological processes'? To be sure, some biological requirements serve as qualifications for partaking in such activities as voting, playing games, or setting up legal systems. But the particular forms that these activities take are matters of decision, conscious adoption of conventions, and human life in its essentials could go on even without these activities.

Even if the contrast is accepted, someone might challenge us and ask why the semantic component of a natural language should not be viewed as subject to biological constraints. To be sure, the view I am advancing implies that there will be biological constraints on the semantics of natural languages; but these constraints are very different from the ones one would expect on the syntactic component. The constraints on the semantics might involve limitations on what would be semantic primitives across natural languages, and constraints on the complexity of expressions that could be understood by a human organism under normal circumstances.

Thus the conception of language as a biological phenomenon leaves us with a strong conceptual bias in favour of the partial independence of syntax from semantics. For the biological constraints on the semantics will have to do with the limits on expressibility, while the biological constraints on the syntax have to do with the forms of the well formed sequences, and thus the forms of processing and parsing.

It should be emphasized that the thesis advanced here does not compare natural languages to biological organs like that of an arm or limb, but to biological processes like breathing and perception. In accounting for the nature of biological units, we can rely on the Aristotelian tetrachotomy of structure, constituent, function, and agentive factors. It makes a difference, however, whether we talk about an organ or a process. With regards to organs such as the eye or ear, we can specify the dominant function; with regards to processes, there might be no specification of dominant function, not because biological processes have no functions, but because they fulfill so many functional specifications. Thus when we look at natural languages, we try to determine what the basic constituents – phonological, semantic, and syntactic parts – are, and also what the essential structures are that link these constituents into larger units. We also look for a partial characterization of the possible agents; i.e. the human minds that can understand and produce language. But there are no reliable guide-posts for selecting something as the 'dominant function' of these processes; nor is there a need for this.

What emerges from these considerations is a pluralistic view of language understanding. Instead of viewing the process as having a single uniform structure – as the models that appeal to mathematicians do – we think of language understanding as involving parallel processing by a number of partially independent mechanisms. What may seem, however, from a mathematical point of view more cumbersome and less appealing, may seem from a biological point of view more natural, and more readily subject to causal explanation. The dependence of syntactic and semantic organization on elements outside the domain of the linguist and philosopher does not make syntactic and semantic analysis impossible; it merely makes it more challenging, and a task that can be ultimately correlated with research in other disciplines concerned with the explanation of human language use and understanding.[6]

Stanford University

NOTES

[*] Earlier versions of this paper were subjected to useful criticisms by my colleagues, Tom Wasow and Dick Oehrle; also by audiences at the University of Texas and the University of California, San Diego. The penultimate version was presented in the fall of 1977 as one of the Bar Hillel Memorial lectures in Tel Aviv. I am delighted to be able to dedicate the final version to my friend and colleague, Jaakko Hintikka, with whom we have been sharing philosophical ideas for more than two decades.

[1] Alfred Tarski, 'The Concept of Truth in Formalized Languages'. First published in Polish in 1933. For English version see Alfred Tarski, *Logic, Semantics, Meta-Mathematics*, Clarendon Press, Oxford, 1956, pp. 152–278.

[2] Asa Kasher, 'The Proper Treatment of Montague-Grammars in Natural Logic and Linguistics', *Theoretical Linguistics* 2 (1975), 133–145.

[3] 'Verifiability', *Proceedings of the Aristotelian Society*, Suppl. vol. 19 (1945).

[4] Richard Montague, 'English as a Formal Language'. First published in 1970; reprinted in *Logic and Philosophy for Linguists*, ed. J. M. E. Moravcsik, Mouton, The Hague, 1974, pp. 112–113.

[5] Joan Bresnan, 'On the Form and Functioning of Transformations', *Linguistic Inquiry* 7 (1976), 304.

[6] Both the notions of the partial independence of syntax and that of a natural language as a biological phenomenon have been developed by Noam Chomsky. I have been stimulated by his proposals, though I am not sure that he would agree with my formulations and applications of these concepts.

AVISHAI MARGALIT

SENSE AND SCIENCE

1. INTRODUCTION

Meaning, thought Bloomfield, is an important matter. Yet the study of meaning, according to him ([2], p. 139), is in poor shape. Its development he took to depend on the development of science: while only science can guarantee 'accurate definitions' of linguistic terms, the state of science is lamentably such that it cannot actually provide such definitions to most linguistic expressions. There are, however, some words whose definitions are in a happier state. Thus, he says,

We can define the names of minerals, for example, in terms of chemistry and mineralogy, as when we say that the ordinary meaning of the English word 'salt' is 'sodium chloride (NaCl)', and we can define the names of plants or animals by means of the technical terms of botany or zoology, but we have no precise way of defining words like 'love' or 'hate', which concern situations which have not been accurately classified – and these latter are in the great majority. ([2], p. 139.)

Judging by current philosophical literature which deals with the semantics of words, expressions like 'love' or 'hate', if indeed they are the majority, are certainly a silent majority, while natural-kind words constitute a very vocal minority. I do not wish, however, to advocate a policy of benign neglect toward this minority (if indeed it is a minority). On the contrary, I believe that the semantics of such words in ordinary language has not as yet been settled, and that a better understanding of what is involved in the assignment of meanings to them is likely to shed light on the nature of language, as well as on further philosophical questions. My concern, then, is with the question of the extent to which science does, or ought to, determine the meaning of descriptive terms (or, more specifically, of natural-kind words) in ordinary language. I hesitate to state that the problems that worry me are the same as those to which Bloomfield addressed himself. After all, Bloomfield shared the ideal (or fantasy) of the language of 'unified science', and probably hoped that the

17

E. Saarinen, R. Hilpinen, I. Niiniluoto, and M. Provence Hintikka (eds.), Essays in Honour of Jaakko Hintikka, 17–47. All Rights Reserved.
Copyright © 1979 by D. Reidel Publishing Company, Dordrecht, Holland.

semantics of ordinary-language words would one day be replaced by the semantics of that kin language, the language of unified science. His own contribution to this project was in clarifying "the linguistic aspects of science" (see [3], p. 219), whereas I shall be concerned with clarifying the scientific aspects of language. By this I do not of course mean the scientification of linguistics but rather the role of science in assigning meanings to everyday expressions. Bloomfield states that "Language plays a very important part in science" ([3], p. 219). I would like to examine the converse, that is, whether science plays, or ought to play, an important part in assigning meanings to terms in natural language.

Even while hoping to replace natural language by a precise and systematic language of science, we saw that Bloomfield actually maintained that 'sodium chloride (NaCl)' is the meaning, and even the ordinary meaning, of 'salt' in *English*. I shall now focus the discussion on this, and similar, claims.

2. THREE IMMEDIATE STRICTURES

Only a minority of English speakers know that 'salt' is NaCl. If Lot were told that his wife, who has turned into a pillar of salt (*Genesis* XIX), has in fact become a pillar of sodium chloride, he would perhaps have found this one more reason to drown his sorrow in wine. Even the wildest stretching of the concept of knowing (or, for that matter, even of *implicitly* knowing) would not enable us to maintain that ordinary English speakers who use the word 'salt' and who sprinkle salt on their salad know that this is sodium chloride. The importance of this fact seemed to Mill sufficient to rule out statements of the sort that the meaning of 'salt' is sodium chloride:

I cannot think we ought to say that the meaning of a word includes matters of fact which are unknown to every person who uses the word unless he has learnt them by special study of a particular department of Nature; or that because a few persons are aware of these matters of fact. . . . I hold that (special scientific connotations apart) a name means, or connotes, only the properties which it is a mark of in the general mind: and that in the case of any additional properties, however uniformly found to accompany these, it remains possible that a thing which did not possess the properties might still be thought entitled to the name. ([17], ch. VIII, pp. 141–142.)

So, since 'sodium chloride' is not a mark "in the general mind" it cannot, according to Mill, be the meaning of 'salt' in English, while surface features like 'white crystalline', 'briny taste' etc., can.

So far, Mill. Chomsky, on the other hand, tells us ([4], p. 143ff) of a scientist S who is capable of studying man as a creature of nature. This scientist, Chomsky suggests, is bound to distinguish three cognitive systems in humans: grammar, common sense, and knowledge of physics. Moreover, he would undoubtedly notice the 'striking differences' among these systems. Among them:

Grammar and common sense are acquired by virtually everyone, effortlessly, rapidly, in a uniform manner, merely by living in a community under minimal conditions of interaction, exposure, and care.... Knowledge of physics, on the other hand, is acquired selectively and often painfully, through generations of labor and careful experiment, with the intervention of individual genius and generally through careful instruction. ([4], p. 144.)

He also points out that in distinction from grammar and common sense, the knowledge of physics is conscious.

The way language (including the dictionary) develops and is acquired is, according to Chomsky, that typical of the systems of grammar and common sense. So, so far as natural language is concerned, sense is a matter of common sense, not of science. Or again, if the concept of meaning (or sense) is to be explanatory of the way language is acquired, understood, and used for purposes of communication, it just cannot be the case that the meanings of a large number of words in the language are known (if at all) to but a small portion of the users of that language.

The best expression of this latter view is to be found in Locke ([14], Ch. VI). Here it is further augmented by the contention that natural languages have in fact preceded science.

But supposing that the *real* essences of substances were discoverable by those that would severely apply themselves to that inquiry, yet we could not think that the ranking of things under general names was regulated by those internal real constitutions, or anything else but their *obvious* appearances: since languages, in all countries, have been established long before science. ([14], ch. VI, p. 75.)

That is, even if science does provide us with the "real essence" of salt, i.e. with its physico-chemical internal structure, the fact that ever since Adam people have used the word 'melach' ('salt') indicates that its meaning has to be determined by what Locke describes as its "superficial" features. This, however, does not exhaust Locke's views on this subject. For instance, in distinction from Chomsky, he is worried by the *lack* of uniformity in our common sensical concepts. It is in fact this lack of uniformity that motivated him to postulate real essences, whether they be known or not ([14], p. 95). The fact that natural-kind words were in use before there was science (Locke), the fact that scientific information, when it is known, is known but to few (Mill), and the fact that the way language is mastered is so radically different from the way science is mastered (Chomsky), all of these seem to come close to demolishing the view that science has any bearing on the meanings of words in ordinary language. However, none of these can be taken as a knock-down argument in this first round, and I believe that the contest between science and common sense over which of them determines meanings is won by the latter (if at all) only on points. After all, the most sophisticated philosophers today agree with Bloomfield. Thus, Kripke would maintain that the meaning of 'salt' is NaCl, and Putnam that NaCl is the predominant sense of 'salt' – provided, of course, that NaCl is indeed the internal structure of salt.

3. WHICH SCIENCE?

The qualification 'provided that NaCl is indeed the internal structure of salt' (at the end of the last section) is important. It directs attention to the question of which is the science that is to determine the meaning of our words. Quite obviously it is science in some objective sense, as distinct from the current scientific knowledge. To be sure, it is the current scientific knowledge that we go by, because this is the best we have. But if it were to turn out in the future science that the internal structure of salt is *not* NaCl, then the meaning of 'salt' is not NaCl even today, in spite of the fact that today we believe that it is.

Leech criticizes Bloomfield's doctrine that science is to provide information pertinent to meaning precisely on the ground that science changes and advances all the time so that scientific definitions aren't constant, as well as on the ground that at any given moment there is more than one theory to explain the phenomena within some given range ([13], p. 2ff). This criticism, however, even if justified given Bloomfield's epistemic approach to 'science', does not apply to Putnam and Kripke: it does not apply where one has in mind the *true science* rather than the *believed science*. But even though Putnam and Kripke are immune to Leech's claim about the underdetermination of scientific theories, they are still vulnerable to Quine's claim about the indeterminacy of translation ([20], Ch. 2; [21]) for, given Quine's arguments, even relative to the true science of the universe the meanings of words in our language are not uniquely determined.

I hope I shall be excused for being determined not to deal with Quine's intriguing thesis of the indeterminacy of translation in this essay. Note, incidentally, that to the extent that this thesis holds with respect to science as supplying meanings for words it certainly holds with respect to common sense, so that it applies either to both or to neither of our contending views and can hence be ignored. I shall therefore pay my due respects by saying that the whole discussion here, concerned as it is with the notion of meaning, takes place in the shadow of the indeterminacy thesis which is aimed at undermining this very notion.

4. WHEREFORE SCIENTIFIC MEANING?

Who needs non-epistemic semantic notions? The scientific approach to semantics justifies the need for such notions as follows. In understanding the nature of natural language we are interested not only in such questions as how is it acquired, how is it understood, or how is it possible to achieve communication with it, but also in the question of how is it possible to talk (truthfully) about the world. It stands to reason that this last question is not unconnected with the former ones. But while the former questions admit of – indeed call for – answers couched in epistemic notions, matters are different with respect to the latter. If the sentence:

"He [Elisha] went forth unto the spring of the waters and cast the salt in there." (II Kings, 2, 21)

is true, then Elisha indeed cast *salt* – and not anything *else*, salt-like as it may be. And this is so only if it was NaCl. To be sure, the men of the city who obliged Elisha by putting salt in the new cruse did not *know* that salt is sodium chloride, but the question whether or not it was salt that they put there turns on the question whether or not it was sodium chloride that they put there.

The point of bringing these examples up is to demonstrate that the investigation of the scientific meaning of 'salt' is part of the investigation of the truth conditions of sentences in which 'salt' occurs in a non-empty way. The more general claim is this. Even though we have gotten used to focus on the logical form of sentences when investigating their truth conditions, this is perforce only a partial inquiry. It is of course amply justified to separate, when dealing with the truth conditions of sentences, the task of investigating their logical form from that of explicating the descriptive terms occurring in them. The former task requires a degree of precision which is not greater than that manifested by the sentence concerned. Thus, consider:

'The Dead Sea contains tons of salt' is true if and only if the Dead Sea contains tons of salt.

The right-hand side of this equivalence does not – and is not supposed to – contain the chemical formula of salt: the right-hand side of this equivalence need not be clearer and more precise than its left-hand side. This is evident in the case of the homophonic equivalence, and it holds true in the case of the heterophonic equivalences as well (see Davidson [5]). The latter task, on the other hand, deals with the descriptive terms occurring in the sentence and may be viewed as aiming at providing the *conditions of reference* of the sentence constituents (words). This task is inseparable from the more general enterprise of determining the truth conditions of the sentence – even if, for reasons of division of labor, more attention has been paid to problems on the level of its logical form.

This is where I see the importance of the attempt to identify the meaning of natural-kind words with their scientific meaning. In determining the extension of 'salt', say – which is part of the task of determining the reference conditions of the sentence constituents, which is, in turn, part of the task of determining its truth conditions – the chemical structure of salt is crucial. The reasoning here is clear. Our ability to say something true about the world depends not just on what we *believe* about the world, but also on its actual structure. In order to explain the success of our language in saying truthful things about salt we need information about the constitution of salt – just as we need such information when we want to explain, for instance, why by mixing salt with water we lower the freezing point of water. The chemical structure of salt is, therefore, essential for determining the truth conditions of sentences where 'salt' occurs in a non-empty way.

Summing up: one of the enterprises undertaken by the theory of language is to explain the connection between language and the world. The explanation of this connection requires that we understand both sides of it, i.e., language and the world. The world, however, is what *science* tells us that is the world. This is why it is hardly surprising that we need scientific information in order to assign meanings to some of our terms –

even if this information is (as yet) unavailable, or known only in part, or known only to few. One central reason, therefore, for identifying the meaning of a natural-kind word with its scientific meaning is the latter's contribution to the truth conditions.

Another reason for this identification is that only the scientific meaning of 'salt', in distinction from any qualitative lexical 'definition' of it, adequately conveys that which remains invariant in all of the true counterfactual conditionals about salt. In the background here stands a conception, central to the theory of meaning and shared by authors who disagree practically about everything else (like Husserl and Carnap, for example), according to which meaning is that which remains fixed under descriptions of possible worlds alternative to our own. So, on this conception, if the internal features of salt fulfil this condition, then they are the meaning of 'salt' rather than the external features (say, whiteness) that can be denied of salt in counterfactual conditionals while still talking about salt.

5. INTERIM SUMMARY AND A NEW OPENING

If science provides meanings to natural-kind words in natural language, then it is hard to see how such a concept of meaning can have an explanatory role regarding the ways in which language is acquired, understood and used for communication. These latter facts would then require an explanation by a different conceptual mechanism. On the other hand, the commonsensical notion of meaning, capable as it may be to account for these facts, cannot apparently satisfy us with respect to the truth conditions of sentences as well as to the 'aboutness' of counterfactual conditionals the way meanings rendered scientifically can. I take it that the success of each of these positions is to be measured according to its ability to explain *well* that which it is supposed to explain, and also to account for that which it is *prima facie* not specifically addressed to.

Many contrasting pairs of notions have cropped up so far; not all of them have been labelled. Among them: the contrast between science and common sense as providers of meanings, the contrast between a realistic and an idealistic approach to semantics, the contrast between internalism and surfacism – that is, between meaning in terms of information on the surface features and meaning in terms of theoretical information on the internal structure (this latter contrasting pair is perhaps one of the legitimate heirs of the phenomenalism-physicalism dichotomy), and more. The sides of these contrasts do not completely overlap, nor should they be expected to.

My aim here, however, is not to embark upon a 'doctrinaire' discussion, that is, to present pairs of contrasting notions such as those just cited and to discuss them on an abstract and principled level. Rather, I would like to take up some aspects of what Kripke gives us as his semantic 'picture', and especially to discuss some of his examples. I am referring to the view according to which it is *theoretical identities*, such as that of water with H_2O and of heat with the motion of molecules, that determine the meanings of the pertinent terms ('water', 'heat'). Kripke uses the word 'picture' so as to avoid having to present a *theory*, presumed to be couched in necessary and sufficient conditions. I judge this to be a mark of sober-mindedness rather than of cowardliness. At the same time, of course, it is hard to criticize a 'picture', since it is not all that clear what would count as counter-examples to it: no counter-example can refute a picture. It is *tact* rather than *logic* that would determine what sort of argument or example damages the picture, and what leaves it intact.

The brush strokes of Putnam's and Kripke's semantic picture are largely their examples. Its evaluation, therefore, is to depend to a large extent on the weight of these examples – more so, at any rate, than on the programmatic proposals. Examples without a doctrine are blind, just as a doctrine without examples is empty: both are necessary. However, their relative weight differs in various philosophical systems. I am distinguishing, then, between philosophical systems characterizable by the 'i.e.' operator, and philosophical pictures characterizable by the 'e.g.' operator. The former are to be appraised with the aid of rules (such as of consistency) and of refuting counter-examples, whereas for the appraisal of the latter, in which examples are crucial, what is needed is what Kant calls the faculty of judgement. My argument throughout this paper will be that the examples of natural-kind words are after all more compatible with the picture of commonsensical semantics than with the picture – currently in vogue – of the scientific semantics.

A semantic picture rival to that of Putnam and Kripke is to be found, in my view, in the writings of Austin: with him, indeed, it is appropriate to talk of a picture rather than of a semantic system or theory. I would want to show that Austin's picture does more justice to natural language than does Putnam's and Kripke's. My discussion, however, is not going to be textual, even though I shall of course refer to the writings of the main protagonists: I shall concentrate on the contrast between the two *approaches* to semantics – the commonsensical *vs.* the scientific. And a final remark: when I talk of the common-sense semantics I do not wish to be understood as implying that the other, scientific one is devoid of common sense. On the contrary, I think that there is plenty of common sense in Putnam's and Kripke's claims and examples. This terminology is only intended to emphasize the contrast between the semantics whose basis of information is primarily common sense and that whose basis of information is primarily science.

6. DIVISION OF LINGUISTIC LABOR

One of the difficulties besetting the 'scientific' semantics was, it will be recalled, that the scientific meaning of natural-kind words is known to few only, if indeed it is known at all. This refers not just to words which are rarely met with, such as 'blenny', 'bract', 'brachiopoda' or 'blastula', but to familiar words used daily such as 'dalia', 'diamond' or 'dog'. What is known to few only is information about the internal or genetic structure of the things designated by these words. How, then, is it possible to assign meanings that are known to and recognized by a tiny minority of the speakers of the language? This is the difficulty which the scientific semantics, in distinction from the common-sense semantics, has to face.

Wittgenstein has proposed a possible direction:

'We are quite sure of it' does not mean just that every single person is certain of it, but that we belong to a community which is bound together by science and education. ([24], Sec. 295.)

Putnam has provided an interesting interpretation to this idea of our being bound together by science and education. The bind, according to him, finds its expression in a division of linguistic labor. Thus the meaning (i.e. the extension) of, say, 'diamond' is determined by an expert – just as the polishing of a diamond is done by an expert. There obtains, further, a relation of dependence between the few who know and the many who don't:

Every linguistic community exemplifies the sort of division of linguistic labor just described: that is, possesses at least some terms whose associated 'criteria' are known only to a subset of the speakers who acquire the terms, and whose use by the other speakers depends upon a structured cooperation between them and the speakers in the relevant subset. ([19], p. 228.)

This division of linguistic labor is, then, the scientific semantics' way out of the problem of 'the knowledge by the few.' But this does not exhaust the merits of this proposal. It is also geared, I believe, to overcoming some of the difficulties encountered in the context of the *common-sense* semantics. For even if the information incorporated into the common-sense semantics is suitable for all it is still a mistake to hold that it is equally divided among the speakers of the language. (Try, e.g., to formulate to yourself the difference between 'fruit' and 'vegetable'.) Accordingly, the sociologization of semantics as it is expressed by the notion of the division of linguistic labor is intended to reduce the burden shouldered by semantical representation systems that are supposed to be 'psychologically real'. Put differently, a sociological mechanism for sense and reference partially replaces the psychological mechanism, and at any rate renders it more realistic.

6.1. *Division of Linguistic Labor or a Homogenous Community of Speakers*

Chomsky's starting point (and possibly Frege's too) involves a degree of abstraction that ignores the variance among speakers. That is, he treats language as a language of a homogenous community of speakers. This idealization is of course useful and productive as far as a description of the syntax of the language is concerned: the variance among speakers of a language concerning their mastery of the syntax of the language is relatively small (even non-educated and unintelligent speakers comprehend complex syntactical structures). The variance concerning mastery of the vocabulary of the language, however, is much greater. In particular there is a huge gap between the total vocabulary of a language, which may comprise as many as a million dictionary entries, and the vocabulary of a typical ideolect in that language which comprises some tens of thousands of words. The mastery of a new word in the language is not the same as the mastery of the sense of a 'new' sentence' provided that one is familiar with its constituents: the first case, but not the second, involves the introduction of a new 'primitive' to be learnt and absorbed. Thus, even though there exists a gap between the unlimited number of possible sentences in the language and the (necessarily limited) number of sentences you have uttered or will in fact utter, it is of no consequence to the question of your knowledge of the language when compared with the gap that exists between the number of entries, say, in the O.E.D. and the number of words known to a typical English speaker. All of this may be held against the idealization that regards the (ideal) speaker of the language as one at the disposal of whom is the entire O.E.D.

6.2. *A Scientific Register in Natural Language*

Another merit of the sociological bent given to the notions of sense and reference is that it dispenses with the need to speak of two languages – the language of science as opposed to the ordinary language (Carnap, Stebbing) – or, again, with the need to speak of two worlds – that of science as opposed to that of common sense (Goodman). Instead of dividing language, or the world, or both into two it divides the information about the language and the world among the users of the language.

I myself share the desire to postulate but one world and one language. I wish to argue for it, however, in a different way.

Following the London school of linguistics let us distinguish between the variance in language according to groups of its *users*, i.e. *dialect*, and the variance in language according to types of its *uses*, i.e. *register*. (See [8].) Now I do not propose to regard the language of science as a dialect, in the sense in which Mandarin is a dialect, in spite of the fact that scientists, like Mandarins, form a relatively well-delineated group. The difference between these two is that while the Mandarin uses his tongue for all uses and purposes, the scientist uses his jargon for specific usages only. Thus there is no reason to expect (and I believe no reason to hope) that 'the language of unified science' will one day become standard in the same way in which Mandarin has become standard.

So I regard the 'language of science', and especially those segments of science which are not formalized within axiomatic systems, as a register of the natural language. Within this register it will, I hold, be true to say that 'water' means H_2O (with the appropriate reservations). The prestige accorded to scientists, like the prestige of the Mandarins, may bring it about that *this* meaning becomes the standard meaning of 'water' in English, but this is not the case at present. And I do not see any reason why descriptive semantics (in distinction from normative semantics) should endow this register with a privileged status.

6.3. *Experts and Lexicographers*

When Adam Smith illustrated the idea of the division of labor through the production of nails it was quite clear what is the product, what are the stages of production and what dexterity is called for in each. Matters are different, however, with the division of linguistic labor.

There are experts for diamonds. Elisabeth Taylor has undoubtedly consulted with some of them. But they are experts for diamonds and not for the word 'diamond'. I take it, then, that Putnam considered the division of labor as a junction between two kinds of experts: experts for diamonds as well as experts for (the linguistic usages of) 'diamond', that is – chemists and lexicographers. The lexicographer presents the chemist with a sample of objects typically referred to in the relevant community of speakers as diamonds, and asks him about their internal structure. The latter's answer, let us say pure carbon, with molecules consisting of one carbon atom in the center connected to four equidistant carbon atoms', determines the extension of 'diamond'. For the extension of 'diamond' is the class of all objects having the same constitution as those in the given sample. Hence, the theoretical structural features of diamonds determine the extension of 'diamond' rather than the superficial similarity to the members of the sample, based on features such as brilliance, hardness, crystalline shape or the like. The division of labor, therefore, is this: the lexicographer supplies the sample, and the chemist ascertains the internal structure, thereby determining the extension.

The controversy between the 'scientific' approach and the 'common-sense' approach to meaning is now revealed to focus on the division of authorities: those charged with determining the paradigmatic sample as opposed to those charged with determining its application. On the scientific (internalist) view the lexicographer brings to his meeting with the chemist only *prima facie* diamonds: he collects only what appears on the face of it (literally) to be a diamond. That is, on this view the lexicographer can at best supply a sample of *prima facie* cases, never a sample of paradigmatic cases. It is the chemist who will have the veto power concerning any foils that may have found their way into the sample. Everyone recognizes this prerogative since everyone knows that if a glittering piece of polished glass that passed as a diamond is pronounced by the expert as glass it will not be called 'diamond' any longer and will be removed from the sample as a mistake.

The 'common-sense' approach to meaning, on the other hand, regards the lexicographer as the one who determines the paradigm cases of objects to which the word in question applies, rather than as merely providing the scientist with *prima facie* cases. How is this conception to be squared with the fact that the common-sense lexicographer will undoubtedly accept the verdict of the chemist regarding the glittering piece of glass and hence will reject it from his sample? It seems to me that there is a way for the common-sense semanticist to overcome this difficulty. A diamond's worth is due to its rarity. The appeal to the expert is usually motivated by the desire to guard against fakes, frauds and forgeries. This is just like the case where one consults an expert in order to ascertain the authenticity of a Vermeer painting. In both cases the expert may use X-rays to arrive at his verdict. But it is important to emphasize that the various physical or chemical tests are supposed to serve the economic purpose of determining the degree of rarity of the object. This is why, given the economic purpose and practice, a polished glass is not a 'near-miss' of a diamond even though its resemblance to a diamond may go pretty far, just like a De Hory forgery is not a 'near-miss' of a Vermeer painting.

Let me explain. A diamond is both a mineral and a gem. Its internal structure is crucial for its being the mineral that it is, but possibly not that important for being the gem that it is. To be sure, 'diamond' is a natural-kind word, but that does not in itself imply the priority of the interest according to which it is classified as a *mineral* over the economic interest according to which it is classified as a *gem*. That these two types of classificatory interest may diverge is attested by cases where a mineral is classified into two different gems even though their internal structure is the same (modulu tiny impurities). Thus for the chemist beryl is a mineral which is the gem emerald when green and aquarium when blue, or again the same internal structure of the corundum underlies both rubies and sapphires. Now if there were a natural mineral as rare as the diamond and whose external features were similar to those of the diamond while its internal structure differed from that of the diamond, it *may* have been classified as the same gem even though it would be a different mineral. I cannot be sure: since the diamond exemplifies *both* of these classificatory dimensions there is no way of telling a priori which would prove dominant had a gem with the described characteristics been discovered. Be that as it may, however, it is clear that the glass crystal is not a 'near-miss' of diamond *either* as a mineral *or* as a gem.

The position I would defend on these matters seems to me to be expressed by Austin's claim to the effect that ordinary language embodies the accumulative experience of generations, and that this experience in turn is bound up with the practical aims of life. Yet, "this is likely enough not to be the best way of arranging things if our interests are more extensive or intellectual than ordinary." ([1], p. 133.)

Putnam has recognized that classifications of objects as well as the senses of the corresponding words may depend upon interests different from the scientific interest. But he wishes all the same to endow the scientific classification as well as the scientific sense with a privileged status (the 'predominant sense', [19], p. 225). My own aim in this essay, however, is to defend Austin's position, or at any rate the semantic picture I believe implicit in it, according to which the scientific sense has *no* priority over the common-sense senses.

7. BAPTISM OF NATURAL KINDS

The division of linguistic labor à la Putnam is synchronic in nature. It relies on a sample-at-a-given-time for the determination of the pertinent extension. Kripke, on the other hand, considers an act of initial baptism to have a special standing, thereby turning the issue of reference determination into a diachronic affair.

It is my view that the adoption of the initial baptism model for the determination of the extension of general names is liable to lead to errors. I would like to take some space to illustrate this point. It is quite easy to fancy that the circumstances of the first occurrence of 'diamond' in the Old Testament in fact constituted the initial act of calling the gem by this name. Well, in Exodus, 28, where the four rows of precious stones decorating the breastplate of the High Priest are described, it says (verse 18): "And the second row shall be an emerald, a sapphire and a diamond." However, on grounds of scientific scholarship, and given that the art of diamond cutting was apparently developed only in the Middle Ages, it is today very much in doubt whether the 'diamond' (in Hebrew 'Yahalom') of the definite description "the third stone in the second row of Aaron's breastplate" is indeed a diamond [6]. At the same time the modern Hebrew usage of 'Yahalom' is, like that of the Middle Ages, in the sense of a diamond. In the light of all this it seems to me to be erroneous to maintain that the extension of 'diamond' in Hebrew (i.e. of 'Yahalom') is the class of all objects whose internal structure is identical to that of the third stone in the second row of Aaron's breastplate.

[Let me add, incidentally, that the diamond example can *not* in my view be explained away on the grounds that the word Yahalom was born again with the rebirth of the Hebrew language – a point that is valid with regards to some biblical words whose original meaning is *not* known and which were coined anew in the Modern Hebrew for modern usages (thus the word 'hashmal', appearing in Ezekiel 1, 4 and translated as 'amber', is today the Hebrew word for *electricity*).]

The upshot, I believe, is that the determination of the extension, and hence of the sense, has to be synchronically rather than diachronically oriented.

Just by way of hinting at some more problems that I have with the act of initial baptism consider the first act of calling something a 'dinosaur', by Sir Richard Owen in 1842. Now Owen of course did not encounter a dinosaur, paradigmatic or otherwise; all that he had to go by were some bones. Suppose it were to have been proved later that Owen was mistaken in his reconstruction, and that the bones in fact belonged to a huge gorilla-like mammal. Would 'dinosaur' have then referred to a mammal? I do not know. But it seems to me that the meaning of 'dinosaur', contrary to Kripke's account, does depend on our *beliefs*, e.g. those concerning the reconstruction of the dinosaur's bones.

8. THE SCIENTIST'S LINGUISTIC HABITS

Let us forget about acts of baptism now and turn to one of Kripke's key passages:

The original concept of cat is: *that kind of thing*, where the kind can be identified by paradigmatic instances. It is not something picked out by any qualitative dictionary definition. ([11], p. 319.)

Now it is quite possible to speak of cats as 'that kind of thing' while pointing to tigers, and lions, and pumas, as well as to leopards, cheetahs, ocelots – and of course to the domestic cat. Was one mistaken in selecting one's sample? Or are all of these borderline cases of 'cat', while the central and paradigmatic cases are the domestic cats? It would certainly be strange and misleading to say that. Moreover, when a zoologist utters, using the scientific register, 'that kind of thing is the largest American cat' while pointing to a jaguar, are we to maintain that his use of 'cat' is a secondary or derivative one? It *will* be considered a secondary sense of 'cat', however, once all the uses of speech are taken into account. Thus, it is not the zoologist's linguistic ways that determine, or ought to determine, the predominent sense of 'cat'.

Or take a different example. The scientific use of 'heat' is in the sense of
the total kinetic energy (in distinction from 'temperature' which
designates the average kinetic energy). Given this use, the scientist may
say, truly and literally (rather than hyperbolically): "There is more heat
in the Dead Sea than in my boiling cup of tea." But this of course does not
accord with the ordinary usage of 'heat' (which corresponds roughly with
that of the scientist's 'temperature').

My claim, then, is that when one speaks of the division of linguistic
labor, the scientist's contribution to it is not his linguistic behavior, or
his scientific register, but rather the information that he is able to provide
concerning the 'essential' (if you will) characteristics of certain objects
sampled for him by the lexicographer.

9. THE TWO FACES OF SIMILARITY

There is an empirical assumption underlying Putnam's or Kripke's
envisaged projection from a sample (of paradigm cases) to the population
(extension). This projection is supposed to be carried out on the basis of
theoretical similarity (identity) among the objects of the sample, and
those of the population at large. Thus Putnam: "The key point is that the
relation same$_L$ is a *theoretical* relation." ([19], p. 225.) This 'is (theoretically)
the same as' picks out different characteristics in different natural kinds:
in living things the genetic origin, in diseases the etiology, and in materials
the molecular structure. And since the usual term contrasted with
'theoretical' is 'observational', the mistaken impression may be formed
that the commonsensical classification to natural kinds is carried out
exclusively on the basis of observational features. This, as I say, is not the
case.

There are two types of features that play a role in the commonsensical
notion of similarity. Using Tversky's happy expression [22] they are the
two 'faces' of similarity. One type of features has to do with their *salience*,
the other with their *diagnostic value*. Salience is determined by perceptual
factors (good form, intensity); the diagnostic value by the degree to which
the pertinent feature is of classificatory and distinctive worth. It is clear,
then, that while perceptual salience is a relatively fixed factor, diagnosticity
is interest-relative and thus largely context-dependent.

When speaking of the commonsensical classification as opposed to the
scientific one I do not mean to suggest that the former is based just on

perceptually salient features and the latter exclusively on diagnostically valuable features. That would be wrong: both faces of similarity are present in the two types of classification. What can be said, though, is that theoretical similarity is based *primarily* on features of a diagnostic value, and relative to a *scientific* interest (that is, with a view to classifying into such groups that will enable adequate systematization and explanatorily powerful empirical and theoretical generalizations). The commonsensical similarity, on the other hand, is indeed based primarily on perceptually salient features, but not on them alone. Diagnostically valuable features do play a role here too, but – and this has to be emphasized – they are generally *practical*-interest relative rather than *scientific*-interest relative.

10. IS THE MEANING OF 'WATER' H_2O?

Let us examine this question in two directions. First, we shall assume one water phenotype to which there corresponds two different genotypes; second, one genotype with two different corresponding phenotypes. We shall be examining the way in which these cases affect the question of the meaning of 'water'. In the next section a similar venture would be undertaken concerning disease words. My aim in all of this is to try to substantiate the claim that in the semantics of natural language information concerning the genotype should *not* be given priority over information about the phenotypes in determining the meaning of a (natural-kind) word. Let me just add, first, that I am of course using the terms 'genotype' and 'phenotype' in a broad sense rather than in the narrow biological sense, and, second, that phenotypical features need not necessarily be *observational* features. Thus, the fact that, say, distilled water is standardly used for medical and industrial purposes (rather than for drinking) will be considered a phenotypical feature of distilled water, while the isotopic structure of heavy water will be considered a genotypical feature of this liquid.

One phenotype, two genotypes. Putnam has proposed the following thought-experiment ([19], p. 223). Suppose somewhere in the galaxy there is a planet similar in all respects to ours. Call it Twin Earth (Huxley would call it Antiterra). Each person on earth has a Doppelgänger there, who is, further, in the same psychological state as his earthly counterpart.

Now, to continue Putnam's piece of science fiction, suppose Rebecca's twin on Twin Earth gives the twin of Abraham's servant a liquid to drink which, although indistinguishable from water in all of its phenotypical features, is not H_2O but some XYZ. Both Rebecca and her Twin-Earthly copy refer to these respective types of liquid as 'water' (let us ignore the extra bit of complication that they speak Hebrew (hence 'Mayim') rather than English), they both believe it's water, they both use it for exactly the same purposes, etc. – and yet there remains the difference in the molecular structure.

Putnam maintains that a liquid which is not H_2O is not water. I agree, of course, that 'water' on Twin Earth is (given our story) not H_2O. But I would still ask: is it not water? Wouldn't we perhaps be licensed to say that the extension of 'water' is any liquid which is *either* H_2O *or* XYZ? I do not think this has to be ruled out.

Consider the possibility that light was discovered to have two un-connected different natures: corpuscules and waves. (It did once seem to be something like that.) What would then have determined the meaning of 'light', the sameness of phenotype or the difference of genotype? It seems to be not at all unreasonable to suppose that the sameness of pheno-type and of practical uses and purposes would outweigh in importance the fact that the genotype was different, insofar as the meaning of 'light' is concerned. This is not to deny, of course, that we have indeed learned to use the ordinary-language 'light' in its physical-scientific sense. Thus we may say that infra-red rays are light which cannot be seen by human eyes, or we are told by Steven Weinberg about an era in which there was light but there was no one to see it, etc. But all of this does not compel us to conclude that the predominant sense of 'light' should be specified in terms of radiant energy as such rather than in terms of that radiant energy that activates our visual sense and enables us to see the body emitting it. On the contrary, the phenomenalistic sense of 'light' un-doubtedly has experiential priority (or 'temporal priority' in Aristotle's terminology) over its physicalistic sense. And as to material priority, this depends on the matter, hence on the interest. And I do not believe that we are to postulate in advance that the scientific interest is the most important one. After all, the scientist can do without the word 'water', and can be satisfied with 'H_2O' and 'XYZ'. Not so the lexicographer. The latter has to account for the meaning of 'water', and therefore has to cover in his sample paradigm cases of the uses of 'water'. This in turn means that his sample has to include both the liquid H_2O and the liquid XYZ.

One genotype, two (or more) phenotypes. This case can be illustrated with the aid of a thought-experiment of Locke. An Englishman raised in Jamaica arrives one day in England and notes during his first night there that the water in his basin freezes. Being familiar neither with ice nor with the word 'ice' he calls it 'hardened water'. Now Locke maintains that when this man points to the ice and calls it '(hardened) water' he is in fact referring to the same "real essence" as when he is pointing at liquid water. He argues, and I think rightly, that "Our distinguishing substances into species by names is not at all founded on their real essences." ([14], Ch. 6, Sec. 20.) In other words, the semantics of names is determined by the phenotypical features of their referents, not by the genotypical ones. And so it is that there is a crucial difference between water and ice – a difference that does indeed warrant their having two names – even though they share in a sense the same genotype. That the difference is crucial is seen when we observe that the truth value of a sentence like "children love to swim in water" changes once 'water' is exchanged for 'ice'.

To be sure, there *is* an 'internal' difference between water and ice, a difference which is blurred when the formula H_2O is used. Water is not *simply* H_2O, but aggregates of molecules denoted by (H_2O); in ice the so-called hydronic bond is stronger. What corresponds rather well with H_2O is in fact steam, or water just about to boil – and these are hardly the paradigmatic cases of 'water'. It is of course tempting to wave all of these 'fine points' and to say that *basically* water is H_2O. (Putnam has managed to insert into the theoretical identity of water the sortal 'that liquid'.) But it is not that simple. The liquid whose constitution is closest to H_2O is distilled water. But to obtain distilled water one goes to the druggist, not to the tap. Distilled water has uses very different from the standard uses of (ordinary) water. It turns out, then, that the meaning of 'water' is determined in no small part by the *impurities* of water, and hence that the 'water' in the expression 'distilled water' is not less syncategorematic than it is in 'heavy water'. (Actually, heavy water is like (ordinary) water in appearances as well as in its chemical characteristics, and differs from it only in some of its physical properties.) With regard to Kripke's example, concerning the alleged Russian discovery of 'polywater': if it turned out that the Russians were right and that indeed there is a liquid identical in internal structure to water but nevertheless differing so much from water, my conjecture is that 'water' in 'polywater' would be taken syncategorematically.

11. DISEASE WORDS

Every one may be affected by disease, but not every one is an expert on diagnosing it. The division of labor involved here is obvious. However, it seems to me that disease words do *not* support the 'scientific semantics' picture. Putnam has made use of the class of disease words in order to establish the internalist semantic picture; he even gave it a prominent role there.

I share the belief that much can be learned about the issues here considered from this group of words. But I doubt whether *what* we learn from it is what Putnam thinks that we do.

On the face of it, it seems that disease words facilitate the upholding of the picture of scientific semantics. The distinction between surface symptoms of a disease and its nature or causes seems all but custom-made for supporting the distinction between superficial (or surface) traits that help determine the reference of a given word on the one hand and the 'deep' traits that provide its meaning on the other: every physician would attest to the importance of the etiology of a disease for its definition. There is a sense in which Wittgenstein shares this view. In the *Blue Book* he says ([23], p. 25): "'A man has angina if this bacillus is found in him' is a tautology or it is a loose way of stating the definition of 'angina'." That is, it is the specific kind of bacillus rather than fever and inflamed throat that constitute the meaning of 'angina': the former is a criterion while the latter but the symptoms.

Putnam and Kripke, needless to say, would not regard Wittgenstein's sentence as a tautology. I doubt that Wittgenstein himself has taken it seriously. It is clear to all that settling the question of which bacilli cause the disease is a matter of empirical discovery, not a linguistic one. Kripke would perhaps maintain here that this identity is necessary while *a posteriori*. But again, it cannot be necessary in any simple way. It may even be possible to induce angina without the presence of the bacteria, just as it is possible to induce tetanus without the tetanus-causing germs themselves but with the appropriate toxin extracted from them. However, all of this may seem unduly pedantic, especially given Wittgenstein's own hedge concerning the 'loose way' in which he states the definition of 'angina'.

I believe, with Austin ([1], p. 77), that with regard to so complex a phenomenon as 'having mumps' it is quite silly to ask which of the causes of the disease is to be identified with the disease itself. It is indeed a mistake, even though perhaps not a silly one, to see the bacilli culture as the angina, and the inflamed throat as an external symptom. Disease, after all, is a significant deviation of the body (or some part thereof) from its normal functioning. The external symptoms constitute the disease – that is, the deviation from normal functioning – just like its other aspects. The question what is 'really' the disease may sound somewhat like the Ryleian question 'where is the parade?' asked after one has seen all the units passing by. To be sure, I am not denying that the bacilli culture has a more significant diagnostic value, with respect to angina, than the inflamed throat. But I still do not see why the meaning of 'angina' is to be identified with that which causes the disease rather than with any of the other symptoms constituting the disease.

Let us look a little closer into the matter. First, a disease may have several different causes. Thus, meningitis has various biological causes (both bacteriological and viral) as well as chemical causes (alcohol). The meaning of 'meningitis' cannot be identified with the presence of meningococcus: this is neither a necessary nor a sufficient condition. Putnam is aware of facts of the meningitis type. He distinguishes, however, between the case where the disease has one 'hidden structure' and the case where it has more than one. In the latter case Putnam concedes that the symptoms rather than the cause may determine the meaning ([19], p. 241). He may also argue that one may be affected by most of the symptoms of measles – like a severe cold, Koplik spots, rash, etc. – without in fact having measles. This may happen, e.g., when one gets a severe cold and develops the other symptoms as a reaction to treatment by penicillin. Without the measles virus, however, this would not be measles. I quite agree that such an accidental occurrence of the measles syndrome would not count as measles. However, it seems to me that were this measles-simulation to occur systematically, it may have come to resemble the meningitis case where the syndrome rather than the etiology determines the disease and hence defines the meaning of the disease name.

The discussion so far seems to warrant the following tentative conclusion. Where there is correspondence between the cause of the disease and its syndrome, it does not really matter which is to be considered its criterion. Where there are several different causes, the syndrome determines. What about the cases where there is one sort of cause with several different syndromes? Here Putnam would argue that it is the cause that determines the predominant sense of the disease word. But is it really so? Let us consider Putnam's own example of a disease, multiple

sclerosis, whose causes are as yet not known. ([19], p. 241.) True, the
causes of this disease are not known, but there are several hypotheses.
The leading one, in fact, suggests that it is caused by the *measles* virus,
which in some cases acts so as to bring about measles, while in others,
multiple sclerosis. Assume, for the sake of the argument, that this conjecture is in fact confirmed, and that 'dormant' puerile measles viruses
cause multiple sclerosis in older age. Would we then say that the extension
of 'measles' is the same as that of 'multiple sclerosis'? Would it really be a
discovery akin to the discovery that the morning star and the evening
star are the same?

I tend to think that not. We do not hesitate to say that measles is
mainly a children's disease, and that the measles that occurs in an adult
person, which is sometimes different in its symptoms from that which
occurs in children, is nevertheless the same disease. Or again, it is quite
common to speak of *polio* both in the cases where the manifestation of the
disease is light *and* in those cases where it is severe and includes paralysis.
But it seems to me that measles and multiple sclerosis are not at all taken
to be on the same continuum of symptoms: they just are two different
diseases.

John Mackie sides with Putnam when he says ([15], p. 99) that by the
name 'measles' we do not refer to the (standard) syndrome of measles
since different people may experience different symptoms of the same
disease. To this I would rejoin that it seems to depend on the degree to
which the symptoms diverge from the standard ones, and, moreover, on
the degree to which those deviant syndromes conform to the recognized
syndrome of *another* disease. Had the measles virus been known to bring
about the syndrome, say, of tuberculosis, the reasons for calling it
'tuberculosis' would in my opinion be as good as those for insisting on
calling it 'measles'.

So: though etiology is undoubtedly important for the definition of
disease words, the external symptoms should by no means be regarded
as of no consequence for the determination of the *meanings* of these words.

12. TRUTH CONDITIONS AND THE SCIENTIFIC MEANING

What is gold is determined by its genotype ('real essence'). *What is called
'gold'* is determined by its phenotype ('nominal essence'). *What is rightly
called 'gold'* is, I believe, the question addressed by Kripke and Putnam,

even though not exactly in this formulation. Their answer, roughly, is that something may rightly be called 'gold' if, and only if, it is gold. And whether or not it is gold is, as said, determined by the genotype. The meaning (or the 'predominant sense') of 'gold' is that which is the ground for rightly calling something 'gold' – in distinction from that which is as a matter of (psychological) fact the ground for our calling various things 'gold'.

What is common in my view to both the common sense and to the scientific approach to meaning is the assumption that the meaning of natural-kind words (as well as that of names in general perhaps) is the information that enables us to project from some paradigmatic cases (a sample) to the extension (the population). The difference between the two approaches can be located in the difference between the projection we actually do and the projection that is in fact justified. The common-sense approach, which is interested in the problem of projection as a psychological problem, uses commonsensical information, while the scientific approach, which attempts to address the question of the *correct* projection, uses scientific information. Put in Kantian terms, the first answers the question *quid facti*, the facts being the observed linguistic behavior, while the latter answers the question *quid juris*, the *jus* being the correspondence between language and the world – i.e., the truth.

I realize, of course, that the question what is rightly called gold is not taken by Kripke and Putnam as an exercise in normative linguistics. They would surely treat it as a different question from, say, the question what is rightly called 'dinner' – the meal in the middle of the day or the one in the evening – when this question is construed as one about a preferred convention (the 'non-U' answer being mid-day, the 'U'-speakers', evening). They would surely say that they are not interested in preferred conventions but in the truth. I am, however, much less sure whether the question what is rightly called 'gold' does not ultimately come down to a problem of normative linguistic, with the S-speakers (i.e. the scientists) replacing the U-speakers. But I'm afraid this is a suspicion rather than an argument.

In the rest of this section I shall be following Putnam's terminology which, following Ziff, speaks of a word's several *senses*, where 'sense' is taken in a non-technical (non-Fregeian) way (like that used in distinguishing 'puppet' *in the sense of* a marionette and 'puppet' *in the sense of* a doll). According to this usage, the element having the atomic number 79 is one sense of 'gold', while a yellow metal (etc.) constitutes another sense of 'gold'. Putnam's view, mentioned earlier, is that the scientific sense is the predominant sense, and the commonsensical one only secondary, or derivative. I shall stick to this terminology, rather than to Kripke's counterpart terminology, which distinguishes between the meaning of a term on the one hand and the fixing of its reference on the other.

I take it that there are quite a number of senses in which both Putnam and Kripke would agree that the priority belongs to the commonsensical sense of a given term (to its 'qualitative definition'). For example, from the point of view of *learning* the term 'gold', the phenotypical (superficial) features have priority over the genotypical (internal) ones: we commonly learn and teach how to use it through explanations that are predominantly phenotypical. At the same time, of course, it may very well be that there is no psychological necessity in the matter. Thus, it may turn out that a term like 'germanium' is (usually) mastered through its genotypical rather than phenotypical features. Nevertheless, as a rough generalization it seems to hold true that from the point of view of mastering natural-kind words in everyday language the commonsensical sense has priority over the scientific one. Another case in point is that the commonsensical sense of such words is more *frequent* than the scientific sense. And frequency of use is, after all, one of the main guidelines of the lexicographer when he comes to grade the priority of the various senses of a given word.

The scientific approach will no doubt reject this last point on the ground that frequency has to do with *usages*, not with *uses*, and while there are no misusages there certainly are misuses. Our problem, they will point out, in determining the predominant sense of a word is to determine at the same time what it is to misuse it – and this cannot be learnt from its usage. More generally, their claim will be that the senses in which the common-sensical sense of a natural-kind word has priority over its scientific sense are cases of what Aristotle calls priority in *knowledge* as distinct from priority in *nature*; or, in other words, cases of priority-of-discovery rather than priority-of-justification. On this view, then, the atomic weight of 197.2 is prior to yellowness, malleability, ductibility, etc., as the sense of 'gold'. It is the predominant sense of 'gold' in that it is the sense that justifies calling something 'gold' rightly. In other words, it is the sense which is relevant to the determination of the truth conditions of sentences in which the word 'gold' occurs in a non-empty way.

I would like to examine this claim, and propose to start by considering the following argument:

The density of water is 1.
Anything whose density is greater than
that of water will sink in water.

Therefore, anything whose density is
greater than 1 will sink in the water
of the Dead Sea. (Compare [18], p. 195.)

The conclusion of course is false, and it is easy to see what has gone wrong. The sense of 'water' in the conclusion is *not* the scientific sense of 'water' as chemically pure. The question arises: if we are to adopt a semantic policy, based on the principle of charity and directing us to take as the predominant sense of a word that sense which maximizes the number of *true sentences* containing this word, then which sense will that be – the scientific or the commonsensical? The answer quite clearly seems to be that if *this* is the test, then the commonsensical sense will emerge as predominant. Now it may be countered by saying that there is no point in adopting a policy that maximizes the number of *true* sentences. Rather, the policy should maximize the number of *interesting* sentences containing the word under consideration. I for my part do not object to this amendment: it seems to me quite right. However, and this is the crux of the matter, I do not see, in the present context, how it is possible to accord to 'interesting' a non-circular account that would support the predominance of the *scientific* sense: to maintain that 'interesting' is, primarily, '*scientifically* interesting' is, I believe, to be guilty of an inappropriate normativism. I'd like to repeat that I concur with Austin's opinion (expressed in the passage quoted on p. 29 above) that it is often *practical* interests that are more important to us in daily life.

12.1. *Hedges and Truth Conditions*

In this section I shall attempt to approach the question of the claim to priority of the scientific sense of a natural-kind word in a somewhat different way.

We are all familiar with such hedge words as 'technically speaking', 'strictly', 'literally', or 'essentially speaking', as well as 'loosely' and 'roughly speaking'. It is my view that, when conjoined with natural-kind words (as in 'this is strictly speaking a cat'), these hedges may be regarded as playing the role of operators which determine the degree of membership in the pertinent natural class. (This evidently presupposes that membership in a natural kind is indeed a matter admitting of degrees, rather than an all-or-nothing affair.) Or again, these operators may be seen as indicating the *distance* of the item in question from some 'central' (i.e. paradigmatic) members of the class – given some geometrical representation of the pertinent natural kind.

Our leading problem may now be formulated thus: Does the scientific sense of a natural-kind word corresponds with the '*strictly* speaking' hedge, or does it rather correspond with the '*technically* speaking' hedge? I shall argue that the latter is the case. I assume that Putnam and Kripke, to the extent that they would accept this way of presenting the issue, would argue that *their* explication of natural-kind words is such that the appropriate hedge-operator to them is the 'strictly speaking' one.

But before arguing my case let me put in a few words of caution and clarification. I am *not* suggesting that their roles as operators can be 'read off' from the ordinary usages of the hedge words I am concerned with. Thus, when one utters 'strictly speaking, he is an idiot' one often uses the hedge as an *intensifier*, meaning to be understood to have said something like 'He is utterly foolish', and *not* meaning to be understood to have used the word 'idiot' in its *scientific* sense, i.e. as referring to one who lacks the ability to develop beyond the mental age of three or four years. (It is when 'idiot' is contrasted with 'imbecile', say, as in 'He is strictly speaking an idiot, not even an imbecile', that the scientific sense is probably referred to.) I am mentioning this in order to draw attention to the fact that there may be differences between the locutionary senses of the hedge words under consideration and their illocutionary senses. Put differently, the pragmatic role of these operators may be at variance with their semantic role. This phenomenon is in fact quite familiar. For instance, it may very well turn out that we often use the expression 'but I *know* it to be so-and-so', or 'I am quite *certain* that such-and-such' precisely when we have some doubts about the matter. Be that as it may, however, I should just like to stress that it is with the *semantic* role of the hedge words that I am concerned, whether or not their pragmatic role converges with it.

Having said all this, let us go back to our problem. Consider the following sentences:

(1)   Technically speaking, a tiger is a cat; but strictly speaking it isn't.

(2)   Strictly speaking ice is not water; but technically speaking it is.

(3)   Technically speaking he is ill, since he has the angina bacilli;
           but strictly speaking he isn't since he shows no symptoms of
           the illness.

It is quite clear that our discussion centers around cases where there is a
divergence between the *typicality conditions* of membership in a natural
kind, as determined by phenotypical features, on the one hand, and the
defining conditions (criteria), usually taken to be determined by genotypical
features, on the other. Following Lakoff [12] I maintain that 'technically
speaking' indicates that the defining criteria hold while the typicality
conditions don't. The use of 'strictly speaking', on the other hand, requires
(or presupposes) that *both* the defining *and* the typicality conditions be met.

As the three examples just cited show, I hold that it is the predominant
sense that is appealed to by the use of the 'strictly speaking' operator and
that the predominant sense is to be equated with the commonsensical
sense. As for the scientific sense, my view is that it contributes to the
precision of a sentence, not to its truth, and, hence, that it is the 'technically
speaking' operator that appeals to it. Once again it is from Austin (see
[7], pp. 167–173) that I borrow this motif of the distinction between truth
and precision. Let me explain.

Consider the pair of sentences:

(4)   She eats a fried egg.
(5)   She eats an egg.

If the first is true, so is the second. The first however is more specific and
more precise (in the sense of being more informative). But it is not more
true than the second. Now ordinary sentences are just as true or as false as
scientific sentences. They are different in that they are usually less precise
than the scientific ones. And like in the case of the meaning of words so
also in the case of the truth of sentences Austin's practical position finds
its expression in the contention that "The statements fit the facts always
more or less loosely, in different ways on different occasions for different
interests and purposes." ([1], pp. 97–98.) I believe that the linguistic
analysis of sentences ought to reflect this contention, at least partially, by
incorporating the hedges into the description of the truth conditions. I
believe, that is, that the 'deep structure' (the 'logical form') of sentences in
ordinary language should contain appropriate hedges, either as operators
on single words or on whole sentences. Thus, to Austin's question whether
it is true that the galaxy is the shape of a fried egg, my answer would be
that it depends which hedge word is represented in the deep structure:
'Roughly speaking, the galaxy is the shape of a fried egg' is true, 'Strictly

speaking, . . .' is not. In a different place Austin questions the truth of the sentence 'Lord Raglan won the battle of Alma.' In my opinion, given that the 'technically speaking' hedge is intended, the sentence is true, since he was at the head of the victorious army. However, it appears that he was quite clearly not the winner, strictly speaking: the *Encyclopedia Britannica* tells us about the battle of Alma that "Generalship was lacking on the allied side, and the victory was attributable to the disciplined nature of the British infantry." (Art. 'Alma'.)

Finally, I should like to make it clear that this appeal to introduce hedges into the logical form of sentences in ordinary language is *not* to be equated with an appeal, à la Lakoff, to introduce hedges as the operators of fuzzy logics. The sentence about the shape of the galaxy is not *roughly true*; rather, *given* that it is in what we might call the roughly-speaking mode, it is *true*.

I believe, then, that Putnam and Kripke's account of the meaning of natural-kind words is not for their uses in the 'strictly speaking' mode of speech but rather for their uses in the 'technically speaking' mode of speech. Had they succeeded in accounting for the 'strictly speaking' mode, they would have won their case, since the strict sense is indeed the predominant sense.

13. MEANING IN COUNTERFACTUALS

The semantic picture of Putnam and Kripke, as has already been shown, is an attempt both to explain the specific contribution of the 'scientific' meaning to the truth conditions of sentences, and to claim that only this meaning is invariant in counterfactual conditionals. However, rather than explain or explicate counterfactuals, Putnam and Kripke rely on them and use them in their attempt to convince us that the 'nominal essences' of a natural-kind object do not convey the meaning of its name. The idea is simple enough. Given any 'surface feature' (say yellowness) it is possible to imagine, counterfactually, that the object in question (gold) lacks it and yet is still called by the same name – but not so when the feature imagined to be lacking is an 'internal', or 'essential' one (having the atomic weight of 197.2).

This technique of discussion appeals to and relies on our intuitions concerning counterfactual states of affairs. I have, elsewhere ([16]), voiced some doubts and reservations about the possibility of arriving at the meaning of words through this technique. I shall not repeat the arguments here: on the contrary, I shall, for the sake of the present argument, assume that this *is* an adequate test for determining meanings. My point here will be that however good this test may be, it still does *not* sort out the 'superficial' features from the 'essential' ones, provided that the former comprise mainly dispositional traits. Consider a given acid. Its internal structure is, say, given by the formula HNO_3. It is given in the appropriate concentration and under the appropriate circumstances for it to turn a litmus paper to red. Now could this solution, in exactly the same conditions, not redden the litmus paper and still be acid (or – and still be rightly called 'acid')? If the answer is that it would *not* be acid, then some of the dispositional characteristics of acid are as essential to it as its internal structure. But then a 'superficial' trait like reddening litmus paper may be sufficient – even according to Putnam and Kripke – for determining the extension of 'acid' and hence its meaning.

Perhaps this suggestion entails some doctrine about necessary causation, but this is not the goods I wish to sell: I refer the interested reader to Harré and Madden ([9]). What I do wish to argue is that the same hypnotical power that is attached to such questions as 'Would it be gold if it did not have the atomic number 79?' is also attached to questions like 'Would it be gold if it were to rust?' There would be no difference in my own readiness to answer these questions – whether positively or negatively; I profess to having no intuitions that would distinguish between them.

A familiar reply to this line of argument is that the superficial dispositional descriptions are but place holders to categorical internal-structure descriptions (i.e., descriptions purged of all dispositional terms). We do, on this view, use such terms as 'soluble', 'brittle', 'irritable', etc., but only in lieu of – and until we are able to apply – what Carnap calls the 'method of structure analysis.' This position, however, is simply untenable. It is just not the case that dispositional terms are disposed of even in the 'deeper' levels of scientific explanations, and there is no reason to expect them to be disposed of in the future science. Thus, Mendeleev's periodic table of the elements essentially involves dispositional terms, since the positions of the elements in the table are determined, among other things, by their dispositions to compound with each other. And so also in the description of the elementary particles, which are distinguished, among other things, by their reactions to various forces applied to them (for example, according to their sensitivity or insensitivity to strong interaction leptons are distinguished from hardons). This applies to many of the other properties as well, such as spin, angular momentum, and more. Dispositional terms, then, are *not* place holders for structural ones. (And even if there is an expectation that they be replaced by terms that are more precise and of more explanatory power it still may turn out that the 'ultimate' terms will be dispositional.)

In any event we are still stuck with the question of what do counterfactual conditionals talk about, according to the commonsense approach to semantics. For Kripke, it will be remembered, the problem of identification across possible worlds is but a pseudo problem: according to him one does not *discover* what 'gold' (say) applies to in these worlds; rather, one *postulates* it. The situation is such that one confronts the pertinent sample, say of bars of gold, and one asks what could happen to *them* in the various possible worlds. That is, one wants to know how different *they* could be and still be bars of gold.

If, on the other hand, one is interested in the *projection* from sample to population, it is *not* possible to stick to one's sample. In this case one does not postulate but rather *discovers* what one talks about.

I, therefore, see the alternative to the Putnam-Kripke position concerning cross-world identification in the view according to which there are different sorts of individuating functions that correspond to a given term. Each of these functions expresses the interest (the 'ideology') for the sake of which the identification is called for. The identification according to 'internal' traits is but one of the various possible functions. I take this to be Hintikka's position [10]. It seems to me that this should be the preferred solution for the semantics of natural language which ought to recognize the variety of classification interests, of which the scientific interest is but one, and not necessarily the most important one.

I am greatly indebted to Edna Ullmann-Margalit who helped me in form and substance.

BIBLIOGRAPHY

[1] Austin, J. L., *Philosophical Papers*, J. O. Urmson and G. J. Warnock (eds.), Oxford 1961.
[2] Bloomfield, L., *Language*, London (1933), 1967.
[3] Bloomfield, L., 'Linguistic Aspects of Science', in *International Encyclopaedia of Unified Science*, vol. 1, O. Neurath, R. Carnap, and C. Morris (eds.) (1938), 1955.
[4] Chomsky, N., *Reflections on Language*, New York 1975.

[5] Davidson, D., 'Truth and Meaning', *Synthese* **17** (1967), 304–323.
[6] *Encyclopaedia Biblica*. Institut Bialik, Hierosolymis, 1965, article 'diamond' (Hebrew).
[7] Furberg, M., *Saying and Meaning: A Main Theme in J. L. Austin's Philosophy*, Oxford (1963), 1971.
[8] Halliday, M. A. K., McIntosh, A., and Strevens, P., 'The Users and Uses of Language', in J. A. Fishman (ed.), *Readings in the Sociology of Language*, The Hague 1970.
[9] Harré, R. and Madden, E. H., *Causal Powers: A Theory of Natural Necessity*, Oxford 1975.
[10] Hintikka, J., 'Semantics for Propositional Attitudes', in Hintikka's *Models for Modalities*, Dordrecht 1969, reprinted in L. Linsky (ed.), *Reference and Modality*, Oxford 1971.
[11] Kripke, S., 'Naming and Necessity', in D. Davidson and G. Harman (eds.), *Semantics of Natural Language*, Dordrecht, Holland, 1972.
[12] Lakoff, G., 'Hedges: A Study of Meaning Criteria and the Logic of Fuzzy Concepts', in *Papers from the 8th Regional Meeting of Chicago Linguistic Society*, 1972.
[13] Leech, G., *Semantics*, London 1974.
[14] Locke, J., *An Essay Concerning Human Understanding*, Book III, New York 1959.
[15] Mackie, J. L., *Problems from Locke*, Oxford 1976.
[16] Margalit, A., 'Open Texture', in A. Margalit (ed.), *Meaning and Use*, Dordrecht, Holland, 1978 (to appear).
[17] Mill, J. S., *A System of Logic*, Book I, Toronto 1937.
[18] Naess, A., *Communication and Argument*, Oslo 1966.
[19] Putnam, H., *Mind, Language and Reality: Philosophical Papers*, vol. 2, Cambridge 1975.
[20] Quine, W. O., *Word and Object*, Cambridge, Mass., 1960.
[21] Quine, W. O., 'Empirically Equivalent Systems of the World', *Erkenntnis* **9** (1975), 313–328.
[22] Tversky, A., 'Features of Similarity', *Psychological Review* **84** (1977), 327–352.
[23] Wittgenstein, L., *The Blue Book*, Oxford 1964.
[24] Wittgenstein, L., *On Certainty*, Oxford 1969.

PATRICK SUPPES

VARIABLE-FREE SEMANTICS FOR NEGATIONS WITH PROSODIC VARIATION*

In several recent publications (Suppes, 1976; Suppes and Macken, 1978) I have argued for a variable-free semantics of quantifiers, attributive adjectives, possessives, and intensifying adverbs. This work is a specialization of my earlier efforts at developing context-free fragments of natural language (1973a, 1973b, 1974). It is also part of my rejection of first-order logic as the appropriate instrument for the analysis of natural language, but I shall not digress here to state in any detail my views on this matter. The central idea is that the syntax of first-order logic is too far removed from that of any natural language, to use it in a sensitive analysis of the meaning of ordinary utterances.

In the present note I restrict myself to negation as it occurs in many simple utterances. I do not claim the analysis is anything like being universally correct, but within the context I have applied it, the intertwining of semantic and prosodic features works out rather well. For other reasons and other interests I have recently become interested in prosody, and I now feel that the ability to incorporate in a faithful way the subtleties of meaning conveyed by various prosodic features is a reasonable semantic demand.

In Section I, I sketch the theory of semantics for generative grammars developed in earlier articles. In Section II, I concentrate on the problems of negation.

I. GENERATIVE GRAMMARS AND THEIR MODEL STRUCTURES

An example will illustrate the framework I have in mind and provide an intuitive introduction to those unfamiliar with the concept of semantic trees for context-free grammars. Consider the tree for the sentence *Some people do not eat some vegetables* shown in Figure 1. On the left of the colon at each node is shown the terminal or nonterminal label. The

49

E. Saarinen, R. Hilpinen, I. Niiniluoto, and M. Provence Hintikka (eds.), Essays in Honour of Jaakko Hintikka, 49–59. All Rights Reserved.
Copyright © 1979 by D. Reidel Publishing Company, Dordrecht, Holland.

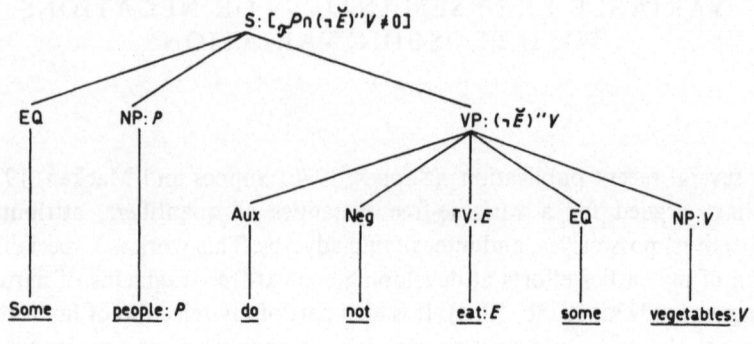

Fig. 1

nonterminal grammatical categories should be obvious: S = sentence, EQ = existential quantifier, NP = noun phrase, Aux = auxiliary, etc. To the right of the colon at a node is shown the denotation of the label if it has one. Thus, in the semantic tree of Figure 1, *people* and its ascendant NP node have the set P of people as denotation; *eat* and its ascendant TV – transitive verb – node have the binary relation E of eating as denotation; *vegetables* and its ascendant NP node have the set V of vegetables as denotation. The VP – verb phrase – node has a denotation composed of set-theoretical operations on E and V. Intuitively the denotation is just the set of people who do not eat some vegetables. The notation $\breve{E}$ is for the *converse* of the relation E; $\neg E$ is the *complement* of the relation $\breve{E}$; and $(\neg\breve{E})''V$ is the *image* of the set V under the relation $\neg\breve{E}$. I take the root of the semantic tree of a sentence to denote the value T or F in a standard Fregean manner. Most of this notation is standard in elementary set theory (see, e.g., Suppes, 1960). Some subtleties about complementation are discussed below.

I give now a quick overview of the relevant formal concepts. First, a structure $G = \langle V, N, P, S\rangle$ is a *phrase-structure grammar* if and only if V and P are finite, nonempty sets, N is a subset of V, S is in N, and $P \subseteq N^* \times V^+$ where N^* is the set of all finite sequences whose terms are elements of N, and V^+ is V^* minus the empty sequence. The grammar G is *context-free* if and only if $P \subseteq N \times V^+$. In the usual terminology, V

is the vocabulary, N is the nonterminal vocabulary, S is the start symbol of derivations or the label of the root of derivation trees of the grammar, and P is the set of production rules. I assume as known the standard definitions of one string of V^* being *G-derivable* from another, the concept of a *derivation tree* of G, and the language $L(G)$ generated by G. (For a detailed treatment of these concepts, see Hopcroft and Ullman, 1969.) A context-free grammar G is *unambiguous* if and only if every terminal string in $L(G)$ has exactly one derivation tree (with respect to G).

Semantics may be introduced in two steps. First the grammar G is extended to a *potentially denoting* grammar by assigning at most one set-theoretical function to each production rule of G. We may show these functions in general by using a notation of square braces; e.g., [NP] is the denotation of NP. In the case of Figure 1,

Production Rule	Semantic Function
$S \rightarrow EQ + NP + VP$	$[S] = [_{\mathscr{F}}[NP] \cap [VP]] \neq 0$
$VP \rightarrow Aux + Neg + TV + EQ + NP$	$[VP] = (\neg[\widetilde{TV}])''[NP],$

where the Frege function $[_{\mathscr{F}}\varphi]$ is defined for any (extensional) sentence φ as follows:

$$[_{\mathscr{F}}\varphi] = \begin{cases} T \text{ if } \varphi \text{ is true (in the given model)} \\ F \text{ otherwise.} \end{cases}$$

And in the case of the other nodes with only one descendant, the semantic function is identity if there is a denotation. For example,

$$TV \rightarrow eat \qquad\qquad [TV] = [eat].$$

The second step is the characterization of model structures. In the general theory of model-theoretic semantics for context-free languages, I use the concept of a *hierarchy* $\mathscr{H}(D)$ of sets built up from a given non-empty domain D by closure under union, subset, and power set 'operations', with T and F excluded from the hierarchy and $T \neq F$. A *model structure* for a given grammar G with terminal vocabulary V_T is a pair $\langle D, v \rangle$ where D is a nonempty set and v is a partial function from $V_T{}^+$ to $\mathscr{H}(D)$. Explicit details are to be found in Suppes (1973a). The treatment here is restricted. First, only terminal words, not terminal phrases, are permitted to denote, so that the domain of the valuation function v is V_T, not $V_T{}^+$.

(The function v remains a partial function because many terminal words – e.g., quantifier words – do not denote.)

The more important restriction is in the hierarchy. In line with my earlier paper (Suppes, 1976), I restrict the model structures to the power set $\mathscr{P}(D)$ of the domain D, i.e., the set of all subsets of D, and the power set of the Cartesian product $D \times D$ – thus, only binary relations are considered. Formally, I define

$$\mathscr{E}(D) = \mathscr{P}(D) \cup P(D \times D),$$

using '$\mathscr{E}$' for extended relation algebras of sets, a terminology introduced in the earlier paper. The valuation function v is then a partial function from V_T to $\mathscr{E}(D)$.

The 'algebraic' operations on elements of $\mathscr{E}(D)$ have mainly already been mentioned: union, intersection, and complementation on arbitrary sets, the converse of relations, the image of a set under a relation: $R''A$. The image of a set under a relation corresponds to function application in Montague grammars. As always, complementation is relative to some given set. From the standpoint of $\mathscr{E}(D)$ the natural set-theoretical choice is $D \cup (D \times D)$, but conceptually this is not very intuitive. For instance, if L is the relation of loving, then $\neg L$ should be the *relation* of not loving, i.e.,

$$\neg L = (D \times D) - L.$$

Consequently, complementation is here taken to mean with respect to $D \times D$ in the case of relations, and with respect to D in the case of sets that are subsets of D. The only point of ambiguity concerns complementation of the empty set or relation, and the context will make clear which is meant.[1]

II. NEGATION

Let the stressed word in a sentence be marked by an accent. Then the following two sentences have closely related meanings in spoken English.

(1)   *John does nót like Mary.*

(2)   *John does not líke Mary.*

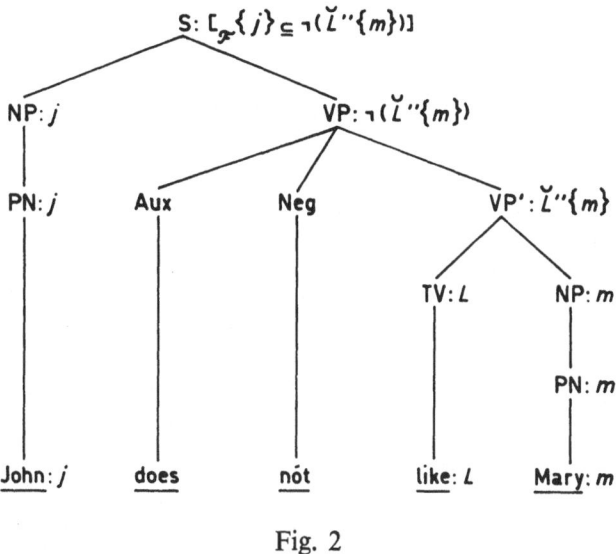

Fig. 2

Yet the semantic trees for (1) and (2) seem different. The critical production rule and the associated semantic function exhibited in Figure 2 is:

(3)   VP → Aux + Neg + VP'   [VP] = ¬[VP'].

The corresponding rule for sentence (2) is close to that used in Figure 1, but for present purposes may be written:

(4)   TV → Aux + Neg + TV'   [TV] = ¬[TV'],

and thus we have the semantic tree as shown in Figure 3. Some readers may be unhappy with the exact form of the production rules (3) and (4). The trees in Figures 2 and 3 might then be thought of as arising from transformations on different phrase-structure rules. This issue is not critical here. Nor is it critical if (3), for example, were replaced by the two rules:

(5) $\begin{cases} \text{VP} \to \text{Neg} + \text{VP}' \\ \text{Neg} \to \text{Aux} + \text{Neg}'. \end{cases}$

As the many types of negation illustrated in Jespersen (1940, Ch. 23) show, there is no point in insisting on the exact form of production rules or of transformations until a very large variety of cases is considered, and my focus in this article is more semantical than syntactical. Admittedly

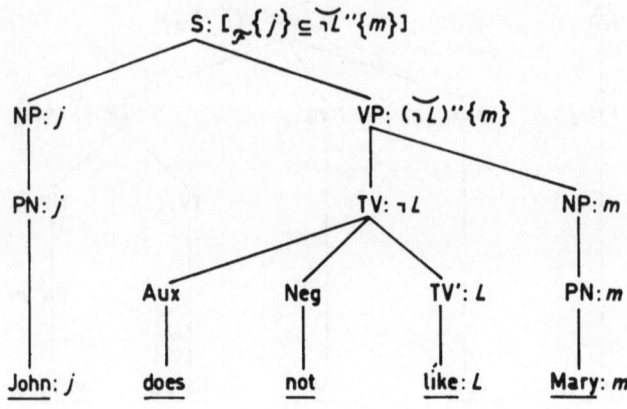

Fig. 3

I have made syntactical choices that facilitate the semantics; this is a strategy that is not fortuitous but one I strongly advocate. The point is that I do not cut much of a syntactical swath in the analysis given here but hope that enough is done to show the virtues of my variable-free semantics in fitting snugly to the surface of some standard ordinary utterances that express negations.

If we look more closely at the meaning of (1) and (2) as expressed in the semantic trees of Figures 2 and 3, we find that they are equivalent in truth-value for all domains, because it is easy to prove for any domain D

(6) $\neg(\breve{L}''\{m\}) = (\overbrace{\neg L})''\{m\}.$

(I also note that

$$\neg(\breve{L}) = (\overbrace{\neg L}),$$

i.e., the order of taking the complement and taking the converse of a relation does not matter.)

Equation (6) would seem to argue that change of stress does not change referential meaning in the use of negation, but the constancy of meaning of (1) and (2) is deceptive. Slightly more complex examples tell a different story. Consider

(7)   *John does nót like some girls.*

(8)   *John does not líke some girls.*

In examining (7) and (8) one criticism of the analysis given above is that the production rules do not seem to be consistent with the general prosodic rule: the last word of an immediate constituent should be stressed. The following six production rules (and their associated semantic functions) do this, and I shall use them in analyzing (7) and (8):

Production Rule[2]	*Semantic Function*
VP → VP′	[VP] = [VP′]
VP → Neg + VP′	[VP] = ¬[VP′]
VP′ → TV + NP	[VP′] = [TV]″[NP]
VP′ → TV + EQ + NP	[VP′] = [TV]″[NP]
TV → Neg + TV̂′	[TV] = ¬[TV′]
Neg → Aux + Nég′	No denotation

The tree for (7) is shown in Figure 4, while the tree for (8) is shown in Figure 5.

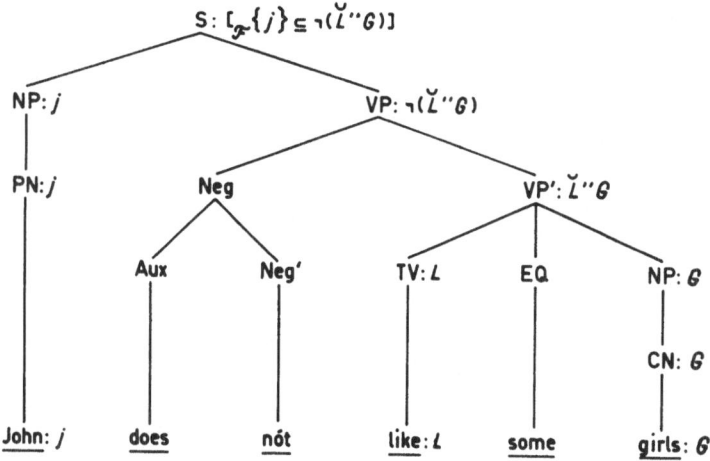

Fig. 4

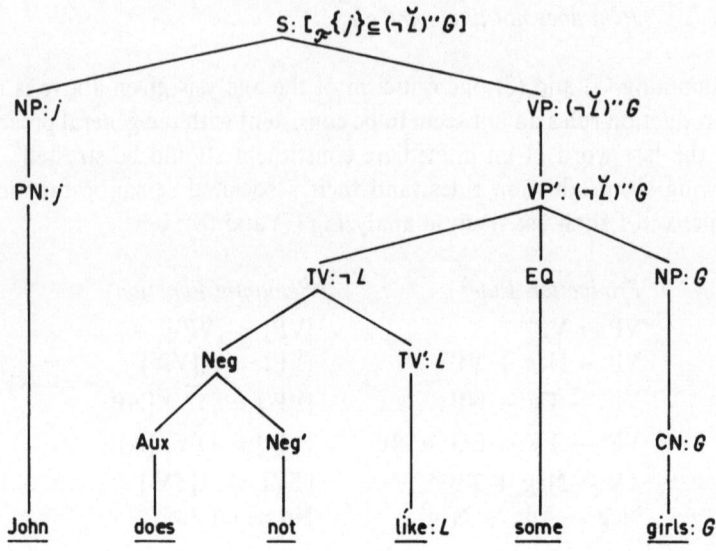

Fig. 5

The following model assigns different truth values to (7) and (8), and thus shows how the difference in stress carries a difference in meaning. The intuitive interpretation of the difference is discussed further below.

Let j = John, m = Mary, and s = Susan, with, of course, all three individuals distinct. So the domain $D = \{j, m, s\}$, and we define G as expected and L in a natural way – so the model is not peculiar:

$$G = \{m, s\}$$
$$L = \{\langle j, m \rangle\}.$$

It is then easy to show that

$$\breve{L}''G = \{j\},$$

so

$$[{}_{\mathscr{F}}\{j\} \subseteq \neg(\breve{L}''G)] = F$$

but

$$(\neg \breve{L})''G = D,$$

so that

$$[_{\mathcal{F}}\{j\} \subseteq (\neg\breve{L})''G] = T,$$

which means that in this model (7) is false and (8) is true.

The intuitive basis of this difference in truth value is brought out by noticing that a somewhat more idiomatic formulation of (7) is:

(7')   *John does nót like girls,*

which is clearly false in the model, since John likes Mary, but (8) is true because he does not like Susan. On the other hand, I emphasize the uniform semantical treatment of the quantifier *some* in both cases – details are to be found in Suppes (1976). The equivalence of (7) and (7') rests on the set-theoretical identity

$$R''A = \bigcup_{a \in A} R''\{a\}$$

for any sets R and A.

Matters are somewhat more complicated in the case of *all*. Corresponding to (7) and (8) consider

(9)   John does nót like all girls.
(10)   John does not líke all girls.

The analysis of *all* in 'object' position in Suppes (1976) is more complicated than it need be. The key to a simpler analysis is to use the set-theoretical analogue of the first-order logic equivalence: $(Ex)\varphi(x)$ iff $\neg(\forall x)\neg\varphi(x)$. Letting UQ stand for universal quantifier, a nonterminal to be rewritten as *all*, we add to the fragment of grammar given above:

$$VP' \to TV + UQ + NP \qquad [VP'] = \neg((\neg[\breve{TV}])''[NP]),$$

where the 'inside' complementation on the right is with respect to $D \times D$ and the 'outside' complementation is with respect to D. Since the trees for (9) and (10) are very similar to those of (7) and (8) respectively, we can go at once to the Frege functions directly. For (9), we have

$$[_{\mathcal{F}}\{j\} \subseteq \neg\neg(\neg\breve{L})''G]$$

and for (10), we have

$$[_{\mathcal{F}}\{j\} \subseteq \neg((\neg\neg\breve{L})''G)],$$

but it is obvious that under the usual rule of 'double negation' for complementation, i.e., for any set A

$$\neg\neg A = A,$$

the Frege function for (9) is the same as that for (8), and the Frege function for (10) is the same as for (7).

In the case of either pair, (7) and (8), or (9) and (10), the difference in meaning within the pair can be seen more clearly by thinking of *not-like*, where the stress is on *like*, as a new transitive verb. So, abstractly, we have:

(8')   *John does* TV *some girls.*
(10')   *John does* TV *all girls.*

In contrast, in the case of (7) or (9), *nót* negates the entire verb phrase, *like some girls* or *like all girls.* Because there is a tendency to insist on the need for presuppositions to analyze explicitly many idiomatic utterances, I emphasize that the semantic distinction made in the present instance requires no use of presuppositions.

The analysis shows how by a prosodic shift in stress we obtain logical equivalence of the existential and universal quantifiers, at least within the restricted framework described. As far as I know, this particular intimate relation between semantics and prosody has not previously been explicitly commented upon. It is a virtue, I believe, of the variable-free semantics I have outlined that the dissection of this relation is relatively simple and straightforward, but the relation itself is justified on independent grounds. Indeed, I consider giving an account of it a reasonable constraint on alternative approaches.

Stanford University

NOTES

* It is a pleasure to contribute this article to the volume dedicated to Jaakko Hintikka on the occasion of his fiftieth birthday. Hintikka has been my part-time colleague at Stanford for many years. I look back with pleasure at our numerous talks about a wide variety of philosophical topics, and especially at our joint seminars, which have ranged from Aristotle to inductive logic. The research reported here has been supported in part by National Science Foundation Grant No. SED77-09698.
[1] For some theoretical purposes it is useful to make the Frege function a Boolean function, and thus to replace the value T by the domain D and the value F by the empty set. This algebraic viewpoint simplifies the argument in Suppes (1976) that quantifier words like *all* and *some* not be part of the syntax of noun phrases, in order to keep the set-theoretical entities that are denoted by words or phrases at the level of elements of $\mathscr{E}(D)$, the extended relation algebra of sets for D.

[2] It is a matter of some controversy to make prosodic features such as primary stress a part of the phrase-structure grammar. Prosodic contours are often added at a later point by many linguists, as Arvin Levine has pointed out to me. For reasons that cannot be gone into here, I much prefer their early introduction on the thesis that the speaker's intention to express a particular meaning should be part of the generation of the utterance from the beginning. Prosodic features that sharply change the meaning, as in some of the examples discussed below, should somehow be marked at an early stage, or at least so I think.

BIBLIOGRAPHY

Hopcroft, J. E. and Ullman, J. D., *Formal Languages and Their Relation to Automata*, Addison-Wesley, New York, 1969.

Jespersen, O., *A Modern English Grammar: On Historical Principles*. V. *Syntax* (vol. 4), Allen and Unwin, London, 1940.

Suppes, P., *Axiomatic Set Theory*, Van Nostrand, New York, 1960.

Suppes, P., 'Semantics of Context-free Fragments of Natural Languages', in K. J. J. Hintikka, J. M. E. Moravcsik and P. Suppes (eds.), *Approaches to Natural Language*, Reidel, Dordrecht, 1973, pp. 370–394. (a)

Suppes, P., 'Congruence of Meaning', *Proceedings and Addresses of the American Philosophical Association* **46** (1973), 21–38. (b)

Suppes, P., 'The Semantics of Children's Language', *American Psychologist* **29** (1974), 103–114.

Suppes, P., 'Elimination of Quantifiers in the Semantics of Natural Language by Use of Extended Relation Algebras', *Revue Internationale de Philosophie* **117–118** (1976), 243–259.

Suppes, P. and Macken, E., 'Steps Toward a Variable-free Semantics of Attributive Adjectives, Possessives, and Intensifying Adverbs', to appear in K. Nelson (ed.), *Children's Language* (vol. 1), Gardner Press, New York, 1978.

LAURI CARLSON AND ALICE TER MEULEN

INFORMATIONAL INDEPENDENCE IN INTENSIONAL CONTEXT

In his paper 'Quantifiers vs. Quantification Theory', Hintikka argues for the existence of partially ordered quantification in English. For instance, he claims that the sentence

(1)   Every writer like some book of his almost as much as every critic dislikes a book he has reviewed

has as one of its possible interpretations that of the partially ordered quantifier sentence

(2) $(x)(Ey)$
        $(z)(Eu)$ $(x$ is a writer $\supset$ (y is a book & x has authored y & (z is a critic $\supset$ (u is a book & z has reviewed u & x likes y almost as much as z dislikes u))))[1]

The force of the partially ordered quantifier prefix is best brought out by a translation of (2) into Skolem function form

(3) $(Ef)(Eg)(x)(z)(x$ is a writer $\supset (f(x)$ is a book & x has authored $f(x))$ & (z is a critic $\supset (g(z)$ is a book & z has reviewed $g(z)$ & x likes $f(x)$ almost as much as z dislikes $g(z)$))))

This reading of (1) is not representable by any first-order sentence (with the same non-logical vocabulary). In particular, it is stronger than any of the possible linear orderings of the four quantifiers of (2) that would extend the partial ordering of (2) to a total one. The branching reading says that each writer has produced at least one book, and on the other hand each critic has reviewed some book or other, which two books the respective men like and dislike to almost the same degree[2]. What is crucial about (2) is that any such book can be found solely on the basis of its writer or reviewer respectively, without anterior information about which (or rather, whose) book it will eventually be compared with.

     We are well aware of the existence of conflicting opinions on Hintikka's thesis. In a recent paper (Hintikka, 1978) a further argument is presented

61

E. Saarinen, R. Hilpinen, I. Niiniluoto, and M. Provence Hintikka (eds.), Essays in Honour of Jaakko Hintikka, 61–72. All Rights Reserved.
Copyright © 1979 by D. Reidel Publishing Company, Dordrecht, Holland.

for the thesis. As we find it particularly persuasive, a slight adaptation of it may be given here. The sentence

(4)   A guy in this city and a girl in every town love each other

implies that in each town there is a girl whom one and the same guy loves. Similarly,

(5)   A guy in every city and a girl in this town love each other

imputes to a particular girl the property of loving several guys in different cities. Now any of the linear representations of

(6)   A guy in every city and a girl in every town love each other

can at most imply one of (4) and (5) only. Either that given any city you can find a guy who has the property described in (4), or that some girl can be found in any town who has the property stated in (5). However, none of the linear readings can capture *both* implications. But presumably both should be implied by (6), if one of them is, given the symmetry of (6) with respect to the simpler sentences (4) and (5). The branching representation does have both implications.

In the remainder of this paper, Hintikka's claim is assumed to hold and our further results are conditional upon it. These results concern the interplay of quantifiers and modalities in sentences that exhibit informational independency. We shall consider a number of sentences with non-linear quantification inside and into an intensional context. It appears that these sentences do not have straightforward translations into any familiar extension of modal logic, including the extension of modal predicate logic with finitely partially ordered quantifiers. The sentences do have representations in fpo. quantification theory with explicit quantification over possible worlds. However, such representations will take us far from the syntactical structure of the English sentences. The sentences therefore present a problem for a translational semantics, in which they are to be represented in an explicit formal language, adhering to strict principles of compositionality (e.g., Montague grammar). In contrast, it will be shown that the sentences are accommodated into the framework of game-theoretical semantics without any additional apparatus.

Our starting point is a wh-question that could be prompted by (1)

(7)   Which book of his does each writer like almost as much as
         every critic dislikes some book he has reviewed?

On Hintikka's theory of the semantics of questions the desideratum – i.e. roughly, the knowledge asked for – is[3]

(8)   I know which book of his each writer likes almost as much as every critic dislikes some book he has reviewed

The main new feature of (8) is that, in virtue of the ordering principles governing the interpretation of English quantifier words, *which* and *each* are understood as *de re* with respect to *know*, whereas *every* and *some* may well be taken as *de dicto* with respect to *know*. In terms of Hintikka's epistemic logic, this means that the first two quantifiers have *know* within their scope, whereas the latter two lie within the scope of *know*[4]. Nevertheless, the informational independency found in (1) is naturally intended to be preserved in (7) and (8). Let us list the putative scope relations between the various operators in (8):

> *which book* is in the scope of *each writer*
> *I know*   –   *which book*
> *every critic* –   *I know*
> *some book* –   *every critic*

Given these scope relations, *some book* should in virtue of the transitivity of scope relations be also in the scope of *each writer*.

However, by arguments analogous to Hintikka's in the aforementioned papers, (8) has also a possible reading where *some book* is taken to be informationally independent of *each writer*. For one thing, (7) seems to be a natural question to ask if someone asserts (1), and therefore it is assumed that the truth of (1) is a presupposition of (7).

It is clear that the informational requirements of (8) cannot be satisfied by any linear ordering of the relevant operators. But what is more problematic is that the required scope relations and independency are not even representable in a combination of modal logic with operators and fpo. quantification theory. For in order to form the desired partially ordered prefix, the quantifiers ought to be brought into a prenex form, but this is blocked by the modal operator. Intuitively we need to be able to prefix the modal operator just to one branch of the fpo. quantifier, so that the quantifiers in the other branch are interpreted as independent of and hence *de re* with respect to the modal operator. However, in order to do this, we would have to change the syntax and semantics of modal operators in a drastic way. The modal operators are then not any longer just sentential operators (objects which prefixed to a formula produce another formula), but they must behave rather like the quantifiers in fpo. quantification theory. In addition we need a way to indicate what part of the formula is affected by the operator, since this does not necessarily coincide with the syntactic position of the operator. We need variable-like modal operators within the formula that are bound by quantifier-like modal operators in the prefix.

We might thus represent (8) as

(9) $(x)(Ey)$ \
         $K_I(z)(Eu)$ / $(x$ is a writer $\supset (y$ is a book & x has authored y & K_I (z is a critic $\supset (u$ is a book & z has reviewed u & x likes y almost as much as z dislikes u))))

We understand the prefixed K_I to bind the later occurrence of K_I in the sense indicated above. The intended meaning of (8) can be brought out by explicit quantification over possible worlds as in

(10) $(x)(Ey)$ \
         $(w)(z)(Eu)$ / $(x$ is a writer at $a \supset (u$ is a book at a & x has authored y at a & $((w$ is an epistemic I-alternative to a & z is a critic at $w) \supset (u$ is a book & z has reviewed u at w & x likes y almost as much as z dislikes u at w))))

Here a is the actual world.

From (10) it is clear that the requirements of the branching reading of (8) are met. The choice of the book for the writer depends only on the writer, whereas the book reviewed by the critic depends on the alternative world. The choice of the book reviewed by the critic remains independent of the particular writer chosen in the actual world.

Of course it is also not problematic to find a second-order representation for (8) using Skolem functions.

(11) $(Ef)(Eg)(w)(x)(z)(x$ is a writer $\supset (f(x)$ is a book & x has authored $f(x)$ & $(w$ is an epistemic I-alternative $\supset (z$ is a critic at $w \supset (u$ is a book at w & z has reviewed u at w & x likes y almost as much as z dislikes u at w)))))

A modification of modal logic along the lines of (9) opens up further problems. For one can strengthen (8) in different ways so as to create further dependencies between the modal operator and the quantifiers.

Consider for instance the following examples

(12)   Every writer knows which book of his he likes almost as much as every critic dislikes a book he has reviewed

and

(13)   Every writer likes a book of his almost as much as every critic may dislike a book he has reviewed

In (12) the modal operator *knows* depends upon the writer in that the epistemic alternativeness relation is a function of the particular writer. In Hintikka's epistemic logic (Hintikka, 1962) this dependency is indicated by indexing the modal operator with a bindable subscript-variable[4]. But (12) cannot be represented by a branching quantifier structure

$$(14) \quad \begin{matrix} (x)(Ey) \\ K_x(z)(Eu) \end{matrix} \Big\rangle \varphi$$

for the subscript x remains unbound in the lower branch. We could, however, represent (12) by a prefix

$$(15) \quad \begin{matrix} (x)(Ey) \\ (v)K_v(z)(Eu) \end{matrix} \Big\rangle ((x = v) \supset \varphi)$$

at the cost of adding one more quantifier and an identity clause in the representation.

Neither can we represent (13) by means of

$$(16) \quad \begin{matrix} (x)(Ey) \\ M(z)(Eu) \end{matrix} \Big\rangle \varphi$$

for the operator M should depend upon the universal quantifier (x). On the other hand, the ordering

$$(17) \quad \begin{matrix} (x)(Ey)M \\ (z)(Eu) \end{matrix} \Big\rangle \varphi$$

will not solve the problem either, since the lower branch will be interpreted *de re* with respect to the modal operator, but the critic and the reviewed book both depend clearly on the interpretation of the modality.

One might think that these examples show that even fpo. quantification theory is not powerful enough to reflect informational independencies between operators in natural language. For partial ordering implies by definition transitivity of scope, but the scope relations required by (8), (12) and (13) appear to exhibit intransitivity.

But again it is not difficult to find partially ordered sentences with explicit quantification over possible worlds that are adequate for representing (12) and (13), just like we did for (8). Such sentences are respectively

(18) $(x)(Ey)$
        $(w)(z)(Eu)$ ⟩ $(x$ is a writer at a ⊃ $(y$ is a book at a & $(w$ is an epistemic x-alternative to a & z is a critic at w ⊃ $(u$ is a book at w & z has reviewed u at w & x likes y almost as much as z dislikes u at $w))))$

(19) $(x)(Ey)(Ew)$
        $(z)(Eu)$ ⟩ $(x$ is a writer at a ⊃ $(y$ is a book at a & x has authored y at a & w is an epistemic alternative to a & $(z$ is a critic at w ⊃ $(u$ is a book at w & z has reviewed u at w & x likes y almost as much as z dislikes u at $w))))$

Moreover, the requirement of transitivity of scope in fpo. quantification theory proves to be vacuous anyway. This follows from results in Walkoe (1970). The only semantically significant part of an ordering of quantifiers is what Walkoe calls their *essential order*. This is the set containing all ordered pairs consisting of a universal quantifier and an existential one, such that the latter is dependent upon the former. In other words, for satisfiability, it suffices to know on which universal quantifier(s) each existential one depends. It is clear that whatever the ordering of the various quantifiers is, the associated essential order is always and vacuously transitive. Walkoe proves that for any ordering of quantifiers there is an equivalent partial ordering, namely its essential order[5].

Furthermore, we know from a result of Enderton[6] that any formula of fpo. quantification theory can be translated into a logically equivalent second-order existential or second-order universal sentence. Hence there are adequate representations of (12) and (13) in second-order logic, using Skolem functions.

(20) $(Ef)(Eg)(w)(x)(z)(x$ is a writer ⊃ $(f(x)$ is a book & x has authored $f(x)$ & $(w$ is an epistemic x-alterna- tive & z is a critic at w ⊃ $(g(z, w)$ is a book at w & z has reviewed $g(z, w)$ at w & x likes $f(x)$ almost as much as z dislikes $g(z, w)$ at $w))))$

represents (12), and (13) is represented by

(21) $(Ef)(Eg)(Eh)(x)(z)(x$ is a writer $\supset$ $(f(x)$ is a book & x has authored $f(x)$ & $g(x)$ is an epistemic alternative & (z is a critic at $g(x)$ $\supset$ $(h(z)$ is a book at $g(x)$ & z has reviewed $h(z)$ at $g(x)$ & x likes $f(x)$ almost as much as z dislikes $h(z)$ at $g(x))))))$

What is shown is that our examples of branching quantification in intensional context do not bring out any new aspects of *informational* independencies. As far as questions of informational independence are concerned, the situation is within limits of fpo. quantification theory.

The source of the trouble with our examples lies rather in the restrictions on the syntax of our formal language. In particular, the notion of scope in modal logic serves several different purposes, which come into conflict in our examples. First of all, scope indicates relations of informational independence. Secondly, the syntactic position of a modal operator marks an argument place, indicating what subformula the modal operator relates to an alternative possible world. These two functions we had to separate in order to capture the informational requirements of (8), (12) and (13). In connection with the latter two we also hit upon a third function: the domain of the interpretation, i.e. the domain of the choice-function connected with each operator or quantifier, may depend on scope relations. For instance, in a sequence consisting of an iterated modality, the second operator introduces alternatives to the worlds over which the first operator ranges. These are not necessarily the same worlds, of course. Similarly, in most treatments of quantified modal logic the range of the quantifier depends on its place in the ordering of the quantifiers and operators. The subscripts of the modal operators bring in a third sense in which different operators may restrict each other's ranges.

In standard modal logic the various functions of scope are connected in a specific way. But this is not at all a conceptual necessity. In fact, there are in the literature on modal logic more indications that the standard conventions of modal logic are not flexible enough to capture some phenomena of natural languages. A case in point are Saarinen's backwards-looking operators[7] that serve among other things to distinguish different functions of scope relations between quantifiers and modal operators. Saarinen shows that there are English sentences that require such an extension of the notion of scope in the logic into which these sentences are to be translated.

In the process of translating our examples to logic, we found it necessary to filter out the informational function of scope relations from the various other functions, which were assembled in the matrix in the form of additional conditions and conjuncts. The disadvantage of this procedure is that it clearly takes us far from the syntactical surface-structure of the English sentences under analysis. The prenex normal form of quantifiers is quite foreign to Engiish. Moreover, in the translations of (8), (12) and (13) to the formulas of fpo. quantification theory or to their corresponding Skolem forms, there is no evident direct correspondence between the syntactic constituents of the English syntax and those of the translations. Any semantical theory that defines exact procedures to translate natural language expressions in some formal language, and adheres to some principle of compositionality will be confronted with an extremely hard case in attempting to account for these sentences. It is as yet far from clear how, if at all, even simple, extensional branching readings of sentences like (1) can be accommodated within a translational framework like Montague grammar[8]. Even if there were an explicit translation procedure for extensional sentences with branching readings within Montague grammar, there would still remain the much more difficult problem of finding an adequate procedure for translating our intensional examples (8), (12) and (13). As we argued, modal verbs can not be translated as operators on sentences but have to be transformed into quantifiers over possible worlds. Otherwise, extensive usage will have to be made of Skolem functions as second-order representations of both quantifiers and modalities. Of course these modifications should be made in such a way that the translation rules do not make any concession to universality of the account of quantifiers. If our claims are correct, it will prove to be problematic for any translational semantics, since the translations cannot be constructed with a one-one mapping onto the syntactic derivation trees of the English sentences, yielding the required branching reading. Obviously, the problem is not that the formal language translated into is not powerful enough – since the intensional logic of Montague's PTQ is of any desired order – but it is rather that the restrictions on the transla-tion procedure make it virtually impossible to construct syntactic trees for surface structures in such a way that the corresponding translations will yield formulas of fpo. quantification theory or the equivalent Skolem forms.

In contrast with these problems for a translational account of our examples, Hintikka's game-theoretical semantics obviates such difficulties by simply skipping the translation stage altogether and directly interpreting the English on its intended models. Let us see how one would treat our examples in game-theoretical semantics. The entire process is rather simple and straightforward. There is no need for any changes in the game-rules or principles as given in Hintikka's papers mentioned in the Bibliography. The game-rules for the interpretation of the original example (1) take care of the interpretation of our examples (8), (12) and (13) as well. The inserted modal moves do not alter the situation significantly. As an example, let us apply the game-rules and principles of Hintikka (1974) to (12). The first rule to apply, according to the principle (O. comm) – a general principle for the order in which quantifiers are to be interpreted – is the rule (G.every) to *every writer*. This gives us

(22)   Henry knows which book of his he likes almost as much as every critic dislikes some book he has reviewed, if Henry is a writer

where *Henry* is the individual chosen by Nature. After detaching the conditional clause, the next rule to apply is (G.know wh)[9]. Myself chooses an individual, say *Sexus*, for *which book*, and the game continues with respect to

(23)   Henry knows that he likes Sexus almost as much as every critic dislikes some book he has reviewed, Sexus is a book and Henry has authored Sexus

After detaching the last two conjuncts, Nature is to choose an epistemic alternative possible world, dependent upon Henry, since the modality is understood as a universal quantification over the epistemic alternatives. There the following sentence is considered

(24)   Henry likes Sexus almost as much as every critic dislikes some book he has reviewed

Now, within this possible world, the next move is Nature's choice of an individual for *every critic*, and the game is continued with

(25)   Henry likes Sexus almost as much as John dislikes some book he has reviewed, if John is a critic

The conditional clause is detached again. Now we come to the crucial move in the game, for here the failure of perfect information enters the picture. We may imagine that Myself here is deputized by an allied player who was not present when the writer was chosen by Nature, so he does not know who was picked or what his name is. This allied Myself-player is to choose a book for the critic, say the book is *Plexus*, then the outcome is

(26)   Henry likes Sexus almost as much as John dislikes Plexus, Plexus is a book and John has reviewed Plexus

Again the last two conjuncts are detached, and we arrive at an atomic sentence that is either true or false according to the states of affairs in that world.

If we consider the strategies Myself has at his disposal, we notice that as far as quantifier moves are concerned, they can be represented by just such Skolem functions as f and g in (20). In other respects the game also yields exactly the same truth-conditions as either (18) or (20).

As is clear from this exposition, the problems which started this paper are due to the syntactical restrictions of the formal languages that are used to capture the meaning of natural languages. As a consequence, serious problems arise for translational semantics, because the required translation cannot be obtained in a straightforward and direct way from the syntactic surface-structure of the English sentence together with its derivation tree.

Game-theoretical semantics, on the contrary, fits the syntactic surface-structure of the English sentences without friction. The interpretive rules apply, according to some general ordering principles, directly to the elements of the sentence and yield the required reading. By obviating the need of an intermediary formal language for the interpretation of natural language, game-theoretical semantics evades the difficult problem of fitting the natural language into the formal framework, or perhaps vice versa. This is not to say that there is no relation between the game-theoretical rules on the one hand and the various parts of the conventional truth-definition for a formal language, on the other hand. It is obvious that some game-rules are direct counterparts in a language of the theory of games of parts of a Tarskian truth definition. But since the game-rules are formulated on constituents of English sentences, it is evident that there are many rules that do not have such a counterpart in the conventional formal languages. It is this fact that makes game-theoretical semantics such a flexible tool for a theory of meaning of natural languages.

Stanford University (A. t. M.)
Academy of Finland (L. C.)

NOTES

[1] We have made some minor changes in the formalization of the matrix of (2), cf. Hintikka (1974), pp. 31–32.

[2] It is interesting to note that an attempt to explicate the intended meaning of (1) in our own words leads us unwittingly to use another form of words which, according to Hintikka, has also an f.p.o. reading. See Hintikka (1974), ex. (46).

[3] See Hintikka (1976), p. 22.

[4] See Hintikka (1962).

[5] See Walkoe (1970), pp. 540–541, Theorems 3.4–3.8 and 4.3.

[6] See Enderton (1970), Theorem 2.

[7] See Saarinen (1977) and (1978). As an example to illustrate the workings of the backwards-looking operators, consider the formula

$$ND_N(Ex)D_{D_N}\varphi$$

Here the N is the modal necessity operator, and D_N, D_{D_N} trace back along the alternativeness relation to the possible world at which the subscripted operator was evaluated. Hence the formula is to be interpreted as: some individual in the actual world satisfies φ in each possible world. One novelty of backwards-looking operators is that they enable us to interpret quantifiers, or, for that matter, operators in general, as informationally independent from an operator in whose scope they are contained. We have to refer the interested reader to Saarinen's publications for a full exposition.

[8] See Gabbay and Moravcsik (1974) and the criticism of that paper by Guenthner and Hoepelman (1975). The promise of the latter to account for branching sentences within Montague grammar remains as yet unfulfilled.

[9] For this rule see Hintikka (1976), p. 116.

BIBLIOGRAPHY

Enderton, Herbert B., 'Finite Partially-Ordered Quantifiers', *Zeitschrift für Mathematische Logik und Grundlagen der Mathematik* **16** (1970), 393–397.

Gabbay, Dov and Moravcsik, Julius, 'Branching Quantifiers, English and Montague-Grammar', *Theoretical Linguistics* **1** (1974), 149–157.

Guenthner, Franz and Hoepelman, Jaap, 'A Note on the Representation of "Branching Quantifiers"', *Theoretical Linguistics* **2** (1975), 285–290.

Hintikka, Jaakko, *Knowledge and Belief*, Cornell University Press, Ithaca, 1962.

Hintikka, Jaakko, 'Quantifiers vs. Quantification Theory', *Linguistic Inquiry* 5 (1974), 153–177.

Hintikka, Jaakko, *The Semantics of Questions and the Questions of Semantics*, Acta Philosophica Fennica vol. 28, no. 4, North-Holland Publishing Company, Amsterdam, 1976.

Hintikka, Jaakko, 'Quantifiers in Natural Language Some Logical Problems, I', in J. Hintikka, I. Niiniluoto and E. Saarinen (eds.), *Essays on Mathematical and Philosophical Logic*, D. Reidel, Dordrecht and Boston, 1978, pp. 295–314.

Montague, Richard, 'The Proper Treatment of Quantification in Ordinary English', reprinted in Thomason, Richmond (ed.), *Formal Philosophy, Selected Papers of R. Montague*, Yale University Press, 1974.

Saarinen, Esa, *Backwards-looking Operators in Intensional Logic and in Philosophical Analysis*, Helsinki University, Ph.D. diss., Helsinki, 1977.

Saarinen, Esa, 'Backwards-Looking Operators in Tense Logic and in Natural Language', in J. Hintikka, I. Niiniluoto and E. Saarinen (eds.), *Essays on Mathematical and Philosophical Logic*, D. Reidel, Dordrecht and Boston, 1978, pp. 341–367.

Walkoe, W. J. Jr., 'Finite Partially-Ordered Quantification', *Journal of Symbolic Logic* 35 (1970), 535–555.

II

MATHEMATICAL LOGIC

DANA SCOTT

A NOTE ON DISTRIBUTIVE NORMAL FORMS

This note should perhaps be called just a 'footnote', since my concern here is in a reformulation of the definition. In a long sequence of papers Hintikka and his coworkers (see the bibliography, which I hope is reasonably complete) have introduced, developed, and applied the idea of this normal form and its *constituents* which are the main ingredient. Usually the description is quite syntactical – since after all these are normal forms of formulae written out in first-order predicate calculus. In the reformulation here the definition will be purely *set theoretical*: the constituents will correspond to certain sets of finite rank ('types' of finite depth) that could be considered quite apart from the usual formal language. However, the translation back to first-order logic is very quick, so not all that much is gained. The exercise of seeing the connection might nevertheless help the reader understand what exactly is being expressed in these normal forms.

It does not seem wrong to say that the idea is independently due to Fraïssé [1955] and Ehrenfeucht [1957] (see also Ehrenfeucht [1961]) whose work was made widely known by the very helpful report in Feferman [1957]. Of course the motivation was quite different, and they did not define normal forms but equivalence relations; Hintikka [1965] makes the situation clear and mentions other work (e.g. Oglesby [1963]). The passage back from equivalences to types is very easy, however; so these are just different aspects of the same basic idea. The recent and readable paper of Flum [1977] should be consulted for connections with model theory and for references to current work. It is too bad that in such a 'canonical' textbook as Chang and Keisler [1973] the notion is relegated to the exercises (pp. 35–6) where it has a good chance of being overlooked. In infinitary logic on the other hand 'back-and-forth' arguments play a bigger rôle than in the model theory of first-order logic. There is no time to discuss this topic here, but the reader may consult the many references given in the bibliography.

The idea was first brought to the attention of the present author by

E. Saarinen, R. Hilpinen, I. Niiniluoto, and M. Provence Hintikka (eds.), Essays in Honour of Jaakko Hintikka, 75–90. All Rights Reserved.
Copyright © 1979 by D. Reidel Publishing Company, Dordrecht, Holland.

William Hanf in conversations in Berkeley in the early 1960's. Hanf had some very original and powerful ideas for proving certain types of theories decidable (and for putting the degree of unsolvability of a first-order theory under control). It took, as I recall, quite a bit of discussion to see that part of his idea just came to the Fraïssé-Ehrenfeucht method. Unfortunately the editors imposed a very strict limitation on size on the papers in the 1963 proceedings 'The Theory of Models', so the nub of Hanf's idea occurs only in a rather condensed form on pp. 141-2 of Hanf [1965]. In Scott [1965] (in the same volume), the definition is presented on p. 340 and reference is made to the work of Karp [1965] (see pp. 408ff.). Thus, this note can also be regarded as a footnote to these 1965 papers.

In Section 1 certain necessary set-theoretical details are explained. The definition itself is given in Section 2, and the basic properties are outlined in Section 3. One application to axiomatic set theory is provided by Section 4, where it is shown that an apparently simple recursive definition is *impossible* because, in view of what we have learned, it would be equivalent to the truth definition.

1. SOME SET-THEORETICAL PRELIMINARIES

The reader is assumed familiar with the usual notions of set theory and model theory. For simplicity a *relational structure* will be taken as a pair $\mathfrak{A} = \langle A, R \rangle$, where A is a non-empty set and $R \subseteq A \times A$ is a *binary* relation. The extension of the definitions to other similarity types will be clear.

Unfortunately, in order to be completely precise, we shall have to recall exactly the definition of *ordered pair*, *relation*, and (*finite*) *sequence*. In the case of *pairs*:

$$(1) \qquad (x, y) = \{\{x\}, \{x, y\}\},$$

as usual. If 0 is the *empty set*, note that $0 \notin (x, y)$ no matter what sets x and y are. Every elementary text proves:

$$(2) \qquad (x, y) = (z, w) \text{ implies } x = z \text{ and } y = w;$$

so (1) is a 'good' (though not unique) definition that reduces pairs to sets. *Cartesian products* are defined by:

(3) $\quad A \times B = \{(x, y) | x \in A \text{ and } y \in B\}$.

Relations are sets of ordered pairs and thus subsets of Cartesian products. As an abbreviation we write 'xRy' for '$(x, y) \in R$' in the usual way.

The *power set* (set of all subsets) is defined by

(4) $\quad \mathbf{P}A = \{X | X \subseteq A\}$;

so, for example, $P(A \times A)$ is the space of all relations defined on A. We shall also want to employ, as a short-hand notation, the idea of the '*proper*' *power set*:

(5) $\quad \mathbb{P}A = \{X \subseteq A | X \neq 0\}$;

that is, the family of all non-empty subsets. We are not concerned here with what *axioms* are required to show that the above statements hold, or that the desired sets exist – all of this is standard in the ZF system.

The *integers* are identified with the finite ordinals: every ordinal *is* the set of all smaller ordinals. The first infinite ordinal is ω, which at the same time is the set of integers. Thus for $m \in \omega$, we can write:

(6) $\quad m = \{0, 1, \ldots, m - 1\}$,

and

(7) $\quad m + 1 = m \cup \{m\}$.

Of course, 0 as an ordinal *is* just the empty set.

A *function* is a relation F such that:

(8) $\quad xFy$ and xFz always imply $y = z$

We write:

(9) $\quad F : A \to B$

to mean that $F \subseteq A \times B$ is a function whose *domain* is all of A (that is, $\forall x \in A \; \exists y \in B \; xFy$). And for $x \in A$, we use the usual *functional notation*: $F(x)$, so that $(x, F(x)) \in F$. The *function space* B^A is the set of all such functions as in (9). In the case of integers $m \in \omega$, the elements of A^m are called *m-termed sequences*. Indeed when $m = 2 = \{0, 1\}$, we find that A^2 and $A \times A$ are in a natural one-one correspondence – but, alas, they are not the *same* sets. So, there is a certain amount of duplication. (It is too bad this cannot be avoided in a simple way!) This duplication requires a separate notation for sequences:

(10) $\quad \langle a_0, a_1, \ldots, a_{m-1} \rangle = \{(0, a_0), (1, a_1), \ldots, (m-1, a_{m-1})\}.$

And we can prove from definition (1):

(11) $\quad \langle a, b \rangle \neq (a, b)$

for all a, b. This does not even require the Axiom of Foundation; however, this axiom would be required to show that, *whatever* the definition of ordered pair, we would have, e.g.:

(12) $\quad \langle 0, 0 \rangle \neq \{\langle 0, 0 \rangle, \langle 1, 0 \rangle\}.$

For finite sequences $a \in A^m$ and $b \in A^k$ we often write a_i for $a(i)$ when $i < m$ (that is, $i \in m$). We will also use the notation:

(13) $\quad |a| = m,$

for the *length* of the sequence a. *Concatenation* of two sequences is defined by:

(14) $\quad a \frown b = \langle a_0, a_1, \ldots, a_{m-1}, b_0, b_1, \ldots, b_{k-1} \rangle.$

Of course, we know that $|a \frown b| = |a| + |b|$ and that definition (14) can be given without the '...' mode of writing. The details are standard. When adding just *one* term onto the end of a sequence, we write for short:

(15) $\quad a * x = a \frown \langle x \rangle.$

The reader should recall that finite sequences of integers are functions like $\pi : m \to k$ and that *permutations* are one-one functions where $m = k$.

The sets of *finite rank* are those with finite transitive closure. Otherwise said, these are the sets generated from 0 by finitely many applications of the binary operation $x, y \mapsto x \cup \{y\}$. Inductively, we can define for $n \in \omega$:

(16) $\quad V_0 = 0; \quad V_{n+1} = \mathbf{P}V_n.$

The union:

(17) $\quad V_\omega = \bigcup_{n \in \omega} V_n$

is the set of all sets of finite rank; while V_n is the set of sets of rank $< n$. We can easily prove:

(18) $\quad V_\omega \times V_\omega \subseteq V_\omega,$ *(closure under pairing)*;

(19) $\quad \bigcup_{m \in \omega} V_\omega^m \subseteq V_\omega$ *(closure under finite sequences)*;

(20) $\displaystyle\bigcup_{y \in V_\omega} y \subseteq V_\omega$ *(closure under membership)*;

(21) $\displaystyle\bigcup_{y \in V_\omega} \mathbf{P}y \subseteq V_\omega$ *(closure under subsets)*;

(22) $\{\mathbf{P}y | y \in V_\omega\} \subseteq V_\omega$ *(closure under powerset)*;

and many other closure properties. For, as is well known, V_ω is a model for all of ZF except for the Axiom of Infinity (indeed, $\omega \notin V_\omega$ even though $\omega \subseteq V_\omega$). In particular V_ω satisfies the Axiom of Foundation because the (finite) ranks are well ordered, and the relation $x \in y$ always implies x has smaller rank than y. Another way to look at V_ω is in terms of the operation $\mathbf{P}_\omega$ of forming the set of all *finite* subsets of a set. We have:

(23) $\qquad V_\omega = \mathbf{P}_\omega V_\omega.$

V_ω is characterized as the *least* solution of Equation (23); and, as is natural, if we assume the Axiom of Foundation overall, V_ω is the *only* solution of (23).

Because the elements of V_ω can be named explicitly in a very simple finite language (using only the symbols: '0', '{', '}', and ','), set construction in V_ω is *recursive*. We can make this explicit by defining a one-one function.

(24) $\qquad \$: \omega \to V_\omega$

as follows. Let $n \in \omega$ be written (uniquely) in binary notation:

(25) $\qquad n = \displaystyle\sum_{i < m} 2^{k_i}$, where $k_0 < k_1 < \ldots < k_{m-1}$

Then we define:

(26) $\qquad \$(n) = \{\$(k_i) | i < m\}.$

Equation (26) is a recursive definition, but it is justified because in (25) the exponents $k_i < n$. In view of (23) it is easy to argue that $\$$ is one-one. It is also easy to see that any number of relationships (for example: $\$(n) \in \$(m)$ or $\$(n) = \mathbf{P}\$(m)$) are recursive. When we say that $A \subseteq V_\omega$ is recursive, we really mean that $\{n \in \omega | \$(n) \in A\}$ is a recursive subset of ω in the usual sense. The inverse function $\# : V_\omega \to \omega$ of course satisfies:

(27) $\qquad \#(x) = \displaystyle\sum_{y \in x} 2^{\#(y)}$

for all $x \in V_\omega$.

2. THE DEFINITION

If $R \subseteq A \times A$ is a relation and $a \in A^m$ is a finite sequence of elements of A, then there is obviously an *induced relation* on the indices of the terms of a:

(1) $R[a] = \{(i, j) \in m \times m \,|\, a_i R a_j\}.$

What occurs on the right-hand side of Equation (1) is a set of finite rank – no matter what the set A is. The only trouble with (1) is that the integer m (note: $m = |a|$) is not uniquely determined. We therefore define for relational systems $\mathfrak{A} = \langle A, R \rangle$:

(2) $\mathfrak{A}[a] = \langle m, R[a] \rangle.$

It would be clear what to do if $\mathfrak{A}$ were of a different similarity type, say $\mathfrak{A} = \langle A, R_0, R_1, \ldots \rangle$ with each $R_i \subseteq A^{k_i}$. Then $R_i[a]$ would have to be redefined so that $R_i[a] \subseteq |a|^{k_i}$ and $\mathfrak{A}[a] = \langle |a|, R_0[a], R_1[a], \ldots \rangle$. As long as the similarity type is finite (a finite number of finitary relations), the induced structure $\mathfrak{A}[a]$ would always be an object in V_ω of finite rank.

If $\mathfrak{A} = \langle A, R \rangle$ and $\mathfrak{B} = \langle B, S \rangle$ are two relational systems, and $a \in A^m$ and $b \in B^k$ are two finite sequences, then we can ask what is the meaning of the equation:

(3) $\mathfrak{A}[a] = \mathfrak{B}[b]$?

Clearly, by definition (2), this implies $m = k$, so the sequences must be of *the same length*. But by (1), it is equally obvious that $a_i R a_j$ holds if and only if $b_i S b_j$. If we introduced the formal language, *then* we could phrase (3) as saying that a and b satisfy *the same atomic formulae* (and hence, *the same quantifier-free formulae*). But there is really no need to bring in the language at this point, for $\mathfrak{A}[a]$ is a finite object containing all the relevant information about quantifier-free formulae. So we can just as well leave the formalism implicit.

Next, given a finite sequence $a \in A^m$ in a relational system $\mathfrak{A} = \langle A, R \rangle$, we can ask how this sequence can be extended from a to $a * x$ (with $x \in A$) or to $a \frown a'$ (with a' also a finite sequence). The totality of possible *types* of extensions gives us information about how a 'sits' in A; whereas $\mathfrak{A}[a]$ only tells us how the a_i relate *among themselves*. To keep this finite, we adopt a step-wise procedure with *types* defined inductively:

(4) $\tau_0^{\mathfrak{A}}[a] = \mathfrak{A}[a];$

(5) $\tau_{n+1}^{\mathfrak{A}}[a] = \{\tau_n^{\mathfrak{A}}[a * x] | x \in A\}.$

If we are interested in the ways we can pass from a to $a \frown a'$ with $a' \in A^k$, then we will look at $\tau_k^{\mathfrak{A}}[a]$ passing through $a * a_0'$, $a * a_0' * a_1'$, ..., $a * a_0' * a_1' * \ldots * a_{k-1}'$. We remark at once that the significance of the equation

(6) $\tau_n^{\mathfrak{A}}[a] = \tau_q^{\mathfrak{B}}[b]$

does not immediately spring to mind, because the definition in (4) and (5) is so highly inductive. We will explain all in the next section, however. We call $\tau_n^{\mathfrak{A}}[a]$ the *n-type* of the sequence a *within* the structure $\mathfrak{A}$. Noting that $0 \in A^0$ is the empty sequence, we see that $\{\tau_n^{\mathfrak{A}}[0] | n \in \omega\}$ forms a sequence of (isomorphism) in variants of $\mathfrak{A}$ *alone*; again the explanation will have to wait of why this is interesting.

To sort out the various types it is useful to classify them into certain levels. We make the classification precise by the definition:

(7) $\mathscr{C}_0^m = \{\langle m, r \rangle | r \subseteq m \times m\};$

(8) $\mathscr{C}_{n+1}^m = \mathbb{P}\mathscr{C}_n^{m+1}.$

If the reader prefers:

(9) $\mathscr{C}_n^m = \mathbb{P}^n \mathscr{C}_0^{m+n},$

where $\mathbb{P}^n$ is the *n*-fold iterate of the operation of forming the family of all non-empty subsets. We perceive at once that for all n, m:

(10) $\mathscr{C}_m^n \in V_\omega,$

and we could even calculate the exact rank of $\mathscr{C}_m^n$ as a recursive function of n and m if this were of interest. What is especially important is the inductive verification of:

(11) $\tau_n^{\mathfrak{A}}[a] \in \mathscr{C}_n^m,$

for all n, m, where $a \in A^m$ and $\mathfrak{A} = \langle A, R \rangle$ is any relational system. To prove this we of course make use of the fact that $A \neq 0$, since, by (5), that is needed to show that $\tau_{n+1}^{\mathfrak{A}}[a]$ is a non-empty set. But otherwise (11) is a direct consequence of the definitions. What is shown by (11) is that given n and m, there are only *finitely many* possible values of $\tau_n^{\mathfrak{A}}[a]$ – even if the set A is infinite.

Following Hintikka, we call the elements of $\mathscr{C}_n^m$ *constituents of depth n*. Since no formal language is involved and since the definitions (4), (5) and (7), (8) are purely set theoretical, the correspondence to Hintikka's original definition may not be so immediately clear; but we shall return to this question. Before doing so, however, it may be useful to illustrate another related definition done in the same style.

In Rantala [1975] the so-called *urn models* are introduced. These are relational systems $\mathfrak{A} = \langle A, R \rangle$ *together with* a sequence of sets $\mathscr{D}_i \subseteq A^i$ such that

(12) $\mathscr{D}_i = \{a \in A^i | a * x \in \mathscr{D}_{i+1} \text{ for some } x \in A\}$

That is, $\mathscr{D}_i$ is the 'projection' of $\mathscr{D}_{i+1}$ along the last coordinate. Rantala sets $\mathscr{D}_0 = A$, which seems irrelevant since $\mathfrak{A}$ already determines A. We shall instead require:

(13) $\mathscr{D}_0 = A^0 = \{0\}$,

which forces all the $\mathscr{D}_i$ to be *non-empty* in view of (12). Writing $\mathscr{D} = \{\mathscr{D}_i | i \in \omega\}$ for short, Rantala's definition (see pp. 463–65 of Rantala [1975]) becomes in our terms:

(14) $\tau_0^{\mathfrak{A},\mathscr{D}}[a] = \mathfrak{A}[a]$;

(15) $\tau_{n+1}^{\mathfrak{A},\mathscr{D}}[a] = \{\tau_n^{\mathfrak{A},\mathscr{D}}[a * x] | a * x \in \mathscr{D}_{m+1}\}$;

where $a \in \mathscr{D}_m$. The effect is to restrict the selection of the extensions $a * x$ ('out of the urn') by a sequence of restrictions *depending* on the previous choice of a. The point of condition (12) is that a selection of x must always be possible, provided $a \in \mathscr{D}_m$ has been following the given selection laws. Again we show inductively that:

(16) $\tau_n^{\mathfrak{A},\mathscr{D}}[a] \in \mathscr{C}_n^m$,

assuming that $a \in \mathscr{D}_m$.

It is not all that clear to me that Rantala has answered, with the introduction of the urn models, the leading question of his paper: "... it has been claimed by certain philosophers that there is an intimate connection between contradiction and change. Can we make any model-theoretical (semantical) sense of such a claim?" (But see the further exposition in Rantala [1977].) Nevertheless the definition is straightforward and natural, and it merits further investigation. There must certainly be some connections with another kind of non-standard semantics: the *chain models* of Karp (see the presentation of Cunningham [1975]), though these were introduced for the purposes of infinitary languages.

3. BASIC PROPERTIES

We prove a sequence of elementary facts that show how the types and constituents behave and how they are connected with first-order logic.

(1)   *The classes $\mathscr{C}_n^m$ are pairwise disjoint.*

Proof: We first have to show that if $C \in \mathscr{C}_n^m$ then $0 \neq C$, $0 \notin C$, and finally $\{0\} \notin C$. Now the first statement is obvious since nothing of the form $\langle m, r \rangle \in \mathscr{C}_0^m$ is ever 0 and, when $n > 0$, $0 \notin \mathscr{C}_n^m$ by definition. But also because $\langle m, r \rangle = \{(0, m), (1, r)\}$, we can never have $0 \notin \langle m, r \rangle$; and, when $n > 0$, we find $0 \notin C$ since otherwise $0 \in \mathscr{C}_{n-1}^{m+1}$. Similarly, since $0 \notin (a, b) = \{\{a\}, \{a, b\}\}$, we see $\{0\} \notin \langle m, r \rangle$; and, when $n > 0$, if $\{0\} \in C \in \mathscr{C}_n^m$, then $\{0\} \in \mathscr{C}_{n-1}^{m+1}$. But, as we have just seen, no such set can contain 0 as an element.

So now for the proof of (1), suppose $C \in \mathscr{C}_n^m \cap \mathscr{C}_q^p$. We take the case $n = 0$ first. Thus $C = \langle m, r \rangle$ for some $r \subseteq m \times m$. If $q = 0$, then $\langle m, r \rangle = \langle p, s \rangle$ (for a suitable s), and $m = p$. So $(m, n) = (p, q)$.

If $g > 0$, then $\langle m, r \rangle \subseteq \mathscr{C}_{q-1}^{p+1}$. But $(0, m) \in \langle m, r \rangle$, so $(0, m) \in \mathscr{C}_{q-1}^{p+1}$. Since all elements of $\mathscr{C}_0^{p+1}$ have the form $\langle p + 1, s \rangle$, we see by (11) of Section 1 that $q - 1 > 0$. Therefore $(0, m) \subseteq \mathscr{C}_{q-2}^{p+2}$. But $(0, m) = \{\{0\}, \{0, m\}\}$ and we just ruled out $\{0\} \in \mathscr{C}_{q-2}^{p+2}$. Thus, this case is impossible, and we are finished with the case $n = 0$. (By symmetry the case $q = 0$ is finished, too.)

Assume now that $n > 0$, and, by way of an induction hypothesis, that we have proved what we want for $n - 1$. Recall $0 \neq C$ and by definition $C \subseteq \mathscr{C}_{n-1}^{m+1}$ and $C \subseteq \mathscr{C}_{q-1}^{p+1}$. But then $\mathscr{C}_{n-1}^{m+1} \cap \mathscr{C}_{q-1}^{p+1} \neq 0$; hence, by hypothesis, $(m + 1, n - 1) = (p + 1, q - 1)$, and so $(m, n) = (p, q)$. □

As a direct corollary, in view of Section 2 (11), we prove:

(2)   *If $\mathfrak{A}, \mathfrak{B}$ are relational structures and a, b finite sequences, then the equation*

$$\tau_n^{\mathfrak{A}}[a] = \tau_q^{\mathfrak{B}}[b]$$
always implies $|a| = |b|$ and $n = q$. □

(And the same holds for the types in the urn models.)

Having shown that the types of different levels cannot be confused, we turn to the questions of whether the equality of types at one level implies the equality at other levels. There are two main results. The first might be called the *selection principle*.

(3)   If $\mathfrak{A} = \langle A, R \rangle$ and $\mathfrak{B} = \langle B, S \rangle$ are relational structures and $a \in A^k$ and $b \in B^k$, and if $\pi : m \to k$ is any mapping on integers, then the equation

$$\tau_n^{\mathfrak{A}}[a] = \tau_n^{\mathfrak{B}}[b]$$

always implies

$$\tau_n^{\mathfrak{A}}[a \circ \pi] = \tau_n^{\mathfrak{B}}[b \circ \pi].$$

Proof: Again we use induction on n. For $n = 0$ if we think of $\tau_n^{\mathfrak{A}}[a] = \langle k, r \rangle$, where

$$r = \{(i, j) \in k \cdot \times k \,|\, a_i R a_j\};$$

then, by the same token $\tau_n^{\mathfrak{A}}[a \circ \pi] = \langle m, r' \rangle$, where

$$r' = \{(i, j) \in m \times m \,|\, a_{\pi(i)} R a_{\pi(j)}\}.$$

But the transformation from r to r' depends *only* on the mapping π:

$$r' = \{(i, j) \in m \times m \,|\, (\pi(i), \pi(j)) \in r\}.$$

Now, since the passage from $\tau_n^{\mathfrak{B}}[b]$ to $\tau_n^{\mathfrak{B}}[b \circ \pi]$ is by the *same* transformation, equality must be preserved.

Next we assume the result for n (and for *all* a, b, π) and proceed to $n + 1$. Suppose $\pi : m \to k$ and $\tau_{n+1}^{\mathfrak{A}}[a] = \tau_{n+1}^{\mathfrak{A}}[b]$. Assume $C \in \tau_{n+1}^{\mathfrak{A}}[a \circ \pi]$. By definition $C = \tau_n^{\mathfrak{A}}[a \circ \pi * x]$ for some $x \in A$. Define $\pi' : m + 1 \to k + 1$, where $\pi'(i) = \pi(i)$ for $i < m$, and $\pi'(m) = k$. Then we can conclude $a \circ \pi * x = (a * x) \circ \pi'$. But $\tau_n^{\mathfrak{A}}[a * x] \in \tau_{n+1}^{\mathfrak{A}}[a]$. So $\tau_n^{\mathfrak{A}}[a * x] \in \tau_{n+1}^{\mathfrak{B}}[b]$. Again by definition $\tau_n^{\mathfrak{A}}[a * x] = \tau_n^{\mathfrak{B}}[b * y]$ for some $y \in B$. By hypothesis $\tau_n^{\mathfrak{A}}[(a * x) \circ \pi'] = \tau_n^{\mathfrak{B}}[(b * y) \circ \pi'] = \tau_n^{\mathfrak{B}}[b \circ \pi * y] \in \tau_{n+1}^{\mathfrak{B}}[b \circ \pi]$. Thus we have proved $\tau_{n+1}^{\mathfrak{A}}[a \circ \pi] \subseteq \tau_{n+1}^{\mathfrak{B}}[b \circ \pi]$. The converse inclusion follows by the same argument, and the case $n + 1$ is established.   □

Statement (3) above includes not only *permutations* (one-one π) but also *deletion* and *repetition* of terms – but all of this is done at the same type level. There is a companion result for passing from higher (better: deeper) types to lower types.

(4)   The equation $\tau_q^{\mathfrak{A}}[a] = \tau_q^{\mathfrak{B}}[b]$ always implies

$$\tau_n^{\mathfrak{A}}[a] = \tau_n^{\mathfrak{B}}[b], \text{ provided } n \leqslant q.$$

Proof: It is sufficient to take the case where $q = n + 1$. Suppose $C \in \tau_{n+1}^{\mathfrak{A}}[a]$. Then $C = \tau_n^{\mathfrak{A}}[a * x]$ for some $x \in A$. But $C \in \tau_{n+1}^{\mathfrak{B}}[b]$ also. Hence, $C = \tau_n^{\mathfrak{B}}[b * y]$ for some $y \in B$. Thus by deletion $\tau_n^{\mathfrak{A}}[a] = \tau_n^{\mathfrak{B}}[b]$ by virtue of (3). □

By set theory we know that the subsets $\Phi \subseteq \mathscr{C}_n^m$ form a *Boolean algebra*. Hence, the predicates (of m-termed sequences) defined by $\tau_n^{\mathfrak{A}}[a] \in \Phi$ obviously are closed under the Boolean operations. For *quantification*, however, we have to increase the depth of the types.

(5)   *If* $\Phi \subseteq \mathscr{C}_n^{m+1}$, *then* $\mathbb{P}\Phi \subseteq \mathscr{C}_{n+1}^m$ *and*
        $\tau_{n+1}^{\mathfrak{A}}[a] \in \mathbb{P}\Phi$ *iff* $\forall x \in A.\ \tau_n^{\mathfrak{A}}[a * x] \in \Phi$.

Proof: That $\mathbb{P}\Phi$ is contained in $\mathscr{C}_{n+1}^m$ follows directly from the definition. The sides of the biconditional are both just rewordings of the statement:

$$\{\tau_n^{\mathfrak{A}}[a * x] \,|\, x \in A\} \subseteq \Phi. \qquad \square$$

The result corresponding to the existential quantifier can be derived from (5) by taking complements. More directly we can argue:

(6) $\tau_{n+1}^{\mathfrak{A}}[a] \cap \Phi \neq 0$ *iff* $\exists x \in A.\ \tau_n^{\mathfrak{A}}[a * x] \in \Phi$.

We could define an operator

$$\mathbb{Q}_n^m : \mathbb{P}\mathscr{C}_n^{+m1} \to \mathbb{P}\mathscr{C}_{n+1}^m$$

by the equation:

$$\mathbb{Q}_n^m \Phi = \{C \in \mathscr{C}_{n+1}^m \,|\, C \cap \Phi \neq 0\}.$$

Then in (6) we could write $\tau_{n+1}^{\mathfrak{A}}[a] \in \mathbb{Q}_n^m \Phi$ on the left-hand side. (In (5) no m, n was needed on $\mathbb{P}$ because *sub*sets are already restricted as to membership.)

Noting that every subset $\Phi \subseteq \mathscr{C}_0^m$ corresponds to a *quantifier-free* formula, and conversely every quantifier free formula φ corresponds to the set of structures $\langle m, r \rangle \in \mathscr{C}_0^m$ which *satisfy* φ with respect to the identity function $a(i) = i$ $(i < m)$ (here the free variables are called: $v_0, v_1, \ldots, v_{m-1}$), then we see from (5) and (6) that every *quantifier* formula Ψ (say, a formula in prenex form with n quantifiers and with $m + n$ variables altogether) corresponds to a set formed from $\Phi \subseteq \mathscr{C}_0^{m+n}$ by applying n times the $\mathbb{P}$, $\mathbb{Q}$-operators giving in the end a set $\Psi \subseteq \mathscr{C}_n^m$. Then: *a sequence* $a \in A^m$ satisfies Ψ in $\mathfrak{A}$ *if and only if* $\tau_n^{\mathfrak{A}}[a] \in \Psi$.

Without going into the (standard) details of the definition of *formula* and *satisfaction* we can argue the converse intuitively as well. Suppose $\Psi \subseteq \mathscr{C}_{n+1}^m$ is any subset of the indicated set of constituents. We ask for the 'meaning' of $\tau_{n+1}^{\mathfrak{A}}[a] \in \Psi$. To write the meaning out formally, we have to take the *disjunction* over all $C \in \Psi$ of the formulae corresponding to the statements $\tau_{n+1}^{\mathfrak{A}}[a] = C$. Since C is a non-empty set of finite rank, suppose

$C = \{C'_0, \ldots, C'_k\}$ and (inductively) that Ψ'_i is the formula (with $m + 1$ free variables) that we have already found to correspond to the statements $\tau^{\mathfrak{B}}_n[a * x] = C'_i$. Thinking what it means for the two sets $\tau^{\mathfrak{A}}_{n+1}[a]$ and C to be equal (and looking back at the definition Section 2 (5)), we write the desired formula (with m free variables $v_0, \ldots, v_{m-1}$) as:

(7) $\forall v_m \bigvee_{i \leqslant k} \Psi'_i \wedge \bigwedge_{i \leqslant k} \exists v_m \Psi'_i].$

The first part says that every element of $\tau^{\mathfrak{A}}_{n+1}[a]$ belongs to C and the second part that the converse holds. But this is *exactly* Hintikka's distributive normal form. Hence we have shown that not only do the subsets $\Psi \subseteq \mathscr{C}^m_n$ represent every first-order definable condition, but also, when they are 'read', the formula that results *is* in normal form. (Thus, not surprisingly, *every* formula has a distributive normal form.) It is also easy (if we had time for more details about first-order languages) to relate the n in $\mathscr{C}^m_n$ to the 'depth' or 'layers' of quantifiers.

In the last section we left unanswered the question of the significance of the equation $\tau^{\mathfrak{A}}_n[a] = \tau^{\mathfrak{B}}_n[b]$, where $\mathfrak{A}$ and $\mathfrak{B}$ are two structures with domains A and B, respectively, and $a \in A^m$ and $b \in B^m$. From what we have just seen, since the equation of course is equivalent to saying that the types belong to the *same* subsets of $\mathscr{C}^m_n$, we can assert that the equation is equivalent to having the two sequences satisfy the *same* formulae of quantifier depth $\leqslant n$. We should also note here that the relation $\tau^{\mathfrak{A}}_n[a] = \tau^{\mathfrak{B}}_n[b]$ can easily be defined independently by the usual back-and-forth idea of Fraïssé-Ehrenfeucht. The only (slight) advantage of the types is that the definition shows that the types are *finite* and it makes (for the author at least) the connection with distributive normal forms seen very simple.

We also wanted to discuss the collection of types $\tau^{\mathfrak{A}}_n[0]$ for $n \in \omega$, where 0 is the empty sequence. (The point of 0 is having *no* free variables.) Thus:

(8) $\{\tau^{\mathfrak{A}}_n[0] | n \in \omega\} = \{\tau^{\mathfrak{B}}_n[0] | n \in \omega\}$ *iff the two structures* $\mathfrak{A}$ *and* $\mathfrak{B}$ *are elementarily equivalent.*

(That is, $\mathfrak{A}$ and $\mathfrak{B}$ should satisfy the same first-order sentences.) Recalling our remarks at the end of the first section, we could also prove:

(9)   *The structure* $\mathfrak{A}$ *has a decidable first-order theory iff* $\{\tau^{\mathfrak{A}}_n[0] | n \in \omega\}$ *is recursive subset of* V_ω.

These are in no way new results, but we emphasize them here because the definitions are really very elementary, purely set theoretical, and done without the aid of any formal syntax. If you like, all syntax has been (rather efficiently) coded up into the sets of finite rank. This is not surprising, but it is somewhat surprising that the definitions (1)–(2) and (4)–(5) of Section 2 contain the whole story of satisfaction of formulae, since in the usual presentations this seems much more complicated.

Hintikka and his coworkers have certainly commented extensively on the question as regards distributive normal forms, but what the author would like to see spelled out is the development of *formal proof* in the same set-theoretical framework. We have defined *truth* and *satisfaction* (in an equivalent form), but what we wish is an easy way of seeing which constituents $C \in \mathscr{C}_n^m$ are *inconsistent* (that is, $C \neq \tau_n^{\mathfrak{A}}[a]$ for all $\mathfrak{A}$ and all a). The collection of such constituents is of course recursively enumerable. The question (which in view of Hintikka's work is hardly a research problem) is *what is the simplest definition of this enumeration.* Of course some definition is possible because we can translate everything into formulae and use the rules of predicate calculus.

Another thing the author would like to see is a similar presentation of the results of Rantala [1975], since he shows that inconsistent constituents can become 'consistent' when we change the notion of model to the so-called urn models. There should be a very direct definition in set-theoretical terms of 'proper' constituents. This seems like an interesting notion that ought to have further applications to some kind of theory of 'finite approximations' to truth of formulae. One would like to see inconsistency in the ordinary sense as the result of 'divergent' sequences of approximations. Perhaps the answer is already contained in the references cited, and the author is unable to recognize the obvious.

4. AN APPLICATION

As we have been so set theoretical throughout, it will not come as a surprise that our application is to (formal) set theory. It is well known that in a system like ZF *recursive definitions can be made explicit.* Usually the definitions concern relations or functions on given sets, and there is not much problem in making them explicit. In the case of a *proper class* like the ordinals, recursive definitions are also desired; and we could, for example, define ordinal addition by the recursion:

(1) $\alpha + \beta = \alpha \cup \{\alpha + \eta \mid \eta < \beta\},$

where we recall that $\alpha = \{\xi \mid \xi < \alpha\}$ and that the less-than relation is thus the same as membership (among ordinals).

Let us try a somewhat similar definition of 'types', where now sequences a will not be restricted to $a \in A^m$ but these will be arbitrary sequences of sets. The given relation will now be replaced by the membership relation. We can write:

(2) $\tau_0[a] = \{(i, j) \in |a| \times |a| \mid a_i \in a_j\},$

(3) $\tau_{n+1}[a] = \{\tau_n[a * x] \mid x \text{ arbitrary}\}.$

It looks like the right-hand side of (3) will give a proper class, but this is not so since we can prove (as before) that $\tau_n[a] \in \mathscr{C}_n^m$, where $m = |a|$. In other words, the types are always sets belonging to the *set* V_ω. But even though *this* is not the trouble, there *is* something wrong with (2)–(3). What?

If, in ZF, there were an explicit definition of $\tau_n[a]$ with *variable n* and a (here n varies over ω and a over *all* finite sequences of sets); then, by what we remarked in the last section, we could define *truth* for set theory within set theory. This contradicts Tarski's theorem. Therefore, there is no explicit definition of $\tau_n[a]$. We could certainly add (2) and (3) as *axioms* (with τ as a new primitive), but this would be an extension of the theory in which the formal consistency of ZF is provable (provided the new symbol is allowed in all axiom schemata). So now we know *why* something is wrong with the recursive definition but still not *what*.

Obviously there is nothing wrong with (2) because this is, as it stands, an explicit definition. Similarly, for any fixed number $0, 1, 2, \ldots, n$ of indices, we could define $\tau_0[a], \tau_1[a], \ldots, \tau_n[a]$ explicitly. The difficulty is with a *variable n*. In (3) to obtain $\tau_{n+1}[a]$ we have to know $\tau_n[a * x]$ for *all* x. Even if in the end we know only a set's worth of information, we have to search through a class's worth of possible contributors to make sure that nothing has been left out. More precisely, define a relation:

(4) $(q, b) \ll (n, a)$ iff $n = q + 1$ and $b = a * x$ for some x.

To know $\tau_{n+1}[a]$ we have to know the value of τ for all $(q, b) \ll (n + 1, a)$.

The point is that $\ll$ is a *well-founded* relation and such relations are good for recursive definitions. On well-founded relations one can do inductive proofs. Thus, by an induction, *if* we had an operation τ satisfying (2)–(3), *then* it is uniquely determined. It is the existence that is the rub.

In the case of $+$ in the ordinals all is well because $\{\eta \mid \eta < \beta\}$ is a *set* (in fact, β). In the case of types $\{(q, b) \mid (q, b) \ll (n + 1, a)\}$ is *never* a set. An analysis of the proof about recursive definitions shows that a *good* well-founded relation must have all initial segments being sets – then the elimination of recursive definitions can be carried out in ZF. That is precisely what we do *not* have with $\ll$. Even though the operation τ of (2)–(3) has a clear intuitive meaning, it simply cannot be defined in ZF.

Merton College, Oxford

BIBLIOGRAPHY

PAPERS ON DISTRIBUTIVE NORMAL FORMS BY JAAKKO HINTIKKA AND COLLABORATORS:

[1953] 'Distributive Normal Forms in the Calculus of Predicates', *Acta Philosophica Fennica* **6** (1953).

[1964] 'Distributive Normal Forms and Deductive Interpolation', *Zeitschrift für Mathematische Logik und Grundlagen der Mathematik* **10** (1964), 185–191.

[1965] 'Distributive Normal Forms in First-Order Logic', in: *Formal Systems and Recursive Functions, Proceedings of the Eighth Logic Colloquium*, Oxford, July 1963 (J. N. Crossley and M. A. E. Dummett, Eds.), North-Holland, Amsterdam, 1965, pp. 47–90. Also reprinted as Chapter XI in: *Logic, Language Games, and Information: Kantian Themes in the Philosophy of Logic*, Oxford University Press, 1973, pp. 242–286.

[1970] 'Surface Information and Depth Information', in: *Information and Inference* (J. Hintikka and P. Suppes, eds), D. Reidel, Dordrecht, 1970, pp. 263–297.

[1970] (with Raimo Tuomela) 'Towards a General Theory of Auxiliary Concepts and Definability in First-Order Theories', in: *loc. cit.*, pp. 298–330.

[1972] 'Constituents and Finite Identifiability', *Journal of Philosophical Logic* **1** (1972), 45–52.

[1973] 'Surface Semantics: Definition and Its Motivation', in: *Truth, Syntax and Modality* (H. Leblanc, ed.), North-Holland, Amsterdam, 1973, pp. 128–147.

[1973] (with Ilkka Niiniluoto) 'On the Surface Semantics of Quantificational Proof Procedures', *Ajatus* (Yearbook of the Philosophical Society of Finland) **35** (1973), 197–215.

[1975] (with Veikko Rantala) 'Systematizing Definability Theory', in: *Proceedings of the third Scandinavian Logic Symposium* (S. Kanger, ed.), North-Holland, Amsterdam, 1975, pp. 40–62.

[1976] (with Veikko Rantala) 'A New Approach to Infinitary Languages', *Annals of Mathematical Logic* **10** (1976), 95–115.

PAPERS BY OTHER AUTHORS

[1955] R. Fraïssé, 'Sur quelques classifications des relations, basées sur des isomorphismes restreints', *Publ. Sci. de l'Universite d'Alger, Serie A (Mathematiques)*,

Part I: **2** (1955), 15–60; Part II: **2** (1955), 273–295.

[1957] A. Ehrenfeucht, 'Applications of Games to some Problems of Mathematical Logic', *Bull. de l'Acad. Pol. Sci.* **5** (1957), 35–37.

[1957] S. Feferman, 'Some Recent Work of Ehrenfeucht and Fraïssé', Summaries of talks presented at the Summer Institute of Symbolic Logic, Cornell (1957), pp. 201–209.

[1961] A. Ehrenfeucht, 'An Application of Games to the Completeness Problem for Formalized Theories', *Fundamenta Mathematicae* **49** (1961), 129–141.

[1963] F. C. Oglesby, 'An Examination of a Decision Procedure', *Memoirs of the Amer. Math. Soc.* **44** (1963).

[1965] W. Hanf, 'Model-Theoretic Methods in the Study of Elementary Logic, in: *The Theory of Models* (J. W. Addison *et al.*, eds), North-Holland, Amsterdam 1965, pp. 132–145.

[1965] C. R. Karp, 'Finite-Quantifier Equivalence', *ibid.*, pp. 407–412.

[1965] D. Scott, 'Logic with Denumerably Long Formulas and· Finite Strings of Quantifiers', *ibid.*, pp. 329–341.

[1973] C. C. Chang and H. J. Keisler, *Model Theory*, North-Holland, Amsterdam 1973.

[1973] V. Rantala, *On the Theory of Definability in First-Order Logic*, Reports from the Institute of Philosophy, University of Helsinki, No. 2, 1973.

[1975] V. Rantala, 'Urn Models: a New Kind of Non-Standard Model for First-Order Logic', *Journal of Philosophical Logic* **4** (1975), 455–474.

[1975] E. Cunningham, 'Chain Models: Applications of Consistency Properties and Back-and-Forth Techniques in Infinite-Quantifier Languages', in: *Infinitary Logic: in Memoriam Carol Karp* (D. W. Kueker, ed.), *Lecture Notes in Mathematics*, Springer-Verlag **492** (1975), 125–142.

[1977] J. Flum, 'Distributive Normal Forms', in: *Proceedings of the Symposiums on Mathematical Logic in Oulu 1974 and in Helsinki 1975* (S. Miettinen and J. Väänänen, eds), Helsinki 1977, pp. 71–76.

[1977] V. Rantala, 'Aspects of Definability', *Acta Philosophica Fennica* **29**, nos. 2–3 (1977).

JENS ERIK FENSTAD

ON THE METAPHYSICS OF THE REAL LINE

In his somewhat unorthodox but stimulating collection of papers on the philosophy of mathematics [6] Jaakko Hintikka has included a paper on non-standard analysis, '*The Metaphysics of the Calculus*', by Abraham Robinson, the creator of that subject.

Hintikka makes the following comment on this paper ([6], p. 3):

The inevitable presence of non-standard models can occasionally be turned into a blessing, however, as demonstrated by Abraham Robinson's non-standard models for analysis. These models serve to vindicate some of the locutions and ideas of the old 'metaphysics of the calculus' which meanwhile have been relegated to the status of hopelessly loose heuristic ideas. Even infinitesimals, the bugbear of every introductory calculus course, can thus be made perfectly respectable. Here is one of several recent developments that demonstrates strikingly the relevance of current work to the traditional issues in the philosophy of mathematics.

We shall elaborate on this remark. Already the many and varied contributions of Abraham Robinson demonstrated convincingly that the non-standard method not only gives us a useful new technique in mathematics, but is also – and this is a direct heritage of the heroic century of the calculus – a powerful source of intuition.

Techniques and intuitions are important enough, but we shall venture one step beyond and propose, perhaps seriously, that non-standard methods may suggest a new idea of the real line. And this will be our 'metaphysics': the real line is not a point set.

1. A COMMUTATIVE DIAGRAM: THE METHOD EXPLAINED

Excellent introductions to non-standard analysis exist, the reader may consult Stroyan and Luxemburg [11] or M. Davis [4]. He should also not fail to look up the recent address of E. Nelson [9].

What we have learned is that the complete ordered field $\mathbb{R}$ of *real numbers* has a (not unique) proper ordered field extension $*\mathbb{R}$, the *hyper-*

E. Saarinen, R. Hilpinen, I. Niiniluoto, and M. Provence Hintikka (eds.), Essays in Honour of Jaakko Hintikka, 91–99. All Rights Reserved.

reals. And every function f or relation S over $\mathbb{R}$ has a natural extension *f, *S over *$\mathbb{R}$.

The strength of the method lies in the fact that the imbedding *: $\mathbb{R} \to$ *$\mathbb{R}$ is elementary, i.e. if ϕ is any sentence in the language of $\mathbb{R}$, then ϕ is true in $\mathbb{R}$ iff it is true in *$\mathbb{R}$ (the *transfer principle*).

This will suffice for our exposition. We shall forego that in the extension *$\mathbb{R}$ we have carefully to distinguish between *standard, internal,* and *external* objects, and that the transfer principle is not the only general principle in play.

To illustrate, and also for purposes of later reference, we shall give a simple – and very standard example of a non-standard characterization of a familiar concept.

A sequence $\{s_n | n \in \mathbb{N}\}$ is a map $s: \mathbb{N} \to \mathbb{R}$, and as such has a natural extension *$s:$ *$\mathbb{N} \to$ *$\mathbb{R}$. It is immediate that *$\mathbb{N} \backslash \mathbb{N} \neq \varnothing$; we also write

$$s_\omega = {}^*s(\omega)$$

for $\omega \in {}^*\mathbb{N} \backslash \mathbb{N}$.

Let us recall a few concepts concerning *$\mathbb{R}$: An element $x \in {}^*\mathbb{R}$ is called *infinitesimal* if $|x| < r$, for all $r \in \mathbb{R}^+$. x is called *finite* if $|x| < r$, for some $r \in \mathbb{R}^+$, otherwise *infinite*. We write $x \approx y$ if $|x - y|$ is infinitesimal.

Every finite $x \in {}^*\mathbb{R}$ is infinitely close to some (unique) $r \in \mathbb{R}$, called the *standard part* of x; in symbols, $r = \mathrm{st}(x)$.

Example. The sequence $\{s_n | n \in \mathbb{N}\}$ converges iff $s_\omega \approx s_\lambda$ for all $\omega, \lambda \in \mathbb{N} \backslash \mathbb{N}$.

And in this case $\lim s_n = \mathrm{st}(s_\omega)$, for any $\omega \in {}^*\mathbb{N} \backslash \mathbb{N}$.

The proof is simple, but reveals some of the characteristic aspects of the method. If the sequence converges, then $|s_\omega - s_\lambda| < \varepsilon$ for all infinite ω and λ, for this inequality will hold from a certain standard n_ε. Hence, $s_\omega \approx s_\lambda$.

Conversely, given $s_\omega \approx s_\lambda$ for all infinite ω, λ. Then if ε is specified, the following sentence is true in *$\mathbb{R}$:

$$\exists y \in N \forall z, w \in N[z, w > y \to |s_z - s_w| < \varepsilon]$$

By transfer the sentence is also true in $\mathbb{R}$, i.e. the sequence converges.

Let us look at the situation in more generality. Given a mathematical structure E we know of two useful extensions, the completion $\hat{E}$, and the non-standard extension *E. In *E we can distinguish a set of *bounded* (prenearstandard) points, B_E (which is defined using the uniformity or norm on E, see [5] for details). We have the following diagram

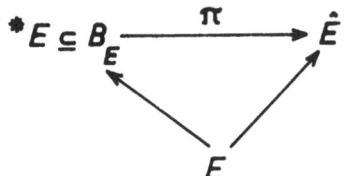

There is a map $\pi : B_E \to \hat{E}$ (onto, but not 1-1) such that whenever F is complete and $f : E \to F$ has an extension to a map $\hat{f} : \hat{E} \to F$, then

$$\text{st}(*f(x)) = \hat{f}(\pi(x))$$

for all $x \in B_E$. This diagram explains the successful use in many situations of non-standard methods.

Let us return to the sequence example. On $\mathbb{N}$ consider the uniformity giving the one-point compactification as its completion, $\hat{\mathbb{N}} = \mathbb{N} \cup \{\infty\}$.

Note that a map $s : \mathbb{N} \to \mathbb{R}$ is uniformly continuous with respect to this uniformity iff it can be extended to a map $\hat{s} : \hat{\mathbb{N}} \to \mathbb{R}$ iff $\lim s(n)$ exists, i.e. the sequence s_n converges. And if this is so, $\lim s(n) = \hat{s}(\infty)$.

Let $*\mathbb{N}$ be a non-standard extension, in this case $B_{\mathbb{N}} = *\mathbb{N}$. Looking back to the diagram we see that $\pi(\omega) = \infty$, for all $\omega \in *\mathbb{N} \backslash \mathbb{N}$. Hence

$$\begin{aligned}
\text{st}(s_\omega) = \text{st}(*s(\omega)) &= \hat{s}(\pi(\omega)) \\
&= \hat{s}(\infty) = \lim s(n).
\end{aligned}$$

And we have convergence iff $s_\omega \approx s_\lambda$, for all infinite ω, λ.

This example is perhaps a curiosity. But what is trivial in a simple setting may yet be important in general. Mathematical structures E have in general both a topologic and an algebraic structure. By the transfer principle the algebra extends to $*E$. Let $E = \mathbb{Q}$, then $*\mathbb{Q}$ is a field in which the finite elements, $\mathbb{Q}_f$, is a subring containing the infinitesimals, $\mathbb{Q}_i$, as a maximal ideal. Hence $\mathbb{Q}_f / \mathbb{Q}_i$ is a field, in fact, it is $\mathbb{R}$.

And, what is well known, the algebra of E does not always extend to $\hat{E}$. The map $\pi : *E \to \hat{E}$ 'confuses' points that from the algebraic point of view should be kept distinct. It is this richness of $*E$ which adds power and intuition to the non-standard method. We give two examples.

Distributions [11]. We work in $*C^\infty(\mathbb{R})$. We thus need no special theory of distributions, but construct the functions we need – e.g. the δ-function – as $*C^\infty$ functions:

Let

$$c(t) = \begin{cases} \exp\left(-1/(1 - |t|^2)\right), & |t| < 1, \\ 0, & \text{otherwise,} \end{cases}$$

i.e. $c(t)$ is a so-called Cauchy 'flat' function.

Let ε be a positive infinitesimal and define $d(t) = c(\frac{t}{\varepsilon})$. Note that d is $*C^\infty$ and has $[-\varepsilon, \varepsilon]$ as carrier. Let $k = \int_{-\infty}^{+\infty} d(t)dt$. Our δ-function will then be

$$\delta_\varepsilon(t) = \frac{1}{k} \cdot d(t) = \frac{1}{k} \cdot c\frac{(t)}{\varepsilon}.$$

We see that δ is positive and in $*C^\infty$. Furthermore, $\int_{-\infty}^{+\infty} \delta(t)dt = 1$.

We shall later on need the following result

LEMMA [11]. Let μ be a *-Borel measure with a finite total variation on a standard bounded interval. The δ-mollification of μ

$$\delta_\varepsilon * \mu(x) = \int_{-\infty}^{+\infty} \delta(x - t)d\mu(t)$$

is a finite distribution (i.e. a $*C^\infty$ function) and for all appropriate φ

$$\langle \varphi, \delta * d \rangle \approx \int_{-\infty}^{+\infty} \varphi(t)d\mu(t)$$

Brownian motion [2]. There is a general measure-theoretic construction due to P. Loeb [8]. Let $\langle X, \mathscr{A}, \nu \rangle$ be an (internal) probability space, i.e. ν is finitely additive (hence *-finite, but not σ-additive). The standard part of ν, st(ν), can be extended to a σ-additive measure $L(\nu)$ on the completion $L(\mathscr{A})$ of the smallest σ-algebra containing $\mathscr{A}$.

This gives us a standard measure space $\langle X, L(\mathscr{A}), L(\nu) \rangle$, however, in many cases on some unusual underlying space. Brownian motion as a *-finite random walk gives an example. This is due to R. Anderson [2].

Let $\Omega = \{-1, 1\}^\eta$, $\eta \in *\mathbb{N} \backslash \mathbb{N}$. Let $\mathscr{A}$ be the algebra of all internal subsets of Ω, and for $A \in \mathscr{A}$, set

$$\nu(A) = \frac{|A|}{2^\eta}$$

where $|A|$ is the (internal) cardinality of A, i.e. we have the 'usual' counting measure.

On the standard space $\langle \Omega, L(\mathscr{A}), L(\nu) \rangle$ we introduce a 'finite' random walk

$$\chi(t, \, \omega) = \eta^{-1/2} \sum_{k=1}^{[\eta t]} \omega_k + \text{'linear interpolation'.}$$

If $\beta(t, \, \omega)$ is the standard part of $\chi(t, \, \omega)$, then β is a standard Brownian motion – a fact which is almost trivial to verify from the explicit construction of $\chi(t, \, \omega)$. And stochastic integration, Ito's lemma, flows in an equally simple and intuitive way.

The reader should observe that in both examples we have used the algebraic richness of the non-standard extension.

2. SINGULAR PERTURBATIONS

We shall illustrate the method by giving some results from a forthcoming paper, *Singular perturbations and non-standard analysis*, [1].

The first one stems from a problem in quantum mechanics and was first discussed in a non-standard setting by Nelson [9]. He introduces the problem in the following way ([9], p. 1186):

Let Δ be the Laplace operator on $L^2(\mathbb{R}^n)$ with the usual domain which makes it a selfadjoint operator and let $H_0 = -\Delta$.

The one-parameter unitary group $t \rightarrow e^{-it \, H_0}$ describes the motion of a free Schrödinger particle in $\mathbb{R}^n$. The question is whether a Schrödinger particle can feel the effect of a force concentrated at a single point, the origin. Let V be a bounded real measurable function (whose bound may be unlimited [i.e. infinite]) which vanishes outside an infinitesimal neighborhood of the origin. We will also use the letters V to denote the bounded selfadjoint operator of multiplication by the function V. Then $H = H_0 + V$ is also a selfadjoint operator and $t \rightarrow e^{-it \, H}$ is a one-parameter unitary group describing the motion of a Schrödinger particle which feels the effect of a force concentrated in an infinitesimal neighborhood of the origin. However, it may be near-trivial.

Indeed it will be if $n \geq 4$. If $n = 3$, interesting things start to happen. But first we have to make our statements precise. We introduce a topology on the set of all self-adjoint operators by requiring that a net H_α converges to H iff $e^{-it \, H_\alpha}\psi \rightarrow e^{-it \, H}\psi$ for all ψ in the underlying Hilbert-space and all reals t. Having a notion of topology, we have the notion of the *monad* of a standard point. Hence we know what it means for an operator H to be infinitely close to the standard operator H_0.

We have the following result which gives a complete description for $n = 3$.

Consider the family of operators

$$H_\alpha = H_0 + \lambda_\varepsilon(\alpha) \cdot \chi_\varepsilon$$

on the non-standard Hilbertspace $L^2(\mathbb{R}^3)$, where

$$\lambda_\varepsilon(\alpha) = -(k + \tfrac{1}{2})^2 \cdot \frac{\pi^2}{\varepsilon^2} + \frac{\pi}{\varepsilon} \cdot \alpha + \beta,$$

with α, $\beta \in \mathbb{R}$, $k \in \mathbb{N}$, ε a fixed positive infinitesimal, and χ_ε the charac-teristic function of the ball $|x| \leq \varepsilon$ in $*\mathbb{R}^3$. As α runs through $\mathbb{R}$ we get a parametrization of *all* self-adjoint extensions of the restriction of H_0 to $C_0^\infty(\mathbb{R}^3 \setminus \{0\})$.

In [1] there is also a discussion of perturbations of the Laplacian of the form

$$-\Delta - \lambda \cdot \sum_{i=1}^n \delta(x - x_i),$$

where $\delta(x - x_i)$ are the δ-measures at the points x_i. Of course, classically this is but a formal expression. Nonstandardly it makes perfectly good sense.

Sturm-Liouville theory is behind these applications. And the above study lead in [1] to a discussion of certain singular perturbations in the general Sturm-Liouville theory. Here we have an operator

$$A(u) = -(pu')' + qu$$

on the standard Hilbertspace $L^2([a, b])$, where p, q are smooth real functions, $p \geq \varepsilon > 0$ and $q \geq 0$. The Sturm-Liouville theory tells us that A has a discrete spectrum $0 < \lambda_0 < \lambda_1 < \ldots$ such that

$$\sum_{i=0}^\infty \lambda_i^{-1} < \infty.$$

Moreover, if $Au_n = \lambda_n u_n$, then u_n has $n + 2$ zeros in $[a, b]$ and the zeros of u_{n+1} separates the zeros of u_n.

There is an interest in the operator A for certain 'singular' functions q. But the problem has been to make sense of the operator when singularities are introduced. We offer one example [1].

Let μ be a bounded positive Borelmeasure on $[a, b]$. Let δ_ε be a delta-function, and consider the δ-mollification

$$q = \delta_\varepsilon * \mu.$$

q is a smooth (i.e. $*C^\infty$) function in $*L^2([a, b])$, and we have in this space a nice (non-standard) operator to which we by transfer can apply the usual Sturm-Liouville theory.

But this may only be a curiosity unless one can prove that the operator (in an appropriate sense) is nearstandard. And this can be done, and hence we obtain an extension of the usual Sturm-Liouville theory to include certain forms of singularities (see [1] for precise statements and details).

Notice at this point an analogy with applications of distribution theory. In both cases we have a given mathematical structure E, in both cases we introduce 'ideal' elements, in non-standard theory an extension $*E$, in distribution theory a completion $\hat{E}$. And in both cases we have to 'pull our results back' in the form of a regularity argument; in non-standard theory we often have to verify that the end product of a construction is near-standard, in distribution theory that a solution is actually a function. And this part of the argument is typically the difficult part. The general theory gives a systematic and convenient setting, which, however, is of no little importance.

But the examples above also point to the difference between the methods. There is an extra algebraic and combinatorial richness in the non-standard approach which is lost in the distribution-theoretic framework. Sometimes, but not necessarily always, this extra richness can be put to good use!

3. A PHILOSOPHIC POSTSCRIPT

We made a suggestion in the introduction: the real line is not a point-set.

Take any axiomatization of elementary geometry, e.g. the paper by A. Tarski in Hintikka [6], or the axioms for the affine plane in Artin's book [3].

In Artin's exposition there are two basic categories of objects, *lines* and *points*. Any point lies on a line, but a line is not a point-set. Axioms with a 'true' geometric content allow us to introduce coordinates from a field k. But only the *ad hoc* Archimedian axiom tells us that the field of an ordered geometry is isomorphic to a subfield of the reals.

It is thus consistent with geometric intuitions, and it may even be true, that the 'real' line is not a point-set, but supports a point-set which could be richer than the set of reals, viz. the set of hyperreals.

And since the real line is not a point-set, there are no grounds for assuming that there is a unique enrichment of the reals on the line. New points can be created. (Note here a weak analogy to constructive mathematics, the constructive continuum is not the set of constructive points.)

In any particular argument we can therefore choose an extension *ℝ of ℝ on the real line. And we have a freedom of choice, different *ℝ can be created for different purposes. Examples exist in current non-standard practice. The method is – we claim – powerful and a rich course for intuitive insight.

Is this necessary? Not in a certain sense. G. Kreisel has pointed out in [7] that current axiomatizations of non-standard analysis are conservative extensions of formal systems for classical standard analysis; see also Nelson [9]. But in a similar sense, since the reals are constructed out of the rationals, the real numbers are also not necessary. In fact, only the set of natural numbers remains together with the usual construction principles of set theory. And this is the traditional cantorian view: Every argument in mathematics is an argument in axiomatic set theory.

Contra this we can insist that the real line is not a point-set.

A final comment should be added. Different extensions *ℝ of ℝ can be created. This is a first degree of freedom. But recent research on the foundation of set theory tells us that ℝ itself is not unique. This is a second degree of freedom from which we cannot escape. Add e.g. the axiom of constructibility and we get a minimal version of ℝ. Add other axioms and non-constructible reals are forced on us. But the standard natural numbers and the geometric real line remain the only objects determined by our intuition.

University of Oslo

BIBLIOGRAPHY

[1] Albeverio, S., Fenstad, J. E. and Høegh-Krohn, R., 'Singular Perturbations and Nonstandard Analysis', to appear.
[2] Anderson, R. M., 'A Non-Standard Representation for Brownian Motion and Ito Integration', *Israel J. Math.* **25** (1976), 15–46.
[3] Artin, E., *Geometric Algebra*, Interscience, New York 1957.
[4] Davis, M., *Applied Nonstandard Analysis*, J. Wiley & Sons, New York, 1977.
[5] Fenstad, J. E. and Nyberg, A., 'Standard and Nonstandard Methods in Uniform Topology', in Gandy and Yates (eds.), *Logic Colloquium '69*, North-Holland, Amsterdam, 1971, pp. 353–359.

[6] Hintikka, J. (ed.), *The Philosophy of Mathematics*, Oxford University Press, Oxford 1969.
[7] Kreisel, G., 'Axiomatizations of Non-Standard Analysis Which Are Conservative Extensions of Formal Systems for Classical Standard Analysis', in W. A. J. Luxemburg (ed.), *Applications of Model Theory to Algebra, Analysis and Probability*, Holt, New York, 1969.
[8] Loeb, P. A., 'Conversion from Non-Standard to Standard Measure Spaces and Applications in Probability Theory', *Trans. Amer. Math. Soc.* **221** (1975), 113–122.
[9] Nelson, E., 'Internal Set Theory: a New Approach to Non-Standard Analysis', *Bull. Amer. Math. Soc.* **83** (1977), 1165–1198.
[10] Robinson, A., 'The Metaphysics of the Calculus', in Hintikka [6], pp. 153–163.
[11] Stroyan, K. D. and Luxemburg, W. A. J., *Introduction to the Theory of Infinitesimals*, Academic Press, New York, 1976.
[12] Tarski, A., 'What is Elementary Geometry?', in Hintikka [6], pp. 164–177.

JUHA OIKKONEN

A GENERALIZATION OF THE INFINITELY DEEP LANGUAGES OF HINTIKKA AND RANTALA

Jaakko Hintikka and Veikko Rantala define in [2] infinitary languages in which sentences may be of infinite depth. Recall that the syntactic structure of the sentences of ordinary finitary as well as infinitary logic may be described in terms of certain trees. For instance the sentence $\forall x \exists y (R(x, y) \lor S(x))$ can be described by the tree:

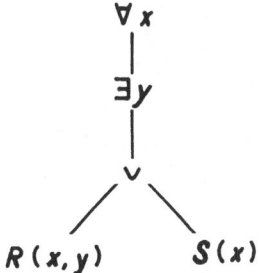

All these trees have only finite branches. This follows from the inductive definition of the notion of a formula. The same property is a necessary requirement for the usual inductive definition of satisfaction. But if satisfaction is defined in game-theoretical terms, then the finiteness of the branches of the syntactical trees is needed no more.

On the other hand one of the basic notions in the theory of Hintikka's distributive normal forms for predicate calculus is that of a constituent. In [6] Veikko Rantala noticed that in order to be able to generalize constituents for infinitary languages, one needs new kinds of formal languages where the syntactical trees of sentences may have infinite branches.

These two remarks led Hintikka and Rantala to define in [2] infinitely deep languages $N_{\kappa\alpha}$, where the sentences are defined in terms of certain (well-founded) trees, which may have infinite branches, and where the satisfaction relation is defined game-theoretically. The new languages are studied besides [2] in Karttunen [3], Rantala [7], and in Oikkonen [5]. For example, [3] gives a back-and-forth characterization for elementary equivalence in these languages.

101

E. Saarinen, R. Hilpinen, I. Niiniluoto, and M. Provence Hintikka (eds.), Essays in Honour of Jaakko Hintikka, 101–112. All Rights Reserved.

The syntactic trees of the infinitely deep languages may have branches with no last element. This fact leads in [2] to several different possible satisfaction relations. In [5] we discussed the relations between infinitely deep languages with several different (abstractly described) satisfaction relations, certain game-quantifier languages, and ordinary infinitary languages with infinitary extra predicates. We showed that, in a sense, all of these have the same expressive power.

In [4] Keisler considers general linearly ordered quantifiers. The aim of the present paper is to consider a notion of infinitely deep languages where general linearly ordered quantifiers are expressible. We show that the results of [5] concerning infinitely deep languages can be naturally generalized to cover the present generalized infinitely deep languages. It then follows from the results of Shelah [8] that the generalized infinitely deep languages can be reduced like the ordinary ones to game-quantifier languages with well-ordered quantifier strings.

This paper is written as a continuation to our [5]. So we refer to [5] for unexplained notation and notions. This is done to spare space, and the energy of the reader who wants to see the general ideas.

1. DEFINITIONS

A *vocabulary* is a set of finitary predicate and function symbols and constant symbols. Vocabularies are denoted by $\mathfrak{L}$, $\mathfrak{K}$, ... Let $\mathfrak{L}$ be a vocabulary. $\mathfrak{L}$ structures are conceived as pairs $\mathfrak{M} = (M, f_{\mathfrak{M}})$, where M is the *domain* of $\mathfrak{M}$, and $f_{\mathfrak{M}}$ is the *interpretation* of $\mathfrak{L}$ in $\mathfrak{M}$. Let $\mathfrak{K} \subseteq \mathfrak{L}$, and let $\mathfrak{M}$ be an $\mathfrak{L}$ structure. $\mathfrak{M} \restriction \mathfrak{K}$ denotes the $\mathfrak{K}$ structure $(M, f_{\mathfrak{M}} \restriction \mathfrak{K})$, which is called the $\mathfrak{K}$ *restriction* of $\mathfrak{M}$. Let $U \in \mathfrak{L} \backslash \mathfrak{K}$ be a unary predicate symbol such that $f_{\mathfrak{M}}(U) \neq \varnothing$. Then $(\mathfrak{M} \restriction \mathfrak{K})^U$ denotes the substructure of $\mathfrak{M} \restriction \mathfrak{K}$ generated by $f_{\mathfrak{M}}(U)$, and it is called the $\mathfrak{K}$, U *restriction* of $\mathfrak{M}$. Conversely, $\mathfrak{M}$ is an *expansion* of $\mathfrak{M} \restriction \mathfrak{K}$, and an *extended expansion* of $(\mathfrak{M} \restriction \mathfrak{K})^U$.

An $\mathfrak{L}$ *logic* is a pair of classes $(\mathfrak{L}^*, \vDash^*)$ where the elements of $\mathfrak{L}^*$ are called *sentences* and $\vDash^*$ is a relation, the *satisfaction relation*, between sentences and $\mathfrak{L}$ structures, if all sentences $\varphi \in \mathfrak{L}^*$ and all $\mathfrak{L}$ structures $\mathfrak{M}$ and $\mathfrak{N}$ satisfy the following *isomorphism condition*:

$$\text{if } \mathfrak{M} \vDash^* \varphi \text{ and } \mathfrak{M} \simeq \mathfrak{N}, \text{ then } \mathfrak{N} \vDash^* \varphi.$$

An operation * which assigns to each vocabulary $\mathfrak{L}$ an $\mathfrak{L}$ logic $(\mathfrak{L}^*, \vDash_{\mathfrak{L}}^*)$ is a *system of logics* if the following requirements are satisfied for all vocabularies $\mathfrak{L}$ and $\mathfrak{K}$ with $\mathfrak{L} \subseteq \mathfrak{K}$:

(i) $\mathfrak{L}^* \subseteq \mathfrak{K}^*$,

(ii)   all $\mathfrak{K}$ structures $\mathfrak{M}$ and all sentences $\varphi \in \mathfrak{L}^*$ satisfy $\mathfrak{M} \vDash_{\mathfrak{K}}^* \varphi$ if and only if $(\mathfrak{M} \restriction \mathfrak{L}) \vDash_{\mathfrak{L}}^* \varphi$.

Requirement (ii) implies that the subscript $\mathfrak{L}$ may be omitted from $\vDash_{\mathfrak{L}}^*$ when * is a system of logics. A system of logics * is *absolute* (in a set parameter a) if, roughly said, there is a Σ_1 definition for the notion of a sentence and a Δ_1 definition for the satisfaction relation. The notion of an absolute logic or system of logics was originally meant in Barwise [1] to be an abstract version of the informal notion of a first-order logic. For details, we refer to Barwise [1] or to Oikkonen [5], where Barwise's original definition is slightly generalized.

Let * be a system of logics and let $\mathfrak{L}$ be a vocabulary. A class $\mathbf{C}$ of $\mathfrak{L}$ structures is a *model class* of $(\mathfrak{L}^*, \vDash^*)$ if there is some $\varphi \in \mathfrak{L}^*$ such that

$$\mathbf{C} = \{\mathfrak{M} | \mathfrak{M} \vDash^* \varphi\}.$$

$\mathbf{C}$ is a PC *class* of $(\mathfrak{L}^*, \vDash^*)$ if there is a model class $\mathbf{D}$ of some $(\mathfrak{K}^*, \vDash^*)$ with $\mathfrak{L} \subseteq \mathfrak{K}$ such that

$$\mathbf{C} = \{\mathfrak{M} \restriction \mathfrak{L} | \mathfrak{M} \in \mathbf{D}\}.$$

Similarly, $\mathbf{C}$ is an RPC *class* of $(\mathfrak{L}^*, \vDash^*)$ if there is a model class $\mathbf{D}$ of some $(\mathfrak{K}^*, \vDash^*)$ together with a unary predicate symbol $U \in \mathfrak{K} \backslash \mathfrak{L}$ such that

$$\mathbf{C} = \{(\mathfrak{M} \restriction \mathfrak{L})^U | \mathfrak{M} \in \mathbf{D}\}.$$

Moreover, $\mathbf{C}$ is c(R)PC if $-\mathbf{C}$, the complement of $\mathbf{C}$ in the class of all $\mathfrak{L}$ structures, is (R)PC; $\mathbf{C}$ is d(R)PC if both $\mathbf{C}$ and $-\mathbf{C}$ are (R)PC.

Keisler [4] considers the logic $\mathfrak{L}(\infty)$ which is that strengthening of $\mathfrak{L}_{\infty\omega}$ where arbitrary *linearily ordered quantifiers* $(T, U, <)$ are allowed. In $(T, U, <)$ $<$ is a linear ordering of $T \cup U$, $T \cap U = \phi$ and the variables in T are thought to be universally quantified and those in U to be existentially quantified. We denote here by $\mathfrak{L}^-(\infty)$ that fragment of Keisler's $\mathfrak{L}(\infty)$ where $\neg\varphi$ is allowed only if $\varphi \in \mathfrak{L}_{\infty\omega}$.

It turned out in [5] that in certain connections the infinitely deep languages can be quite well generalized to a form where atomic formulas are replaced by arbitrary sentences of absolute logics. We study similar questions here, and so we present the definition in the abstract form.

Another change which we shall do to the original way of defining infinitely deep languages is the omission of negations: Under certain weak requirements on the satisfaction relation, the negation symbols can be 'pushed down' in infinitely deep sentences so that they only appear in front of atomic formulas. See Karttunen [3]. This means that if the atomic formulas are replaced by atomic formulas and their negations, then the infinitely deep languages can be defined without using negations in the syntactical trees at all. This situation is simply generalized here, when we define abstract infinitely deep languages without negations. If one wants to consider negations, then one can refer to our [5] and make the required changes to the present presentation.

A *generalized tree* is a partially ordered set $\mathfrak{T} = (T, <)$ where $\{t_0 \in T | t_0 \leq t\}$ is linearly ordered for all $t \in T$. A *branch* b of a generalized tree $\mathfrak{T}$ is a maximal linearly ordered subset of T. By Zorn's lemma, every point t of a generalized tree is contained in at least one branch.

Let * be an absolute system of logics, and let $\mathfrak{L}$ be a vocabulary. Let $\mathbf{C}$ be a proper class of constant symbols such that $\mathfrak{L} \cap \mathbf{C} = \varnothing$. $\langle \mathfrak{L}^* \rangle^\infty$, is the collection of all pairs $\theta = (\mathfrak{T}, f)$, the *sentences* of $\langle \mathfrak{L}^* \rangle^\infty$, where $\mathfrak{T} = (T, <)$ is a generalized tree and f is a function from T into the collection

$$\{\bigvee, \bigwedge\} \cup \{Qc | c \in \mathbf{C} \text{ and } Q \text{ is } \exists \text{ or } \forall\} \cup \bigcup_{\substack{c \subseteq \mathbf{C} \\ \text{a set}}} (\mathfrak{L} \cup C)^*$$

such that the following requirements are satisfied:

(i) if b_0 and b_1 are distinct branches of $\mathfrak{T}$ then $b_0 \cap b_1$ has a greatest element,

(ii) if t is not maximal then $f(t) \in \{\bigvee, \bigwedge\} \cup \{Qc | c \in \mathbf{C} \text{ and } Q \text{ is } \exists \text{ or } \forall\}$,

(iii) $f(t) \in \{\bigvee, \bigwedge\}$ if and only if there are $t_0, t_1 > t$ such that $\{t_2 | t < t_2 \leq t_0\} \cap \{t_2 | t < t_2 \leq t_1\} = \varnothing$,

(iv) if $f(t)$ is $\exists c$ or $\forall c$ then there is no such $t_0 < t$ that $f(t_0)$ is $\exists c$ or $\forall c$,

(v) if t is maximal, then $f(t) \in (\mathfrak{L} \cup C)^*$ where $C = \{c \in \mathbf{C} | \text{ there is some } t_0 < t \text{ such that } f(t_0) \text{ is } \exists c \text{ or } \forall c\}$.

The satisfaction relation for $\langle \mathfrak{L}^* \rangle^\infty$ is going to be defined game-theoretically somewhat similarly to what is done in [2] for ordinary infinitely deep languages. But there are some essential changes; so we shall discuss certain details. Let $\theta = (\mathfrak{T}, f)$ be a sentence of $\langle \mathfrak{L}^* \rangle^\infty$, and let $\mathfrak{M}$ be an $\mathfrak{L}$ structure. The *semantic game* $G(\theta, \mathfrak{M})$ has the players

∃ and ∀. If $t \in T$ is non-maximal and $f(t)$ is $\bigvee$ or of the form $\exists c$, then player ∃ has a *move* at t; if t is non-maximal and $f(t)$ is $\bigwedge$ or of the form $\forall c$, then player ∀ has a *move* at t.

Let $X \subseteq T$. An *interpretation* of X to M is a function **a** which assigns an element $\mathbf{a}(t) \in M$ to each such $t \in X$ that $f(t)$ is of the form $\exists c$ or $\forall c$. If **a** is an interpretation of $X \subseteq T$ to M and if $Y \subseteq X$, then the restriction $\mathbf{a} \upharpoonright Y$ is defined in the natural way. Especially, if $t \in X$, then $\mathbf{a} \upharpoonright t$ denotes $\mathbf{a} \upharpoonright \{t_0 \in X | t_0 \leq t\}$.

A *strategy* of ∃ in $G(\theta, \mathfrak{M})$ is a function s which assigns to each pair $(t, \mathbf{a})$ where ∃ has a move at t and **a** is an interpretation of $\{t_0 | t_0 \prec t\}$ to M, a *choice* $s(t, \mathbf{a})$ of ∃ at t. If $f(t)$ is of the form $\exists c$, then $s(t, \mathbf{a}) \in M$; if $f(t)$ is $\bigvee$, then $s(t, \mathbf{a})$ is a non-empty family of branches of $\mathfrak{X}$ with the following properties:

    (i) if $b_0, b_1 \in s(t, \mathbf{a})$, then there is some $t_0 \in b_0 \cap b_1$ such that $t < t_0$,

    (ii) if $b_0 \in s(t, \mathbf{a})$ and if b_1 is such that $b_0 \cap b_1$ contains some $t_0 > t$, then $b_1 \in s(t, \mathbf{a})$.

The strategies of ∃ in $G(\theta, \mathfrak{M})$ are denoted by s, possibly with subscripts, and the set of all strategies of ∃ is denoted by $S = S(\theta, \mathfrak{M})$. The *strategies* and *choices* of player ∀ are defined in a similar way. The strategies of ∀ are denoted by p, possibly with subscripts, and the set of all strategies of ∀ is denoted by $P = P(\theta, \mathfrak{M})$.

A *play* of $G(\theta, \mathfrak{M})$ is a pair $(b, \mathbf{a})$ where b is a branch of $\mathfrak{X}$ and **a** interprets b to M. The notion of a strategy is essentially weaker in the present context than in the usual one where all branches are required to be well-ordered. In the usual case one can prove by an inductive argument that a pair of strategies $(s, p) \in S \times P$ determines a unique play, but in the present situation this is not the case. This will be seen below in an example. First a definition: A play $(b, \mathbf{a})$ is *consistent* with $(s, p) \in S \times P$ if all $t \in b$ satisfy:

    (i) if $f(t)$ is $\bigvee$, then $b \in s(t, \mathbf{a} \upharpoonright t)$, if $f(t)$ is $\bigwedge$, then $b \in p(t, \mathbf{a} \upharpoonright t)$,

    (ii) if $f(t)$ is $\exists c$, then $\mathbf{a}(t) = s(t, \mathbf{a} \upharpoonright t)$, if $f(t)$ is $\forall c$, then $\mathbf{a}(t) = p(t, \mathbf{a} \upharpoonright t)$.

Example: Let $\mathfrak{X} = ((0, 1), <)$ be the open unit interval of the real line with its natural ordering. Let c_t be a constant symbol for each $t \in (0, 1)$.

Let $f(t)$ be $\forall c_t$ if t is rational and $\exists c_t$ if t is irrational. (In other words $\theta = (\mathfrak{T}, f)$ corresponds to the linearily ordered quantifier $((0, 1) \cap Q, (0, 1) \setminus Q, <)$.) θ is a trivial case of a sentence of $\check{\mathbb{Q}}\mathfrak{L}^*\check{\mathbb{Q}}^\infty$, but it is rich enough for the present purposes. Let $M = \{0, 1\}$. Consider the game $G(\theta, M)$. Let $s \in S$ and $p \in P$ be such that for all $\mathbf{a} : (0, 1) \to M$

$$g(t, \mathbf{a}{\restriction}t) = \begin{cases} 0, & \text{if } \mathbf{a}(t_0) = 0 \text{ for all } t_0 \in (0, t) \\ 1, & \text{otherwise} \end{cases}$$

holds for $g = s$ when t is rational, and for $g = p$ when t is irrational. Let $\mathbf{0} : (0, 1) \to M$ and $\mathbf{1} : (0, 1) \to M$ be such that $\mathbf{0}(t) = 0$ and $\mathbf{1}(t) = 1$ for all $t \in (0, 1)$. Then, obviously both the plays $((0, 1), \mathbf{0})$ and $((0, 1), \mathbf{1})$ are consistent with (s, p).

Even though a pair of strategies does not uniquely determine a play of the semantic game, the notion of a winning strategy is still available. Actually, Keisler [4] has the same situation. Let $T_{\theta, \mathfrak{M}}$ be a collection of plays of $G(\theta, \mathfrak{M})$; if $(b, \mathbf{a}) \in T_{\theta, \mathfrak{M}}$, then $\exists$ wins it. A strategy $s \in S$ is a *winning strategy* for $\exists$ w.r.t. $T_{\theta, \mathfrak{M}}$ in $G(\theta, \mathfrak{M})$, if $\exists$ wins every play $(b, \mathbf{a})$ consistent with any (s, p) where $p \in P$.

What is now missing from a satisfaction relation for $\check{\mathbb{Q}}\mathfrak{L}^*\check{\mathbb{Q}}^\infty$ is a uniform and natural definition of $T_{\theta, \mathfrak{M}}$. Consider a play $(b, \mathbf{a})$ of $G(\theta, \mathfrak{M})$. Let $C = \{c \in C \mid f(t) \text{ is } \exists c \text{ or } \forall c \text{ for some } t \in b\}$. If b has a maximal element t then it is natural to set $(b, \mathbf{a}) \in T_{\theta, \mathfrak{M}}$ if and only if $\mathfrak{M}_\mathbf{a} \vDash^* f(t)$, where $\mathfrak{M}_\mathbf{a}$, is that expansion of $\mathfrak{M}$ which is obtained by interpreting the constant symbols of C by means of $\mathbf{a}$. If b has no maximal element, then there is no obvious criterion for $(b, \mathbf{a}) \in T_{\theta, \mathfrak{M}}$. This is the reason for discussing several different satisfaction relations for infinitely deep languages. In the latter case it is natural to think that the relation $(b, \mathbf{a}) \in T_{\theta, \mathfrak{M}}$ depends only on b, not on $\mathbf{a}$, because the play meets no sentence whose truth relative to $\mathbf{a}$ should be tested. (On the other hand the omission of $\mathbf{a}$ is only a notational simplification, because b can obviously be replaced by $(b, \mathbf{a})$ in many places below.) Let $\mathbf{T}(\mathfrak{M}, b, \theta, \mathfrak{L})$ be a relation. It is called a *truth definition* if $\mathbf{T}(\mathfrak{M}, b, \theta, \mathfrak{L})$ and $\mathfrak{M} \simeq \mathfrak{N}$ imply that $\mathbf{T}(\mathfrak{M}, b, \theta, \mathfrak{L})$. The collection $T_{\theta, \mathfrak{M}}$ is *defined* by a truth definition $\mathbf{T}(\mathfrak{M}, b, \theta, \mathfrak{L})$ if the following hold:

(i) if b has a maximal element t, then $(b, \mathbf{a}) \in T_{\theta, \mathfrak{M}}$ if and only if $\mathfrak{M}_\mathbf{a} \vDash^* f(t)$,

(ii) if b has no maximal element, then $(b, \mathbf{a}) \in T_{\theta, \mathfrak{M}}$ if and only if $\mathbf{T}(\mathfrak{M}, b, \theta, \mathfrak{L})$ holds.

This definition yields in an obvious way a satisfaction relation for $\S\mathfrak{L}^*\S^\infty$: Different satisfaction relations of $\S\mathfrak{L}^*\S^\infty$ will be denoted by $\S\vDash^*\S'$, $\S\vDash^*\S''$, ... The satisfaction relation $\S\vDash^*\S'$ is *defined* by the truth relation $T(\mathfrak{M}, b, \theta, \mathfrak{L})$ if all $\theta \in \S\mathfrak{L}^*\S^\infty$ and all $\mathfrak{L}$ structures $\mathfrak{M}$ satisfy the requirement

$$\mathfrak{M}\S\vDash^*\S' \; \theta \text{ if and only if } \exists \text{ has in } G(\theta, \mathfrak{M}) \text{ a winning strategy}$$
w.r.t. the $T_{\theta, \mathfrak{M}}$ defined by $T(\mathfrak{M}, b, \theta, \mathfrak{L})$.

Especially, $\S\vDash^*\S_0$ is the satisfaction relation defined by that T where $T(\mathfrak{M}, b, \theta, \mathfrak{L})$ holds for all $\mathfrak{M}$, b, θ, and $\mathfrak{L}$. In other words, in $\S\vDash^*\S_0$ all branches with no maximal element are thought to be won by $\exists$.

2. TRANSLATIONS INTO GAME QUANTIFIER LANGUAGES

We shall present in this article results analogous to those of Chapter 9 in our [5]. But there is an essential difference between the present situation and that in [5]. There all the branches of the syntactic trees are well-ordered. So a (well-ordered) game quantifier can be used to describe in a way all possible branches, if the quantifier string is long enough. This means that all the branches can be isomorphically embedded to initial segments of the ordering of the game quantifier. In the present context where branches may not be well-ordered this is not the case: Two branches may have completely different linear orderings like those of the natural and the real numbers. In such a case there cannot be any linearly ordered set $\mathfrak{J}$ such that all branches of a generalized tree $\mathfrak{T}$ could be isomorphically embedded to initial segments of $\mathfrak{J}$. To avoid these difficulties we shall need the following two lemmas:

LEMMA A. Let $\mathfrak{T} = (T, <)$ be a generalized tree such that if b_0 and b_1 are its distinct branches, then $b_0 \cap b_1 = \{t_0 | t_0 \leq t\}$ for some $t \in T$. Then there are a generalized tree $\mathfrak{T}' = (T', <')$, an isomorphical embedding $g: \mathfrak{T} \to \mathfrak{T}'$, a linearly ordered set $\mathfrak{J} = (I, \prec)$ and a mapping $h : T' \to I$ such that the following requirements hold:

(i) if b_0 and b_1 are distinct branches of $\mathfrak{T}'$, then $b_0 \cap b_1 = \{t' \in T' | t' \leq' g(t)\}$ for some $t \in T$, and moreover, for every branch b of $\mathfrak{T}'$ such that $f(t) \in b$, the set $\{t' \in T' | f(t) <' t'\}$ has a smallest element,

(ii) $f(T)$ is cofinal in T' in the sense that for all $t' \in T'$ there is $t \in T$ such that $t' \leq' f(t)$,

(iii) if t_0, $t_1 \in T'$ are such that $t_0 <' t_1$ then $h(t_0) \prec h(t_1)$, and for every

$t \in T'$ the set $\{h(t_0)|t_0 <' t\}$ is an initial segment of $\mathfrak{I}$,
  (iv) $\operatorname{card}(T) = \operatorname{card}(T')$.

Proof: We shall use a simple compactness argument: Let U and V be unary predicate symbols, P and Q binary predicate symbols and F be a unary function symbol. Let $\check{t}$ be a constant symbol for each $t \in T$. Let Φ be the set of the following statements in predicate calculus:

  (1) the diagram of $(\mathfrak{X}, t)_{t \in T}$,
  (2) $U(\check{t})$ (for every $t \in T$),
  (3) 'U and P form a generalized tree, and $P \subseteq U \times U$',
  (4) 'Q is a linear ordering of V, and $Q \subseteq V \times V$',
  (5) $\forall x(\check{t}Px \land xP\check{t}_0 \rightarrow \neg xP\check{t}_1)$ (for all t, t_0, $t_1 \in T$ such that $t < t_0$, t_1, but no $t_2 \in T$ satisfies $t < t_2 \leq t_0$ and $t_2 < t_1$).
  (6) $\forall x(\check{t}Px \rightarrow \exists y(\check{t}Py \land yPx \land \forall z(zPy \rightarrow (zP\check{t} \lor z = \check{t}))))$ (for all $t \in T$ such that there are distinct branches b_0 and b_1 of $\mathfrak{X}$ such that $b_0 \cap b_1 = \{t_0|t_0 \leq t\}$),
  (7) $\forall xy(xPy \rightarrow H(x)QH(y)) \land \forall xy\exists z(yQH(x) \rightarrow y = H(z) \land zPx)$.

Here statements (5) and (6) correspond to part (i) of the claim, and statement (7) corresponds to part (iii). By using well-founded trees, it is easy to show that every finite subset of Φ is satisfiable. Then the Compactness Theorem implies that Φ has a model, and by the downward Löwenheim–Skolem Theorem, there exists one of the same cardinality as $\mathfrak{X}$. Let $\mathfrak{M}$ be such a model, and let $\mathfrak{X}^* = (U^{\mathfrak{M}}, P^{\mathfrak{M}}) = (T^*, <^*)$, $\mathfrak{I} = (V^{\mathfrak{M}}, Q^{\mathfrak{M}}) = (I, '\prec)$ and $h = H^{\mathfrak{M}} \restriction T$. Let $g : t \mapsto \check{t}^{\mathfrak{M}}$ be the natural embedding of $\mathfrak{X}$ to $\mathfrak{X}^*$. Let

$$T' = \{t_0 \in T^*|\text{there is some } t \in T \text{ such that } t_0 <^* g(t)\},$$

and $<' = <^* \restriction T'$. Then $\mathfrak{X}' = (T', <')$, g, $\mathfrak{I}$ and $h = h^* \restriction T'$ satisfy the claims. ∎

A *normalized* sentence of $\mathring{\mathfrak{L}}^*\mathring{\mathfrak{V}}^\infty$ is a quadruple $\theta = (\mathfrak{X}, \mathfrak{I}, h, f)$ where $(\mathfrak{X}, f) \in \mathring{\mathfrak{L}}^*\mathring{\mathfrak{V}}^\infty$ is such that if b_0 and b_1 are branches of such that $b_0 \cap b_1 = \{t_0|t_0 < t\}$, then for all branches b of $\mathfrak{X}$ with $t \in b$, there is a smallest element in $\{t_0 \in b|t < t_0\}$, $\mathfrak{I} = (I, \prec)$ is a linearily ordered set and $h : T \rightarrow I$ is a function such that $t_0 < t_1$ implies $h(t_0) \prec h(t_1)$, and for all $t \in T$ the set $\{h(t_0)|t_0 < t\}$ is an initial segment of $\mathfrak{I}$. Normalized sentences will be treated like ordinary sentences of $\mathring{\mathfrak{L}}^*\mathring{\mathfrak{V}}^\infty$.

Remark: If θ is a normalized sentence of $\mathring{\mathfrak{L}}^*\mathring{\mathfrak{V}}^\infty$ then the choices of disjuncts and conjuncts in $G(\theta, \mathfrak{M})$ at some stage $(t, \mathbf{a})$ of the game can be reduced to the choices of immediate successors of t. Moreover, the stages of $G(\theta, \mathfrak{M})$ can naturally be indexed $\mathfrak{F}$. A truth definition $T(\mathfrak{M}, b, \theta, \mathfrak{L})$ is *propositional* if for all sentences $\theta_0 = (\mathfrak{T}_0, f_0)$ and $\theta_1 = (\mathfrak{T}_1, f_1)$ such that $\mathfrak{T}_0 \subseteq \mathfrak{T}_1, f_0 = f_1 \restriction T_0$ and in $T_1 \backslash T_0$ f_1 obtains only values of the forms $\exists c$ and $\forall c$, then

$$T(\mathfrak{M}, b, \theta_1, \mathfrak{L}) \text{ and } T(\mathfrak{M}, b \cap T_0, \theta_0, \mathfrak{L})$$

are equivalent for all branches b of $\mathfrak{T}_1$ and all $\mathfrak{L}$ structures $\mathfrak{M}$. A satisfaction relation $\langle\!\vdash\!\ast\!\langle\!\rangle'$ is *propositional*, if it is defined by some propositional truth definition. The idea behind propositional truth definitions is that they, in a sense, make only requirements of the choices of disjuncts and conjuncts on plays corresponding to a branch with no maximal element. For example, all the satisfaction relations defined in [2] for ordinary infinitely deep languages have such a property.

LEMMA B. Let $\theta \in \langle\!\mathfrak{L}\!\ast\!\langle\!\rangle^\infty$ and let $\langle\!\vdash\!\ast\!\langle\!\rangle'$ be a propositional satisfaction relation for $\langle\!\mathfrak{L}\!\ast\!\langle\!\rangle^\infty$. Then there is a normalized sentence θ' of $\langle\!\mathfrak{L}\!\ast\!\langle\!\rangle^\infty$ such that all $\mathfrak{L}$ structures $\mathfrak{M}$ satisfy

$$\mathfrak{M}\langle\!\vdash\!\ast\!\langle\!\rangle' \; \theta \text{ if and only if } \mathfrak{M}\langle\!\vdash\!\ast\!\langle\!\rangle' \; \theta'.$$

Proof: Define $\theta' = (\mathfrak{T}', \mathfrak{F}, h, f')$ as follows: Let $\mathfrak{T}', \mathfrak{F}, g$ and h be like in Lemma A. We may suppose that $\mathfrak{T} \subseteq \mathfrak{T}'$, and that g is the canonical embedding. Then a branch b of $\mathfrak{T}$ has a maximal element if and only if the corresponding (unique) branch of $\mathfrak{T}'$ has a maximal element; moreover, if t is the maximal element of b then t is also the maximal element of the corresponding branch of $\mathfrak{T}'$. By Lemma A, $f'(t')$ must be of the form $\exists c$ or $\forall c$ for $t' \in T' \backslash T$. Let f' be any such extension of f.

Let $\mathfrak{M}$ be an $\mathfrak{L}$ structure. Consider the games $G = G(\theta, \mathfrak{M})$ and $G' = G(\theta', \mathfrak{M})$. Let the sets of strategies of players $\exists$ and $\forall$ in these games be S and S', and P and P', respectively.

Suppose that $s' \in S'$ is a winning strategy for $\exists$ in G'. Let c be any function which interprets $T' \backslash T$ to M. Define $s \in S$ by setting $s(t, \mathbf{a}) = s'(t, \mathbf{a} \cup (\mathbf{c} \restriction t))$ where the restriction $\mathbf{c} \restriction t$ is defined in the sense of θ'. Let $p \in P$ and let $(b, \mathbf{a})$ be a play of G consistent with (s, p). Let $(b', \mathbf{a}')$ be a play of G' extending $(b, \mathbf{a})$ in the natural way. Obviously, there is some $p' \in P'$ such that $(b', \mathbf{a})$ is consistent with (s', p'). So $\exists$ wins $(b', \mathbf{a}')$, because s' is a winning strategy for $\exists$. But $\langle\!\vdash\!\ast\!\langle\!\rangle'$ is propositional. Hence $\exists$ must win

(b, **a**) in G. (Recall that b has a maximal element if and only if b' has one, and if such elements exist, then they are identical. Thus, if b has a maximal element then ∃'s winning (b, **a**) or (b', **a**$'$) both mean the satisfaction of the same sentence in (essentially) the same model. And if b has no maximal element, then the required equivalence follows directly from propositionality.) It follows that s is a winning strategy for ∃ in G. By a similar argument one can show that if ∃ has a winning strategy in G, then ∃ has one also in G'.

We are now able to present our main result concerning the relations of $\emptyset\mathfrak{L}*\emptyset^{\infty}$ to $\mathfrak{L}^{-}(\infty)$. Because the proofs are very similar to those in Chapter 9 of [5], we present our results in a single theorem.

THEOREM. Let * be an absolute system of logics, and let $\mathfrak{L}$ be a vocabulary. Let $\theta = (\mathfrak{X}, \mathfrak{J}, h, f)$ be a normalized sentence of $\emptyset\mathfrak{L}*\emptyset^{\infty}$, and let $\emptyset\vDash*\emptyset'$ be a satisfaction relation for $\emptyset\mathfrak{L}*\emptyset^{\infty}$.

(i) If $\emptyset\vDash*\emptyset'$ is $\emptyset\vDash*\emptyset_0$, then there are sentences φ_0, $\varphi_1 \in \mathfrak{K}^{-}(\infty)$ for certain $\mathfrak{K} \subseteq \mathfrak{L}$ of the form

$$\langle \forall v_{i0} \exists v_{i1} | i \in I \rangle \bigwedge_{i \in I} \sigma_i$$

such that the model class of θ is the dRPC class of $\mathfrak{L}^{-}(\infty)$ defined by φ_0 and φ_1 (and some unary predicate symbol U). Here $\langle \forall v_{i0} \exists v_{i1} | i \in I \rangle$ indicates an alternating linearly ordered quantifier of order type $2 \cdot \mathfrak{J}$.

(ii) If $\emptyset\vDash*\emptyset'$ is definable by a truth definition $\mathbf{T}(\mathfrak{M}, b, \theta, \mathfrak{L})$ which is (set theoretically) $\varDelta_1$, then the model class of θ is a dRPC class of $\mathfrak{L}^{-}(\infty)$. Moreover $\langle \forall v_{i0} \exists v_{i1} | i \in I \rangle$ is the only type of linearly ordered quantifier needed.

(iii) If $\emptyset\vDash*\emptyset'$ is definable by a truth definition $\mathbf{T}(b, \theta, \mathfrak{L})$ (which does not depend on $\mathfrak{M}$), then the model class of θ is a dRPC class of $\mathfrak{L}^{-}(\infty)$. Again $\langle \forall v_{i0} \exists v_{i1} | i \in I \rangle$ is the only type of linearly ordered quantifier needed.

Proof in outline: The basic idea of the proof is to consider extended expansions $\mathfrak{N}$ of an $\mathfrak{L}$ structure $\mathfrak{M}$ which in a sense consist of $\mathfrak{M}$ and a copy of θ. The class of such $\mathfrak{N}$ is definable in $\mathfrak{L}_{\infty\omega}$. The choices in $G(\theta, \mathfrak{M})$ can be considered as choices of elements of such extended expansions $\mathfrak{N}$. (Recall, that since θ is normalized, the choices of disjuncts and conjuncts in $G(\theta, \mathfrak{M})$ correspond to the choices of certain elements of $\mathfrak{X}$; that is, of those of $\mathfrak{N}$.) Therefore, the rules of $G(\theta, \mathfrak{M})$ can be coded on such $\mathfrak{N}$ by a suitable game-quantifier sentence. The linearly ordered quantifier $\langle \forall v_{i0} \exists v_{i1} | i \in I \rangle$ is used to describe different stages of $G(\theta, \mathfrak{M})$. To be more precise, v_{i0} or v_{i1} refers to the choice made at the ith stage of $G(\theta, \mathfrak{M})$, depending on which of the players $\forall$ and $\exists$ has the move.   ■

Remark: This theorem has strengthenings and corollaries anologous to the results of Chapter 9 in [5]:

(i) The assumption that θ is normalized can be abandoned by Lemma B, if $\S\vDash^*\S'$ is supposed to be propositional.

(ii) If the considerations are restricted to structures of cardinality $\lambda \geq \text{card}(\text{TC}(\{\theta, \mathfrak{L}\}))$ (or $\lambda \geq \exp$ (card $(\text{TC}(\{\theta, \mathfrak{L}\})))$) corresponding to cases (ii) and (iii) of the Theorem), then dRPC can be replaced by dPC in the Theorem. (The PC-results of Chapter 9 of [5] should be corrected to a similar form.)

(iii) Part (i) of the Theorem implies that all model classes of $(\S\mathfrak{L}^*\S^\infty, \S\vDash^*\S_0)$ can be represented as RPC classes of $\mathfrak{L}_{\infty\infty}$, if infinitary extra predicates with linearily ordered variable places are allowed, because theorem 2.5 of [5] can easily be generalized to the present context.

(iv) Suppose that $\mathfrak{L}^*$ contains an equivalent of each atomic $\mathfrak{L}$ sentence. Then all the logics $(\S\mathfrak{L}^*\S^\infty, \S\vDash^*\S')$ and $(\S\mathfrak{L}^*\S^\infty, \S\vDash^*\S'')$ have the same RPC and cRPC classes (as $\mathfrak{L}^-(\infty)$), if $\S\vDash^*\S'$ and $\S\vDash^*\S''$ are propositional and both satisfy either the requirement of part (ii) or (iii) of the theorem. If propositionality is abandoned, then the result holds for the fragments of normalized sentences.

(v) All the previous results can be strengthened to concern certain fragments of the infinitely deep languages and $\mathfrak{L}^-(\infty)$.

(vi) It follows from the results of Shelah [8] and the above theorem that the present generalized infinitely deep languages can actually be translated to ordinary game quantifier languages with well-ordered quantifier strings.

University of Helsinki

BIBLIOGRAPHY

[1] Barwise, J., 'Absolute Logics and $L_{\infty\omega}$', *Ann. of Math Logic* **4** (1973), 309–340.
[2] Hintikka, J. and Rantala, V., 'A New Approach to Infinitary Languages', *Ann. of Math. Logic* **10** (1976), 95–115.
[3] Karttunen, M., 'Infinitary Languages $N_{\infty\lambda}$ and Generalized Partial Isomorphisms', in J. Hintikka, I. Niiniluoto, and E. Saarinen (eds.), *Essays on Mathematical and Philosophical Logic*, D. Reidel, Dordrecht and Boston, 1978, 153–168.

[4] Keisler, H. J., 'Formulas with Linearly Ordered Quantifiers', in *The Syntax and Semantics of Infinitary Languages*, Lecture Notes in Mathematics **72**, Springer-Verlag, Berlin-Heidelberg-New York, 1968, pp. 96–130.

[5] Oikkonen, J., 'Second Order Definability, Game Quantifiers and Related Expressions', in *Commentationes Physico-Mathematicae* **48**, no. **1** (1978).

[6] Rantala, V., 'On the Theory of Definability in First-Order Logic', *Reports from the Institute of Philosophy, University of Helsinki*, no. 2, 1973.

[7] Rantala, V., 'Game Theoretical Semantics and Back-and-Forth', in the same volume as [3].

[8] Shelah, S., 'On Languages with Non-Homogeneous Strings of Quantifiers', *Israel Journal of Math.* **8** (1970), 75–79.

III

APPLICATIONS OF FORMAL METHODS

ŁADISLAV TONDL

ON THE POSSIBILITIES OF INFORMATION
EVALUATION OF GRAPHICAL COMMUNICATIONS

The present paper is an attempt to exploit the term 'transmitted information' in the concept introduced by J. Hintikka [1] and his school for evaluating graphical forms of data in association with the digital presentation of these data. The term 'information synonymity', based on the concept of 'transmitted information' and introduced by the author in [5], is used to compare the graphical and digital forms of the same data.

1. INTRODUCTION

The present cybernetic and information equipment and its peripheral devices offer the possibility of processing and displaying a large amount of data. The high speed printers included in this equipment are capable of providing the user with enormous amounts of data which are complete and accurately displayed. However, they do not always allow for easy orientation and selection of the relevant data, essential for a given management problem, or decision, or they do not always clearly indicate the associations among the individual data. In an effort to overcome these difficulties and obstacles of the present computer storing, seeking and processing of large amounts of data, certain trends in theoretical and technical cybernetics are being developed, in particular:
– research into the criteria and means of evaluation of data relevance and the seeking of relevant data;
– solution of problems of admissible data reduction; and
– seeking suitable forms of data aggregation, enabling individual data packages to be interpreted easily and quickly.
In developing these trends the data and data sets are usually considered in their discrete form, i.e., data which can be represented as a sequence of alpha-numeric symbols, as series, matrices, etc. From a linguistic point of view this is a question of certain texts operating with verbal elements, whether the verbal elements are of natural, formalized or

115

E. Saarinen, R. Hilpinen, I. Niiniluoto, and M. Provence Hintikka (eds.), Essays in Honour of Jaakko Hintikka, 115–134. All Rights Reserved.

algorithmic languages being irrelevant.

However, already at the beginning of the development of information and cybernetic techniques data in non-verbal form were used, i.e., data of pictorial shapes, graphical patterns, etc. A relatively extensive field of research was established, so-called pattern recognition, pattern shapes produced by nature itself, as well as pattern shapes as man-made artifices (e.g., letters of handwriting or other symbols) being involved. However, present-day cybernetic equipment also stimulated another field which operates with pattern and graphical shapes at the output of cybernetic systems. In this way a new trend in cybernetic equipment was constituted, which is usually characterized as computer graphics. The term 'graphical' here is understood in a narrower sense of the word, i.e., as 'pattern', operating with graphical shapes of a non-verbal nature. This includes the most various plotting equipment, X-Y recorders, pattern displays, active and interactive displays, etc.

Pattern shapes and their use in information systems allow for a special method of solving problems of relevant data selection, reduction or aggregation. Pattern shapes represent, in suitable form, aggregated data 'packages' or a compressed form of data sets, which enables quick and effective orientation in a data set, seeking required or sought data and easy orientation in these data, etc. For example, if we were to transform all the data contained in a map into the traditional form of a text, seeking relevant data in the form of this text would be incomparably more tedious and difficult than in the pattern (graphical, in the narrower sense) form. Maps, graphs, networks and other forms of patterns are, therefore, also a text or a data set of their kind which, however, in suitable form enables a quicker, less extensive and usually also more effective interpretation to be made. Figuratively speaking, if the graphical forms of the communication in the given sense are also a text, this text, as compared to the traditional forms of text, may then be read on more than one page simultaneously.

It is, therefore, not by chance that in many fields of human activity, particularly technical fields, all spheres in which maps are used, as well as numerous data which can be localized, etc., it is convenient to operate with data sets and data in graphical form in the way mentioned. Although these sets may be digitized and the transformation of the records into digitized form presents no difficulties, the graphical form retains its advantages, particularly from the users' point of view. These circumstances also stimulated the research into 'graphical languages', 'picture languages', 'two-dimensional languages', etc. A review of the problems of these languages, particularly of their syntax, is given in [2], [3] and [4]. However, this involves a relatively wide range of considerably heterogeneous problems which can be solved depending on the user's requirements and the technical means used. Attempts at constructing a universally conceived 'picture syntax' or a general syntax of graphical languages did not lead to a solution which would be acceptable for all purposes.

An information evaluation of pattern shapes drawing on the basis of the concept of information synonymity introduced in [5] will be described. This will then involve a possible approach to the semantics of pattern shapes founded on transinformation analysis in the sense of [1]. As regards the syntax of pattern shapes, we shall restrict ourselves to the application of the net structure of a data set to the concept of graphical communication as described by J. Bertin [6].

2. GRAPHICAL COMMUNICATION

A communication of this type requires that the information carrier (as opposed to a sequentially organized series of alphanumeric symbols in which the appropriate syntactic and semantic rules are respected) represents pattern or graphical shapes which have a meaning as a whole and which can also be interpreted as a whole. This global interpretation does not, of course, exclude partial interpretations of the individual parts of the shape.

By shape we understand a certain organization of constitutive elements which can be realized by human activity or by technical device, imitating this activity. This usually involves point, line or surface elements which may be combined with one another or combined with others, e.g., alphanumeric elements. In some cases iconic models, operating with elements of three-dimensional space, may come close to pattern shapes.

Pattern or graphical shapes may be considered isolated manifestations of human activity, and also interpreted as such, or as expressions of a certain graphical language which also respect that which is specific for any meaningful use of the language (utterance), i.e. certain syntactic and semantic rules. Between both types of usages of pattern shapes there are, of course, various stepwise transitions. Whereas the first type is more specific for the sphere of plastic arts and painting, in which individual works are involved, the second type is important for the sphere of intellectual and technically oriented communication. This differentiation between the two types of pattern shapes must be considered relative and conditioned, because in interpreting isolated manifestations of human activity, e.g. in the sphere of arts, it is not only possible but also expedient to consider some sort of 'handwriting type', 'language type', etc. However, within the scope of this study we shall deal exclusively with the latter type of graphical communication in which the relations of the usable symbols and appropriate syntactic and also semantic rules are much more essential. This is also so, because in some cases, e.g., with some maps or graphical data pertinent to project documentation, some symbols (operating with point, line or surface elements) and their interpretation are given in the form of a binding norm.

One could discuss the various possibilities of classifying the patterns or graphical shapes. Of quite a different nature are graphs of the distribution of a certain property from the point of view of the distribution of another property (e.g., an age tree of a given population), a graphical representation of the distribution of certain phenomena in space by means of a map, the graphical representation of the association of certain stages or phases in time with the aid of a net, etc. J. Bertin [6] distinguishes between three types of graphical shapes:

  (a) diagrams,
  (b) networks, and
  (c) cartographic patterns.

This classification is undoubtedly rather rough and does not encompass all the possibilities or transitional forms. However, the concept of 'pattern' is always essential (this term corresponds to the terms of the psychology of perception, in particular the terms 'Gestalt', pattern, 'image', etc.). By 'shape' J. Bertin understands a significant perceptible form which we are able to record within a minimal time of perception. One can also understand this as if the perception of shape were capable of cutting down the time required for transmitting information and, simultaneously, allowing for the transmission of a large amount of information over a relatively short period.

The perception of shape in graphical communication provides the possibility of recording and interpreting the essential components of the information being provided, within a relatively short time, and of proceeding from these essential components as required to detailed components. Moreover, the shape expresses a certain complex whole which may be perceived and interpreted as such a whole in its essential components which is a relatively inextensive set of data abundant in information.

If it is possible in graphical communication to express and interpret the shape as a complex whole, further (detailed) data may be added to this whole in the next interpretation of the graphical communication. This then means that the graphical communication can be interpreted as a chain in the data set such that further (detailed) data are subordinated to a superior datum. In the theory of data sets this method of the chaining of the superior datum (expressing the meaning of the shape in complex form) and the detailed data corresponds to data sets with a net structure of the 'master' type.

The superior data (which is denoted by M in the schemas shown below) can be understood as if this data were to express the fundamental characteristic, visually quickly and complexly expressible, of the state of things being represented in a given universe. In graphs this may involve, e.g., the nature of the distribution or occurrence of the represented phenomena, in maps the fundamental characteristics or types of represented objects. The superior data also represents that which the user, who has a sufficient knowledge of the graphical tools used, first 'reads', that from which the reading of the graphical form used begins. From the reading of the superior data (M) one may then proceed to the reading of detailed data ($D_1, D_2, \ldots, D_n$). We speak of the net structure in this connection because one may proceed from any detailed data back to the superior data or to any other detailed data.

This method of interpretation of a graphical communication assumes that the data $M, D_1, D_2, \ldots, D_n$ are understood to be sentences or (relatively small) sets of sentences. This also corresponds to the hypothesis that various forms of graphical communication are a text *sui generis*. Whereas the traditional forms of text, operating with alphanumeric symbols, are usually based on a sequential ordering of the partial data (represented as sentences or sets of sentences), the sequential arrangement of the superior data and the detailed data only represents one of the possibilities of interpreting the graphical communication. The following forms of ordering can be considered as most important of these forms of interpretation (and, therefore, also of digital transcription):

I. One of the possible forms of chaining the superior and detailed data is sequential ordering which can be expressed schematically as shown in Figure 1. The feedback of the data last mentioned in the sequence of

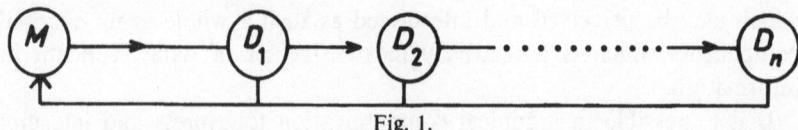

Fig. 1.

detailed data with the superior data is in no way necessary. This method of chaining can be applied always when it is necessary to negotiate the whole ordered sequence of detailed data, e.g., in determining contours of a certain formation on a map. This method is also effective in automated digitizing processes in which, e.g., the digitizer is following the course of a certain isoline, certain contours, etc. In digitizing with the aid of a digitizer it is usually irrelevant whether the detailed data are scanned from D_1 to D_n, or vice versa. This is also true of automated plotting on plotters which in fact represent the sequential ordering of sets of detailed data; indeed, only these data as a whole will provide the overall shape.

II. Another form of chaining the superior data and the detailed data in interpreting a graphical communication is a pattern which enables one to proceed from the superior data to any of the detailed data and back to the former again (Figure 2).

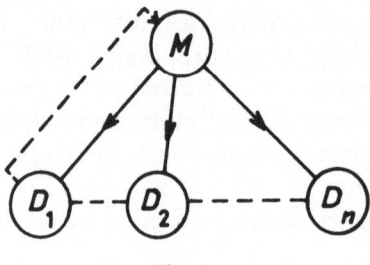

Fig. 2.

The pattern or graphical form, in this case, need not be 'read in all details', i.e., scanning all the detailed data. This involves situations in which only parts of the map, if a map is being read, are essential for the user, in which only a selected part of the pattern or graphical 'text' is sufficient. The essential thing is that this part, i.e., a certain selected subset of the set $\{D_1, D_2, \ldots, D_n\}$, together with the superior data M, is meaningful so that it is capable of providing the user with the information he wants. This method of reading a pattern or graphical form corresponds to, e.g., looking for certain key words in a catalogue or dictionary, to select desired data from a certain register, etc., in cases when it is unnecessary to read the whole catalogue, dictionary or similar text.

III. Another important type of chaining the superior data to detailed data in interpreting a graphical communication is the tree ordering in which one may proceed along the individual branches. This can be expressed schematically as shown in Figure 3.

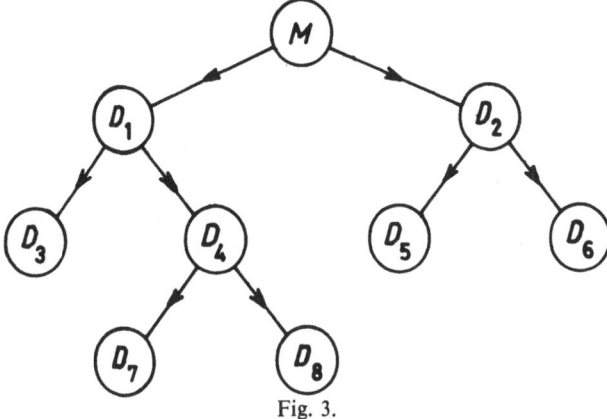

Fig. 3.

The tree graph assumes that the data on the lower levels are specified in turn in the graphical communication. In determining a given object on a map, for example, further characteristics of an object are gradually specified according to the hierarchy of the levels of these characteristics. This specification can also be understood as a gradual elimination of inadequate elements and this, in turn, can be represented by a matrix pattern. The gradual elimination, for example, can be applied in improving the accuracy of the localization of a given object, in the gradual intro-duction of further details for interpreting a pattern, etc.

These three types of interpretation (which do not apparently exhaust all possible types and only represent the most usual types, occurring mainly in the interpretations of maps) may also be considered as a metalanguage transcription of that which is being communicated by graphical or pictorial means. If we consider the communication, expressed by graphical or pictorial means, as a concrete communication in the language of the objects, the metalanguage analysis of this communication can be effected at several levels:

(a) at the description level,

(b) at the syntactic (generative) level, and

(c) at the semantic level.

One may say, therefore, that the mentioned types of interpretation of a graphical communication represent a metalanguage transcription at the

semantic level, the metalanguage being a language of the type of a natural
or formalized language, operating with sequentially ordered alpha-
numerical symbols. In information systems which assume parallelism of
graphical and digital forms of output, this type of metalanguage transcription in a semantic metalanguage may be considered to be transcription
in a translational language, which is the common semantic metalanguage
to both versions of output data.

The description level of the metalanguage transcription of a graphical
communication means the concept of a graphical communication as an
organization of certain fundamental elements. A system of description
attributes or relations, which can be coordinated with the individual
fundamental elements, pairs of these elements, triads, etc., can be
characterized as the forms of this organization. Symbols operating with
- point elements,
- line elements,
- surface elements
or symbols operating with combinations of these elements, are considered
to be fundamental elements. The choice of the fundamental elements and
description attributes and relations depends on the nature of the graphical
language used. If the device in question is, for example, a plotting automatic instrument, the fundamental elements are the symbols which the
given device is capable of plotting. The description attributes are, e.g., the
connection of two point elements with a line section or a curve of given
properties, the properties of line elements, overlapping, contact of
surface elements, etc.

The syntactic level of metalanguage analysis of a graphical communication means the concept of a graphical communication as a
realization of certain syntactic (generative) rules in the system of a certain
graphical language. Some authors, e.g., R. Narasimhan [7], speak in this
connection of the generative level and generative description. This requires
the application of Chomsky's concept of generative grammars to the
sphere of graphical communication. Assuming that the graphical language
(GL) is formed by a system of elementary symbols X, a system of their
attributes and relations P, a system of formation rules F and a system of
transformation rules T, a graphical language can be defined as:

$$GL = (X, P, T, F).$$

The elementary form of generative rules which correspond to this structure
of the graphical language was presented by R. Narasimhan [7].

3. SEMANTICS OF A PATTERN

The semantic analysis of a pattern can be considered from several points of view. For example, one could consider the denotation of a given pattern, i.e., the relation of this pattern to a certain state of affairs in the appropriate universe, the meaning of the given pattern, i.e., the signification invariant which is preserved whatever is the use of the pattern, etc. This method of semantic analysis, which is based on Frege's differentiation of meaning and denotation, however, is closely related to purely verbal communication and is only of limited significance for graphical communication. Below we shall understand by semantics of a pattern an interpretation of the pattern such that is based on the transcription of the given pattern or graphical shape into the semantic metalanguage, which has a verbal character and which can be understood as a sequentially ordered text.

This concept of semantics of pattern is particularly convenient in information systems for which it is expedient to reckon with parallelism of data in graphical and digital forms and in which it is possible to select data of graphical and digital form from the same data base. In these and similar cases, the semantic metalanguage functions as a translational language for the graphical and digital data selection form. Since selection from the data base is possible in graphical and digital form, it is most expedient to compare the advantages and disadvantages of both forms from the point of view of information. If we are going to speak of the information point of view or of the information evaluation of data in graphical and digital form, we shall have in mind the information evaluation based on transinformation analysis.

Transinformation analysis does not assume the information evaluation of a certain datum or data set as such, but a method of information evaluation such that it considers to what extent the given data or data sets contribute to the user's or data addressee's information. This means that in transinformation analysis one takes into account to what extent the user is equipped with information in advance, i.e., for example, a given set of problems, problem situations, etc., as well as to what extent the communicated data increase this *a priori* outfit of the user, or, in other words, to what extent they reduce the original indeterminacy or original entropic level.

From this point of view, one is then able to construct the semantic analysis of data presented in graphical form, as well as data or data sets communicated in digital form. Moreover, one must assume that the data presented in the form of a pattern or graphical shape and data in digital form are synonymous to the user with respect to the same *a priori* outfit. In this case, synonymity should not be understood as absolute synonymity, but as synonymity which can be realized with respect to the user and his *a priori* outfit with information. These conditions are met by the concept of 'information synonymity', introduced and defined in [5].

In information systems one frequently encounters situations in which one has to determine whether the graphical form of data representation and the digital data form are synonymous. For example, the graphical representations of the terrain as a map or a map schema should be informationally synonymous with the set of sentences expressing the primary and derived (extrapolated) data based on the appropriate measurements. Analogously, it is desirable that the graphical representation of the development of certain indices within a certain interval be informationally synonymous with the set of data expressing the magnitude of these indices at the appropriate times within a given interval. In automated information systems or at the outputs of cybernetic devices for processing data one may reckon with sequentially ordered sets of data which are produced by the high speed printers of these devices, as well as with graphical representations of the same data set in the form of a pattern or graphical shape.

It is, therefore, expedient to conceive the semantic analysis of a pattern as a semantic analysis of a text or as text semantics. As in the case of a sequentially ordered text, the interpretation of each individual sentence is usually incomplete if one does not take into account the interpretation of the foregoing sentences of the text; also the interpretation of each individual symbol or symbol complex in the pattern requires the whole pattern to be taken into account. In text semantics it is usually emphasized that the ordered sequence of sentences of the text gradually fixes the universe, the concrete associations, or the so-called semantic localization. A more detailed analysis of text semantics and semantic localization is presented in [8]. If we were to extract a sentence and isolate it from the context of the whole text and if we were unaware of the procedure of semantic localization implemented in the foregoing sentences, we would be unable to grasp the meaning of the sentence within the scope of the text as a whole. This is also true of the individual symbols or symbol complexes within the scope of a pattern in graphical communication. However, there is a very substantial difference here: Whereas in a sequentially ordered text the procedure of semantic localization is usually very tedious and requires that the whole or nearly whole part of the foregoing text be run, in graphical communication the analogous process is usually easier and quicker. Schematically expressed, the semantic localization of a symbol or symbol complex expressing a detailed data D_i is usually effected by reverting to the superior datum M from D_i, i.e., to a datum expressing the fundamental characteristics of the state of affairs in a given universe which can be grasped quickly.

4. INFORMATION SYNONYMITY AND INFORMATION EVALUATION OF A PATTERN

In order to be able to establish the criteria of information evaluation of a pattern, we must assume that the text, represented by a sequentially ordered sequence of sentences, and the text *sui generis*, represented by the pattern, are informationally synonymous (in the sense of [5]). We may consider informationally synonymous such data, data sets, texts or communications characterized in any other way, provided they can be substituted for one another with respect to a problem or complex of problems; the measure of transmitted information (transinformation), which these communications yield with respect to a given problem or complex of problems, being preserved. It should be emphasized that the concept of 'information synonymity', characterized in this manner, is always rendered relative to a certain *a priori* information status, which contains the specification of the given problem or complex of problems, or possibly also that which is already known of the problem.

In the subsequent analysis we shall use, apart from the usual logic symbolics, the following symbols: Pattern will be denoted g, the pattern being considered a text *sui generis* which represents a certain organization (e.g., in the form of given graphical schemas) of the superior data M and detailed data $D_1, D_2, \ldots, D_n$, i.e.,

$$g = F(M, D_1, D_2, \ldots, D_n).$$

The data or data set in digital form will be denoted by s and we shall consider it to be a sequentially ordered text, i.e., a sequence of individual sentences:

$$s = \{S_1, S_2, \ldots, S_m\}.$$

The *a priori* information outfit will be denoted by i, however, i may also be considered to be a class of sentences which specify the given problem or complex of problems, or even other data which are available in advance, i.e.,

$$i = \{I_1, I_2, \ldots, I_0\}.$$

The definition of the term 'information synonymity' is based on the concept of 'transmitted information' (transinformation). The transmitted information, which we shall denote TI, expresses to what extent a new datum or data set s reduces the original uncertainty associated with i. This means that the transmitted information is a function of the original uncertainty associated with i, which we shall denote $U(i)$, and of the conditional uncertainty associated with i under the assumption that also s is available, and this we shall denote $U(i/s)$. This means that

$$TI(s/i) = f[U(i), U(i/s)],$$

$TI(s/i)$ representing the information transmitted by the data or data set s with respect to that to which s is related. Since the concept of 'transmitted information', generally characterized in this manner, is symmetric, it is expedient to introduce the normalized asymmetric measure of transmitted information with respect to i, i.e., $TI_i(s/i)$. In the extensive literature on the semantic theory of information one may find a number of alternative definitions of the measure of transmitted information. The three most important definitions of the normalized measure of transmitted information are given in [10]. Of these three definitions the most convenient for solving the given problem is the measure of transmitted information which is based on using the so-called measure of content of semantic information (cont), introduced by Carnap and Bar-Hillel [9]. According to this definition

$$TI_i(s/i) =_{df} \frac{\text{cont}(i) - \text{cont}(i \to s)}{\text{cont}(i)} = \frac{1 - p(i \vee s)}{1 - p(i)} =$$

$$= 1 - p(s)\frac{1 - p(i/s)}{1 - p(i)} = p(\sim s/\sim i),$$

(the measure defined in this way corresponds to formula 6.11 in [10], where p is the probability measure used). This probability measure may be defined, if so required, as a frequency-constructed probability or a probability measure constructed in another way, e.g., as a so-called subjective probability. (It is assumed, of course, that this satisfies the well-known Kolmogorov axioms of the theory of probability.)

Since the measure of content of the semantic information of a data set i

$$\text{cont}(i) = 1 - p(i),$$

this measure of content may also be used to determine the original and conditional uncertainty. This means that

$$U(i) = \text{cont}(i) = 1 - p(i)$$
$$U(i/s) = p(i \vee s) - p(i) = p(i) - p(i \cdot s).$$

The mentioned measure of transmitted information (in normalized form) may then be defined as

$$TI_i(s/i) =_{df} \frac{U(i) - U(i/s)}{U(i)}$$

The values of the measure of content are between 0 and 1. This implies that $0 \leq TI_i(s/i) \leq 1$, and $TI_i(s/i) = \max TI_i = 1$ if and only if the transmitted information s is sufficiently strong to imply i, i.e. provided i is the logical consequence of s. It also holds that $TI_i(s/i) = \min TI_i = 0$ if and only if i is implied by the negation of s, i.e., provided i is the logical consequence of $\sim s$.

The mentioned measure of transmitted information TI_i is additive with respect to the conjunction of two data or data sets s and s', i.e.

$$TI_i(s.s'/i) = TI_i(s/i) + TI_i(s'/i)$$

provided the condition that the disjunction $s \vee s'$ is logically true, is satisfied.

Concerning the other important properties of the described transmitted information TI_i, one should mention the conditions for information irrelevance. A datum or data set s' is informationally irrelevant with respect to $TI_i(s/i)$ and, therefore s' does not increase the measure of transmitted information $TI_i(s/i)$, i.e.

$$TI_i(s.s'/i) = TI_i(s/i),$$

provided the condition that the disjunction $s' \vee i$ is logically true, is satisfied.

On the ground of the concept of 'transmitted information' it is possible to introduce the concept of 'information synonymity'. Two data or data sets are informationally synonymous with respect to a certain *a priori* outfit if and only if the measure of the transmitted information of both data with respect to *i* is the same. Since the measure of transmitted information can also be the same if it is zero or very small, one must, bear in mind that this measure must attain a certain sufficient level with respect to *i*. This depends on the demands imposed on the quality of the solution of a given problem or complex of problems. If we are to solve urbanistic problems within a limited part of Prague, we require the appropriate maps which determine the planimetry and altimetry of this part. These maps may also be expressed in digital form. In order to be able to compare both forms of expressing the data, the data must be adequate in both cases with respect to the given problem. A general map on a small scale, which depicts the area being considered, and the corresponding digital expression may transmit the same information with respect to the same problem, however, the information is insufficient. We shall consider the transmitted information to be sufficient with respect to *i*, provided it is capable of providing the solution of the given problem with the required quality, of eliminating the original uncertainty in a satisfactory way, of guaranteeing the solution with a risk of inaccuracy which is within certain limits, etc. This means that the sufficient transmitted information is determined conventionally. The transmitted information, provided with respect to *i* is sufficient, if

$$TI_i(s/i) \geq \varepsilon,$$

where ε is defined conventionally with respect to the requirements of the given type.

On the basis of these assumptions we are able to define information synonymity of a pattern *g* and a data set (in digital form) *s* with respect to *i* (denoted by the symbol $\text{Syn}_i\, g, s$) as follows:

$$\text{Syn}_i\, g, s =_{df} [TI_i(g/i) = TI_i(s/i)] \cdot [TI_i(s/i) \geq \varepsilon].$$

This definition expresses the conditions which are imposed on information synonymity: the pattern *g* and the data set *s* are informationally synonymous with respect to *i* if and only if they can be substituted for one another without changing the measure of transmitted information with respect to *i*, and this measure is greater or equal to the sufficient transmitted information.

This definition or information synonymity renders the term 'synonymity of a pattern and data set in digital form' relative, with respect to the *a priori* information outfit, on the one hand, and with respect to the conventionally adopted measure of sufficient transmitted information, on the other. This means, for example, that if *g* and *s* are informationally synonymous with respect to *i*, they need not be synonymous with respect to a different problem and a different *a priori* information outfit. The existing information synonymity may also be disrupted if the conditions imposed on the limits of the sufficient transmitted information are changed.

Since the definition of the term 'information synonymity' is based on the application of the measure of content of the semantic information cont, it can be proved that

$$\text{Syn}_i \, g, s = [p(g \vee i) = p(s \vee i)] \, . \, [p(s \vee i) \leq 1 - \varepsilon/1 - p(i)/].$$

The construction of the probability measure *p* can also be justified on the basis of the so-called 'algebra of events', in which the individual events or states of affairs are considered to be the subsets of that universe to which the patterns or data sets used are related.

The term 'information synonymity' may form the basis for the information evaluation of patterns (rendered relative with respect to a given *a priori* information outfit inclusive of the specification of a given problem or complex of problems), provided these patterns can be digitized. If this transformation, usually executed by means of the appropriate digitization, is reliable enough to enable us to assume information synonymity of the graphical shape and data set in digital form, the information value of the pattern is determined intermediately by means of the information value of the synonymous data set, i.e.,

$$\text{Syn}_i \, g, s \rightarrow TI_i(g/i) = TI_i(s/i).$$

5. INFORMATION SYNONYMITY AND THE TIME FACTOR

If the pattern *g* and the data set *s* are informationally synonymous, i.e., if they transmit the same information with respect to a given *a priori* information outfit and a given problem *i*, this need not mean that the pattern and the data set synonymous to it are equally convenient for the user. From the user's point of view not only the quality and extent of the transmitted information are essential, but also other factors relating to the transmission. Of these factors the most important is the time factor.

The time factor is essential especially from the point of view of using the information in management and decision processes. In these processes the time required for sufficient interpretation of the transmitted data in graphical form and of the data in digital form is also important. One of the specific features of graphical communication is the possibility of recording and interpreting within a relatively short time the essential components of the transmitted information, of obtaining these components relatively quickly and in condensed form, so that they can be gradually supplemented by further detailed data.

If g is the pattern and

$$g = F(M, D_1, D_2, \ldots, D_n),$$

and s is the digitized data set,

$$s = \{S_1, S_2, \ldots, S_m\},$$

the interpretation of both texts takes place over a certain time t. This means that one also has to add to the information evaluation $TI_i(g/i)$ and $TI_i(s/i)$, the comparison of the time necessary for interpreting g, i.e., t_g, with the time required to interpret s, i.e., t_s. If Syn_i g, s and, moreover $t_g < t_s$, it is undoubtedly more convenient for the user to base the decision or control process on the interpretation of the pattern.

The advantages of graphical communication from the point of view of the time factor need not be just limited to the shorter time required to interpret g or s. It is even more important that the interpretation of the pattern or the interpretation of the data set are usually quite different: From a purely intuitive point of view it is already clear that in the gradual reading of a sequentially ordered text it usually takes some time before one 'reaches the essence of the matter', before we 'arrive at the merit of the whole affair', etc. This usually calls for a long series of data to be considered, which can only be read and interpreted gradually in sequence. The interpretation thus proceeds from

$$S_1 \text{ to } S_2, S_3, \ldots, \text{ to } S_m,$$

which presupposes that a series of sentences, which are not always essential for clarifying the merit of the matter, has to be read and interpreted.

On the contrary, in reading and interpreting a graphical form we start with that which we characterized as the superior data, and we select the most convenient sequence of reading and interpretation of other detailed data which, moreover, may practically be reviewed at a glance. Schematically this can be expressed as in Figure 4:

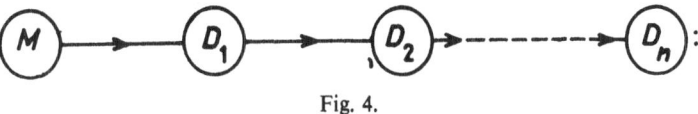

Fig. 4.

many of the detailed data may be omitted in this procedure.

The procedure of interpretation may be expressed as a gradual decreasing of the uncertainty or original entropic level. This also means that the process of transmission of information, the measure of which we expressed in terms of the measure of transmitted information (transinformation) TI_i, always takes place in the form of a sequence of certain steps, i.e., also in time. In the course of the transmission of information, therefore, $TI_i(s/i)$ or $TI_i(g/i)$ are not realized at once, instantaneously, but as a sequence of certain information doses or information quanta.

Since we defined $TI_i(g/i)$ and $TI_i(s/i)$ as

$$\frac{U(i) - U(i/g)}{U(i)} \text{ and } \frac{U(i) - U(i/s)}{U(i)}, \text{ respectively,}$$

$TI_i(g/i)$ or $TI_i(s/i)$ can only be considered as the result which is always obtained at time t_g or t_s. In fact, the measure of original uncertainty or measure of original entropic level decreases gradually, i.e., as a sequence of partial decreases of the original uncertainty, e.g., as the sequence

$$\frac{U(i) - U(i/g_0)}{U(i)}, \frac{U(i) - U(i/g_0 \cdot g_1)}{U(i)}, \ldots,$$
$$\frac{U(i) - U(i/g_0 \cdot g_1 \ldots g_n)}{U(i)}$$

where $g_0, g_1, \ldots, g_n$ are the sequentially read and interpreted elements of the pattern as a whole. The process of sequential reading and interpreting the individual sentences of the data set in digital form can be expressed analogously.

The advantages of graphical communication can be seen particularly in that the mentioned sequence of gradual reducing of the original uncertainty, in the case of graphical communication can be organized expediently enough to guarantee a considerable reduction of the original

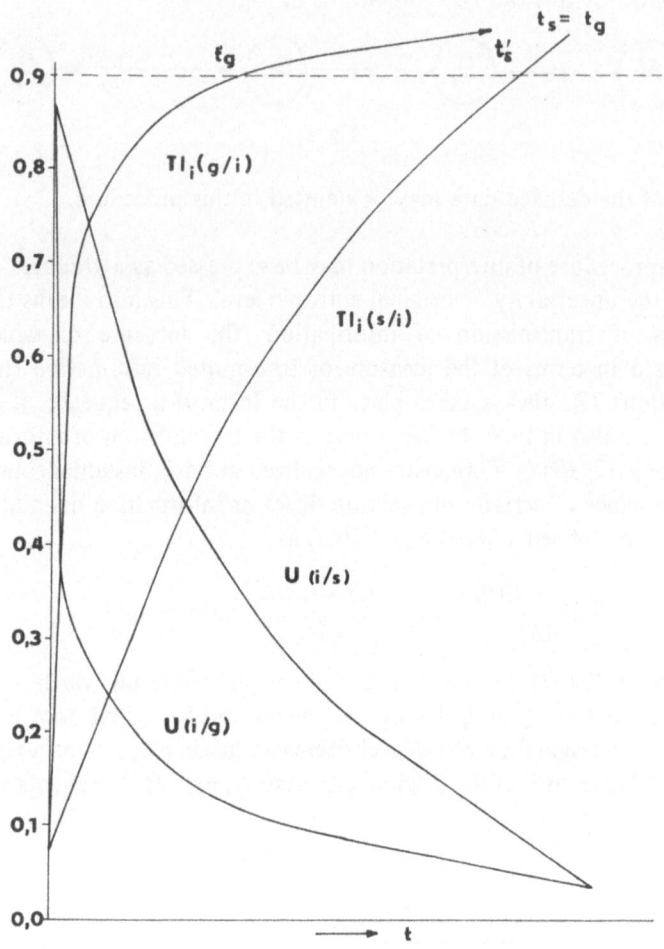

Fig. 5.

uncertainty within the first few steps. This can be expressed schematically in the figure in which the *x*-axis represents the time rhythm and the *y*-axis the measure of uncertainty and its gradual decrease.

The figure indicates that in gradually interpreting a graphical communication the decrease of the original uncertainty is at first relatively rapid, because a considerable decrease is already effected by the superior data. As opposed to this in the gradual interpretation of a data set in digital form and in the form of a sequentially ordered text the decrease of the original uncertainty usually proceeds quite slowly at first. In the figure 5

$$TI_i(g/i) = TI_i(s/i),$$
$$TI_i(s/i) \geq \varepsilon \text{ and } t_g = t_s.$$

If in the same process of gradual reducing of the original uncertainty or original entropic level it is sufficient to achieve that which we have characterized as the sufficient transmitted interpretation, which is conventionally determined by the value ε (under the assumption that $\text{Syn}_i g, s$ and $TI_i(s/i) \geq \varepsilon$), then in the gradual interpretation of a graphical communication it is possible to achieve the corresponding value of the decrease (shown in the given figure on the y-axis by the value $U(i) - U(i/g)$ much earlier, i.e., in time t_g'. This is shorter than the time required to achieve the same value of decrease in the gradual interpretation of a data set in digital form, i.e., time t_s'. This means that the graphical form of communication, should be preferred, even though

$$\text{Syn}_i g, s$$
$$TI_i(g/i) = TI_i(s/i) \geq \varepsilon,$$
$$t_g = t_s, \text{ assuming that}$$
$$t_g' < t_s'.$$

It should be emphasized that the above considerations in no way justify one in drawing the general conclusion about a general priority of the graphical form of communication over digital forms. This priority is only justified in some fields of human activity, particularly in those where one works with maps, graphical documentation in the fields of machine engineering, construction, in some branches of natural and technical sciences, in all the fields in which data localizable in time and space are being considered, etc. The justification of this priority depends on meeting relatively severe conditions, associated with 'information synonymity' and 'sufficient transmitted information'.

Design Institute, Prague

BIBLIOGRAPHY

[1] Hintikka, Jaakko, 'The Varieties of Information and Scientific Explanation', in B. van Rootselaar and J. F. Staal (eds.), *Logic, Methodology and Philosophy of Science* III, North-Holland Publ. Co., Amsterdam, 1968.

[2] Kaneff, S. (ed.), *Picture Language Machines*, Academic Press, London and New York, 1970.
[3] Nake, F., and Rosenfeld, A., *Graphic Languages*, North-Holland Publ. Co., Amsterdam, 1972.
[4] Rosenfeld, A., *Picture Processing by Computer*, Academic Press, New York and London, 1969.
[5] Tondl, Ł., *Problemy semantiki*, Progress, Moscow 1975.
[6] Bertin, J., *Semiologie graphique*, Mouton – Gauthier-Villars, Paris, 1967.
[7] Narasimhan, R., 'Picture Languages' in [2].
[8] Tondl, Ł., *Data a informace*. (Unpublished monograph).
[9] Carnap, R., and Bar-Hillel, Y., 'Semantic Information', in W. Jackson (ed.), *Communication Theory*, London 1953.
[10] Niiniluoto, I., and Tuomela, R., *Theoretical Concepts and Hypothetico-Inductive Inference*, D. Reidel, Publ. Co., Dordrecht and Boston, 1973.

KLEMENS SZANIAWSKI

ON FORMAL ASPECTS OF DISTRIBUTIVE JUSTICE

1. INTRODUCTION

The concept of distributive justice is here considered in an extremely simplified form, viz. relative to the situation when n objects are to be divided among m persons ($1 \leqslant n \leqslant m \geqslant 2$) and no account is taken of factors that could differentiate the persons as far as the division is concerned, such as their merits or needs. By so limiting the concept of justice, we gain the possibility of a formal analysis. On the other hand, the applicability of such a concept in actual social contexts is rather problematic.

In the model under consideration, distributive justice essentially amounts to some form of symmetry. Probably the most natural way to express this symmetry is to postulate that everybody has the same chance of getting what he desires. The problem then arises, whether such postulate is consistent with another natural desideratum, viz. that the division of objects between individuals be Pareto optimal. This question must be answered in the negative. It has been shown[1] that the two postulates are, in general, incompatible.

The present essay is an attempt to push the analysis a little further. In particular, I will try to show that the 'postulate of equal chances' can be replaced by another postulate which does not contradict the Pareto optimality. It thus appears that even in so simple a model, we can have more than one concept of justice. Whether the two concepts are equally satisfactory, is a matter of ethical judgment.

I shall begin by restating the problem and the earlier result. The alternative postulate will then be formulated and compared with the previous one.

2. BASIC CONCEPTS

Let $G = \{A_1, A_2, \ldots, A_n\}$ be a set of objects which are to be distributed

135

E. Saarinen, R. Hilpinen, I. Niiniluoto, and M. Provence Hintikka (eds.), *Essays in Honour of Jaakko Hintikka*, 135–146. *All Rights Reserved.*

among individuals belonging to the set $P = \{1, 2, \ldots, m\}$. The word 'object' is here given a wide interpretation. For instance, the right to sit in the King's presence may also be called object. However, unlike the last-mentioned privilege, elements of G are treated as unique: no object may go to more than one person.

For simplicity, elements of G will be denoted by $A, B, C, \ldots$ It is assumed that there are at most as many objects as persons: $n \leqslant m$. The reason for this restriction will be given below.

Let $\leqslant_k$ be the (weak) preference ordering of the set G by the person k. Strong preference $<_k$ and indifference $=_k$ are defined in the usual way. The set $\{\leqslant_1, \leqslant_2, \ldots, \leqslant_m\}$ is called 'profile of preference orderings'.

The empty set $\varnothing$ is also subject to evaluation in this sense, since one must allow for the possibility of someone's getting no object from the set G: we shall then say that this person gets $\varnothing$. It is postulated that

(1)   for all k and all j: $\varnothing <_k A_j$.

In other words, everybody strongly prefers getting any object to getting nothing. In this sense, elements of G are 'goods'.

By *division* I shall mean any way of attributing objects to individuals, i.e., any function from G to P.

Very little can be said about divisions, before further concepts are introduced. It is, however, clear that in view of $n \leqslant m$ and (1), a division attributing to some person more than one object would violate the principle of justice. A postulate[2] eliminating such divisions will be adopted:

(R)   No person gets more than one object.

From now on, I shall consider only divisions satisfying R.

Further analysis is made possible by introducing randomization. Its role is, of course, to 'smoothe out' inequalities inherent in necessarily discrete divisions. In the simplest case, when two people want to get one indivisible object, the ultimate inequality is unavoidable. Symmetry is then saved by giving them equal chances of getting what they want. It is, indeed, a common practice to 'toss a coin for it'. The definitions below are intended as a generalization of this simple solution.

Any probability distribution on the set D of all divisions will be called *distribution*. Division is, of course, a special case of distribution in this sense.

A *distribution rule* is a function from the set of all profiles of preference orderings to the set of all distributions. In other words, a distribution rule determines, for each profile of preference orderings, the random procedure to be followed in the choice of actual division.

The central problem now is, how to define distribution rules that are ethically satisfactory.

3. EQUAL CHANCES OF SATISFACTION

In the absence of any other criteria differentiating the participants in the division, distributive justice must consist in treating them on an equal basis. Perfect equality in the actual division of goods is, in general, unattainable. We can, however, postulate equality in terms of chances. It is not accidental that, in the trivial example above, *fair* coin was used to decide who will get the desired object. Any other distribution would violate the principle of equal treatment of the two participants.

In order to generalize this idea, let the objects in G be numbered, for each individual, according to their preference-ordering by that individual. (We assume, from now on, that indifference does not occur. The assumption entails no loss of generality, since indifference makes it easier to solve the problem: if someone is indifferent between A and B, his tastes can be accommodated in more ways than when he strongly prefers A to B.) Let $p_i(k)$ denote the probability that individual i gets the k-th object in his hierarchy of preferences. It is natural to adopt the following postulate.

(E)   For all $i, j = 1, \ldots, m$ and all $k = 1, \ldots, n$: $p_i(k) = p_j(k)$.

In other words, the distribution rule satisfying E must be such that everybody has the same probability of getting the k-th object in his preference ordering ($k = 1, \ldots, n$).

Let us see how E works in some specific case. In order to avoid clumsy formulations, I will limit the analysis to the case $m = n$. The limitation leaves essentially unchanged the conceptual structure of the problem.

In view of R, each person gets exactly one object. Any division can, therefore, be represented as a permutation of elements in G, under the interpretation that the i-th element in the permutation is assigned to the i-th person. For $n = 3$, we have 6 possible divisions, an easily manageable number.

In the discussion below, I will use the following examples.

(a)	(b)	(c)	(d)	(e)
1 2 3	1 2 3	1 2 3	1 2 3	1 2 3
A A A	A B C	A A A	A C C	A A C
B B B	. . .	B B C	B B A	B B B
C C C	. . .	C C B	C A B	C C A

Each case represents the structure of preferences in the 3-elementary group. Objects are listed according to their decreasing order of preference in a given person's hierarchy. Thus, in the case (d), the person 3 wants C in the first place, then A, then B.

All possible divisions, together with their probabilities x_i, are:

$$ABC\ x_1 \qquad BCA\ x_4$$
$$ACB\ x_2 \qquad CAB\ x_5$$
$$BAC\ x_3 \qquad CBA\ x_6$$

Given the profile (c) of preference orderings, the postulate E imposes on the distribution $\{x_i\}$ a condition expressed by the following set of equations.

$$\begin{aligned}
x_1 + x_2 &= x_3 + x_4 = x_5 + x_6 \\
(2) \qquad x_3 + x_4 &= x_1 + x_6 = x_1 + x_3 \\
x_5 + x_6 &= x_2 + x_4 = x_2 + x_5
\end{aligned}$$

For instance, the equations in the second line ensure that 1, 2 and 3 have the same probability of getting the object that stands second in their hierarchy of preferences, defined by (c). For 1, this object is B. B is assigned to 1 in the third and fourth division. Therefore, $x_3 + x_4$ is the probability that 1 gets the second object of his preference. By the same reasoning, this probability is $x_1 + x_6$ for 2 and $x_1 + x_3$ for 3.

The set (2) is, of course, redundant and can be reduced to:

$$\begin{aligned}
(3) \qquad x_1 &= x_4 = x_5 \\
x_2 &= x_3 = x_6
\end{aligned}$$

An insight into the meaning of these conditions is obtained when divisions are represented by the ordinal numbers of objects assigned to the respective persons:

$$\begin{aligned}
122\ x_1 \qquad 133\ x_2 \\
(4) \qquad 231\ x_4 \qquad 212\ x_3 \\
313\ x_5 \qquad 321\ x_6
\end{aligned}$$

It is then easy to see that the probabilities of obtaining objects ranking as No 1, No 2 and No 3 are, for 1, 2 and 3:

	1	2	3
No 1	1/3	1/3	1/3
No 2	1/3	1/3	1/3
No 3	1/3	1/3	1/3

(5)

This follows from (3) and (4). In each column of such a table, the probabilities must of course sum up to 1.

A distribution satisfying E exists, for all profiles of preference orderings. It simply is the equidistribution $x_i = 1/n!$. When all divisions are equiprobable the probability that the i-th individual gets the object No 1 in his preference ranking is $(n - 1)!/n! = 1/n$. And the same argument goes for object No 2, etc. Therefore, in tables of the type (5) all the entries are equal to $1/n$, which shows that the postulate E is met by such a distribution.

4. OPTIMALITY

The notation adopted in (4) suggests another desideratum, besides that of distributive justice. When we compare, for instance, the first two divisions (to denote them, I will use the symbols d_i), it is easy to notice that d_1 dominates d_2, in game-theoretic sense. Under d_1, each person gets at least as much as under d_2, while persons 2 and 3 actually get more, since the objects assigned to them rank as second instead of third. Obviously, d_2 is an unsatisfactory way to assign objects to persons, as long as d_1 is possible. According to the principle of Pareto optimality, we ought to exclude such divisions as inadmissible.

Formally, let a division be represented as $k_1 k_2 \ldots k_n$, where k_i is the rank number of the object assigned to the person i. We shall say that $k_1 k_2 \ldots k_n$ dominates $l_1 l_2 \ldots l_n$ when $k_i \leqslant l_i$ for all i and $k_i < l_i$ for some i. The postulate of Pareto optimality can be expressed as:

(O)   If a division is dominated by some other division its probability is zero.

Out of all possible variants of optimality, that of Pareto seems to be the weakest, therefore the least controversial. Also, the requirement above is consistent, in the sense that it can be met for all possible profiles of preference orderings. This follows from the fact that the relation of dominance is transitive. In a finite set of divisions, partially ordered by that relation, there must be at least one non-dominated element.

The postulates E and O, however, turn out to be inconsistent.

5. EQUALITY VERSUS OPTIMALITY

Their inconsistency is exhibited in the example (c), discussed above. By the postulate O, $x_2 = 0$ since d_2 is dominated by d_1, and $x_5 = 0$ since d_5 is dominated by d_3. By the postulate E, $x_2 = x_3 = x_6$ and $x_5 = x_4 = x_1$. It follows that $x_i = 0$ for $i = 1, \ldots, 6$. Which means that a distribution satisfying E and O for the profile of preference orderings described in (c) does not exist. But a distribution rule associates a distribution to *each* profile of preference orderings. Therefore, a distribution rule satisfying both E and O does not exist.

When faced by inconsistency of simple and intuitively convincing requirements, we can look for two types of solution. The first one consists in weakening of one or more of the desiderata. In the present case this seems hardly feasible. The Pareto optimality is already an extremely weak version of optimality. As to the equality of chances, one could of course restrict it to objects ranking as first in the respective hierarchies, but the intuitively convincing character of the postulate E would then be lost.

Another way out is analogous to that followed in the case of the Arrow postulates for democratic welfare function. It was shown that their inconsistency vanishes when the range of their application is suitably limited. The question is, for which profiles of preference orderings the postulates E and O can be jointly satisfied?

It is rather obvious that E and O are consistent when the conflict of interests between the participants in the division is either maximal or minimal. In the first case, represented by example (a), all preference orderings are identical. Domination cannot arise, since in any division, an improvement of someone's position implies a worsening in the position of another person (the situation is somehow analogous to that of zero-sum game, where any arbitration scheme fails because of complete opposition of interests). It follows that a dominated division does not exist and postulate O is vacuously satisfied. In the example (b), there is no conflict of interests: the highest ranking objects are different for each person (the rest of the preference scale is immaterial), therefore the division ABC dominates all other divisions. At the same time it satisfies, by itself, the postulate E, since it is represented by identical numbers 111. Hence, the distribution assigning probability 1 to ABC satisfies both E and O.

The question, whether the disjunction of (a) and (b) is a necessary condition for the consistency of E and O, must be answered in the negative. Cases (a) and (b) are not the only ones for which the two postulates can be reconciled. Case (d) serves as a counter-example. For this structure of preferences, the postulate E requires the following grouping:

$$
(6) \quad
\begin{array}{llll}
d_1 & ACB & 113 & x_1 = 0 \\
d_2 & BAC & 231 & x_2 = 0 \\
d_3 & CBA & 322 & x_3 = 0 \\
\hline
d_4 & ABC & 121 & x_4 = 0.5 \\
d_5 & BCA & 212 & x_5 = 0.5 \\
\hline
d_6 & CAB & 333 & x_6 = 0
\end{array}
$$

By E, we have $x_1 = x_2 = x_3$ and $x_4 = x_5$. By O, $x_3 = 0$ (d_3 dominated by d_4 and d_5) and $x_6 = 0$ (d_6 dominated by any other division). Therefore, the distribution on the right-hand side is implied by E and O. It yields the table of probabilities:

$$
(7) \quad
\begin{array}{c|ccc}
 & 1 & 2 & 3 \\
\hline
\text{No 1} & 0.5 & 0.5 & 0.5 \\
\text{No 2} & 0.5 & 0.5 & 0.5 \\
\text{No 3} & 0 & 0 & 0
\end{array}
$$

We thus see that joint satisfiability of E and O is not limited to the two extreme cases of total conflict and complete lack of conflict. A sufficient and necessary condition is, however, not easy to find.

6. EQUAL CHANCES OF CHOICE

The postulate to treat symmetrically the participants in the division can be interpreted in more ways than one. Instead of speaking in terms of a person getting the k-th object of preference, we could consider the order in which the participants are allowed to make their choice. Equality would then consist in such a randomization that everybody has the same chance of being the k-th to make his choice ($k = 1, \ldots, n - 1$; the n-th person has, of course, no choice but to take the object that is left; if $n < m$, the same goes for the remaining persons: $n + 1, \ldots, m$ who then get $\varnothing$). Formally we have the postulate

(C) $\Pr\{i$ is the k-th to choose$\} = \Pr\{j$ is the k-th to choose$\}$ for all $i \neq j = 1, \ldots, m$ and all $k = 1, \ldots, n - 1$.

This amounts to postulating equiprobability of all possible orders in which people present themselves to make their choice.

The use of the term 'choice' in the wording of postulate C helps to see the rationale of C. The 'decisional' language is, however, inessential, since C can be expressed directly in terms of individual preferences. The procedure is as follows.

For any permutation of participants in the division, we assign to the first element of that permutation the highest ranking object in his hierarchy of preference; out of the remaining objects, we assign to second person the highest ranking object in his hierarchy of preference; and so on, until the set G is exhausted. In this way, we obtain a specific division, determined uniquely by the permutation we started with (the uniqueness depends on the assumption that only strong preferences are allowed). Postulate C demands that all permutations of participants be equiprobable. Therefore, all the divisions generated by them must also be equiprobable. Of course, some divisions appear more than once, because they are generated by more than one permutation. Their probabilities increase accordingly.

Let us see how the postulate C works in the case of example (c), used above to show inconsistency of E and O. Next to the permutations of the elements of P, I list the divisions they generate.

$$
\begin{array}{ll}
& 123\ ABC \quad 231\ BAC \\
(8) \quad & 132\ ABC \quad 312\ BCA \\
& 213\ BAC \quad 321\ CBA
\end{array}
$$

(8) entails the following distribution:

$$
\begin{array}{ll}
& ABC\ 2/6 \quad BCA\ 1/6 \\
(9) \quad & ACB\ 0 \quad\ \ \ CAB\ 0 \\
& BAC\ 2/6 \quad CBA\ 1/6
\end{array}
$$

The table of probabilities, analogous to (5), summarizes this result:

(10)

	1	2	3
No 1	2/6	2/6	4/6
No 2	3/6	3/6	2/6
No 3	1/6	1/6	0

It is easy to see that the two divisions, ACB and CAB, responsible for the conflict of postulates E and O (because they are the dominated ones), now have probability zero, which means that the distribution (9) satisfies Pareto optimality. This is no accident. The procedure employed in the transition from permutations of persons to corresponding divisions automatically ensures Pareto optimality, since in each step an individual is assigned his *best* object, out of the set of objects that are still available. Dominated divisions are thus eliminated and there can be no conflict between postulates C and O.

A comparison of (10) with (5) shows that equality of chances exhibited in (5) is now violated in favor of the person 3. This phenomenon is easily explained by the fact that 3 profits from his somewhat excentric tastes: preferences of 1 and 2 are identical, hence the competition between them is complete. It is also worth noticing that the position of 1 and 2 is also improved when we pass from (5) to (10) Their gain was made possible by the fact that 'absolute equality', postulated in E and implemented in (5), has been given up.

7. THE TWO VERSIONS OF EQUALITY

Let me first point out some formal differences between E and C. The postulate C determines uniquely the distribution, for any profile of preference orderings (provided that only strong preferences are allowed). On the other hand, E specifies a class, usually infinite, of distributions that are satisfactory from this point of view. For instance, there are infinitely many distributions satisfying (3), although they all yield the table (5). It follows that E and C diverge when the structure of preferences is such that the C-distribution does not belong to the E-class of distributions.

It is rather obvious that the two extreme cases (*a*) and (*b*), i.e., total conflict of interests and complete lack of conflict, do ensure the agreement of E and C. They are, however, not necessary for the agreement to take place. If individual preferences exhibit certain symmetry, this is reflected in the C-distribution which, for this reason, satisfies the requirement E. For instance, in the case of $m = 4$, the profile

(11)
$$
\begin{array}{c}
1\ 2\ 3\ 4 \\
A\,A\,C\,C \\
B\,B\,D\,D \\
C\,C\,A\,A \\
D\,DB\,B
\end{array}
$$

yields, by C, equiprobability of the four non-dominated divisions

(12)
$$
\begin{array}{l}
ABCD\ (1212)\ 0.25 \\
ABDC\ (1221)\ 0.25 \\
BACD\ (2112)\ 0.25 \\
BADC\ (2121)\ 0.25
\end{array}
$$

while any other division gets the probability zero. The distribution (12) evidently satisfies E, by granting every person probability 0.5 for the top object and 0.5 for the second best object.

The essential difference between the two requirements consists, of course, in the meaning of equality. According to E, the participants are (probabilistically) equal with respect to the degree to which their wishes are satisfied. According to C, they are given (probabilistically) equal access to the set of objects they value. This is far from being the same thing.

It can be said that E demands more, in the sense that it postulates equal satisfaction of tastes, irrespective of what these tastes are. Postulate C is more formal: in demanding equal access to G, it makes the degree of satisfaction depend, for each individual, upon the configuration of tastes in the group.

For some profiles of preferences, C can yield results which seem rather unintuitive, unless one knows the procedure by which they have been obtained. Thus, for the profile (d), we obtain by C the following probabilities.

(13)
	1	2	3
No 1	4/6	3/6	3/6
No 2	2/6	3/6	2/6
No 3	0	0	1/6

When we compare this with the symmetric table (7), we see that in the change from E to C, the first individual has gained and the third has lost, while the position of the second person has remained unchanged. It is not immediately clear why this should be so.

On the other hand, the desirable property of C is that the distribution it determines is always Pareto-optimal, i.e. dominated divisions are excluded. The C-distribution can be strikingly efficient as shown by the example (*e*). Given such profile of preference orderings, the condition imposed by E defines a family of distributions in terms of the numbers *a* and *b* ($a + b = 1/3$), whereas C eliminates all divisions except those represented by the ordinal numbers 121 and 211.

$$
(14) \quad
\begin{array}{llll}
ABC & a\ 0.5 & ACB & b\ 0 \\
BCA & a\ 0 & BAC & b\ 0.5 \\
CAB & a\ 0 & CBA & b\ 0
\end{array}
$$

The resulting tables of probabilities for E and C are, respectively

(15)

	1	2	3		1	2	3
No 1	1/3	1/3	1/3	No 1	1/2	1/2	1
No 2	1/3	1/3	1/3	No 2	1/2	1/2	0
No 3	1/3	1/3	1/3	No 3	0	0	0

The contrast between fundamental equality and optimal allocation of objects is here apparent.

SUMMARY

There is, of course, much more to the concept of distributive justice than what was discussed above. It is usually assumed that members of the society have a common, or at least prevailing, hierarchy of values. Then the main ingredient of justice is a criterion, according to which the distribution takes place. Perhaps the most difficult problems are due to the conflict of several, equally acceptable, criteria.

In the present analysis, it was assumed that the only factor by which the individuals are differentiated, was the preference ranking of goods. Justice was reduced to egalitarianism. In order to make equality possible, randomization was introduced, in spite of its somewhat artificial character.

It turned out that even in such extremely simple situation there are at least two acceptable concepts of justice. The first one is more radical, in the sense that it postulates equal degree of satisfaction for everybody. It has the drawback of being non-optimal. A more modest requirement is, to give everybody equal opportunity to choose. It implicitly accords privilege to people whose preferences deviate from the prevailing pattern.

Finally, a word of comment on the relation between the problem of distribution rule and that of social welfare function. Similarity of data is obvious: in both cases, the domain of the function is the set of all profiles of preference orderings. Their ranges are different. But the analogy goes further than that. It is easy to notice that, for each individual, the preference ordering of the set G of goods determines uniquely the preference ordering of the set D of all possible divisions. We, therefore, can translate the original problem into the terminology of Arrow. The social welfare function would lead from the set of all profiles of preference orderings of D to D itself. By the impossibility theorem, no such function satisfying the postulates of democracy exists. Unless, of course, the transition from preferences on G to preferences on D restricts the domain of the function in such a way as to make the theorem of Arrow inoperative. This, however, is another problem.

University of Warsaw

NOTES

[1] K. Szaniawski, 'O pojęciu podziału dóbr' ['On the Concept of Distribution of Goods'], *Studia Filozoficzne* 2 (1966). The main result is described in K. Szaniawski, 'Formal Analysis of Evaluative Concepts', *International Social Science Journal* 27 (1975).
[2] The postulate has formerly been expressed in a more general way, viz. that no one gets $k + 2$ (or more) objects while someone else gets k objects. Coupled with the assumption that there can be more objects than persons, this had counterintuitive consequences. Preference ordering of the set G does not uniquely determine preference ordering of the set 2^G. Thus we can have it that both person 1 and person 2 strongly prefer A to $\{B, C, D\}$. Then, under the natural assumption of monotonicity: $X \subset Y \Rightarrow X <_k Y(X$ and Y ranging over $2^G)$, the division d_1 attributing A to 1 and $\{B, C, D\}$ to 2 is more egalitarian than the division d_2 attributing $\{A, B\}$ to 1 and $\{C, D\}$ to 2. But d_1 is prohibited by the above postulate, while d_2 is not.
One way to block this undesirable consequence is to introduce the restriction: $n \leq m$. The danger in unrestricted use of R was pointed out to me by Mogens Blegvad. I want to thank Professor Blegvad for his valuable criticism.

KRISTER SEGERBERG

SOME REFLECTIONS ON METHOD IN THE THEORY OF SOCIAL CHOICE

1. INTRODUCTION

A group of individuals deliberating which of several alternatives to select as the decision of the group as a whole – that is the kind of situation one encounters in *social choice theory*. To find general rules – if such there are – for helping groups to decide wisely is what the theory of social choice is about.

Suppose again that you have a group of individuals who are to divide among themselves a set of items in such a way that each individual gets exactly one item (there are as many individuals as items, say). This problem is sometimes called the assignment problem. The study of general rules for picking good assignments of items to individuals seems to have no standard name, but for the purposes of this paper *assignment theory* will do.

The present paper deals with the relationship between social choice theory and assignment theory. It seems as if this subject was first broached by Peter Gärdenfors. In [1] Gärdenfors discussed solutions to the assignment problem that depend only on what preferences the various individuals in the group hold over the available items. While, as he says, his approach "may bring to mind the theory of social choice", he warns that "the assignment problem has a different logical structure". As this author remarked briefly in [4], this warning could be misleading, for in a clear theoretical sense assignment theory is but a special case of social choice theory; this connection is explained in detail below. The semi-published paper [3] shows that Gärdenfors is aware of this connection but that he regards it without much enthusiasm. However, it seems to this author that the connection has at least methodological or foundational interest. This paper is written in defense of that view.

147

E. Saarinen, R. Hilpinen, I. Niiniluoto, and M. Provence Hintikka (eds.), Essays in Honour of Jaakko Hintikka, 147–159. All Rights Reserved.

2. GENERALITIES

If X is any set, then $|X|$ denotes the cardinality of X.

Let X be a set and R a binary relation in X. R is *transitive* if

$$\forall x, y, z(\langle x, y \rangle \in R \wedge \langle y, z \rangle \in R \cdot \Rightarrow \cdot \langle x, z \rangle \in R),$$

strongly connected if

$$\forall x, y(\langle x, y \rangle \in R \vee \langle y, x \rangle \in R),$$

reflexive if

$$\forall x(\langle x, x \rangle \in R),$$

and *antisymmetric* if

$$\forall x, y(\langle x, y \rangle \in R \wedge \langle y, x \rangle \in R \cdot \Rightarrow \cdot x = y).$$

If R is transitive and strongly connected, then we say that R is a *preference order* (*in* X). Thus preference orders are reflexive but not necessarily antisymmetric.

If R is a preference order in X, then we write, for all $x, y \in X$,

(1) $x \succsim y$ *(rel R)* iff $\langle x, y \rangle \in R$,
(2) $x \succ y$ *(rel R)* iff $\langle x, y \rangle \in R \wedge \langle y, x \rangle \notin R$,
(3) $x \sim y$ *(rel R)* iff $\langle x, y \rangle \in R \wedge \langle y, x \rangle \in R$.

This symbolism may be read as follows, where (1) is a reading of (1'), (2) of (2'), and (3) of (3'):

(1')   'x is at least as good as y, relative to R',
(2')   'x is better than y, relative to R',
(3')   'x and y are equally good, relative to R'.

If it is clear from the context what preference order is referred to, we may omit the suffix '*(rel R)*'.

If R is a preference order in X and $x \in X$, then we define the *rank*, the *positive rank*, and the *negative rank of x in R* – in symbols, *rank* (R, x), *rank*$^+$ (R, x), and *rank*$^-$ (R, x), respectively – as follows:

$$rank\ (R, x) = rank^+\ (R, x) - rank^-\ (R, x),$$
$$rank^+\ (R, x) = |\{y : x \succ y\ (rel\ R)\}|,$$
$$rank^-\ (R, x) = |\{y : y \succ x\ (rel\ R)\}|.$$

3. SOCIAL CHOICE

Let I be a set of *individuals* and A a set of *alternatives*. A *(social choice)* *situation involving I over A* is a function from I to the set of preference orders in A. If s is a situation involving I over A, and if $i \in I$ and a, $b \in A$, then we may say that *in s, i prefers a to b* if and only if $a \succ b$ (*rel* $s(i)$), and that *in s, i is indifferent between a and b* if and only if $a \sim b$ (*rel* $s(i)$).

A *social choice function (for I over A)* is a function assigning to each situation involving I over A a preference order in A. If F is a social choice function for I over A, s is a situation involving I over A, and a, $b \in A$, then we may say that *in s, I as a group prefers a to b under F* if and only if $a \succ b$ (*rel* $F(s)$), and that *in s, I as a group is indifferent between a and b under F* if and only if $a \sim b$ (*rel* $F(s)$). An alternative $a \in A$ is *optimal in s according to F* if, for all $b \in A$, $a \succeq b$ (*rel* $F(s)$).

There are countless social choice functions to choose between. If a particular candidate is attractive but not perfect there are various ways of trying to improve it. One that will play a role in this paper is iteration. This device may be expedient if one seeks a function that is decisive in the sense that it keeps to a minimum the number of couples $\langle a, b \rangle$ such that $a \sim b$ (relative to the preference orders picked by the function). Suppose for a certain situation s involving I over A and alternative $a \in A$ the set

$$[a] = \{b : a \sim b \ (rel \ F(s))\}$$

contains more than one member. In order to obtain a more decisive order one might try to rank the elements of $[a]$. Now even though they come out equally good when F is applied to s, it is conceivable that the elements of $[a]$ become differentiated if F is applied to the *restricted situation* $s{\restriction}[a]$ *involving I over* $[a]$, where, for each $i \in I$,

$$(s{\restriction}[a])(i) = s(i) \cap [a] \times [a].$$

So one may try that. And this procedure may be repeated again and again until no more changes result.

An informal presentation of this idea is found in [2]. The following is a more formal account.

Let F be any social choice function (for some given I and A). For any nonnegative integer n we define the *n'th iteration* F^n of F as follows. First,

$$F^0 = F.$$

Next, suppose that F^n has been defined. Then we stipulate, for any

$a, b \in A$ and any situation s,

$$a \underset{\sim}{\succ} b \; (rel \; F^{n+1}(s)) \text{ iff}$$
either $a \succ b \; (rel \; F^n(s))$
or $a \sim b \; (rel \; F^n(s))$ and $a \underset{\sim}{\succ} b \; (rel \; F(s{\restriction}[a]_n))$,

where

$$[a]_n = \{c : a \sim c \; (rel \; F^n(s))\}.$$

(The fact that F is a social choice function can be used to show that this definition is meaningful.) By the *iterative closure* F^ω of F we mean the function $\bigcap_{n < \omega} F^n$. It is easy to see that if I and A are finite, then there is some $n < \omega$ such that, for all $i > n$, $F^i = F^\omega$.

By the *Borda function*, for any finite fixed I and A, we mean the function B such that, for all situations s involving I over A, and all $a, b \in A$,

$$a \underset{\sim}{\succ} b \; (rel \; B(s)) \text{ iff} \sum_{i \in I} rank \; (s(i), a) \geq \sum_{i \in I} rank \; (s(i), b).$$

For discussion of and intuitive considerations on the Borda function as well as what we call its iterative closure, see [2].

4. ASSIGNMENT

Let I be a finite set of *individuals* and J a set of *items*. For technical reasons we assume, throughout the paper, that $|I| = |J|$ (see [1], p. 332). A(n *assignment*) situation involving I over J is a function from I to the set of preference orders in J. If σ is a situation involving I over J, and if $i \in I$ and $j, k \in J$, then we may say that *in σ, i prefers getting j to getting k* if and only if $j \succ k \; (rel \; \sigma(i))$, and that *in σ, i is indifferent between getting j and getting k* if and only if $j \sim k \; (rel \; \sigma(i))$.

An *assignment function* (*for I over J*) is a function assigning to each situation involving I over J a preference order in the set B of bijections from I to J. (A bijection from I to J is a function from I to J that is one-one and onto; since $|I| = |J|$, this definition is meaningful.) If F is an assignment function for I over J, σ is a situation involving I over J, and $b, c \in B$, then we may say that *in σ, I as a group prefers b to c under F* if and only if $b \succ c \; (rel \; F(\sigma))$, and that *in σ, I as a group is indifferent between b and c under F* if and only if $b \sim c \; (rel \; F(\sigma))$. A bijection $b \in B$ is *optimal* *in σ according to F* if, for all $c \in B$, $b \succeq c$ (rel $F(\sigma)$).

As in the case of social choice above, there are indefinitely many assignment functions. And also here we may encounter the problem of trying to enhance the decisiveness of a particular assignment function by an iterative procedure. Suppose for a certain situation σ involving I over J and a certain member b of the set B of bijections from I to J that the set

$$[b] = \{c : b \sim c \text{ (rel } F(\sigma))\}$$

contains more than one member. In order to obtain a more decisive order one might try to rank the elements of $[b]$. Now even though they come out equally good when F is applied to σ, it is conceivable that the elements of $[b]$ become differentiated if F is applied to the *restricted situation* $\sigma | I_b$ involving I_b over J_b, where

$$I_b = \{i \in I : \exists c, d \in [b](c(i) \neq d(i))\},$$
$$J_b = \{j \in J : \exists i \in I \exists c, d \in [b](j = c(i) \wedge j \neq d(i))\},$$

and, for each $i \in I$,

$$(\sigma | I_b)(i) = \sigma(i) \cap J_b \times J_b.$$

So one may try that. And this procedure may be repeated again and again until no more changes result.

[Digression: It is worth mentioning that of course one must check that the definition of $\sigma | I_b$ is meaningful. Among other things one must show that $|I_b| = |J_b|$. I and J are finite, so the easiest way to do that is perhaps to show that $|I - I_b| = |J - J_b|$, which may be accomplished as follows. For each $i \in I - I_b$ let $\varphi(i)$ denote the unique $j \in J$ such that, for all $c, d \in [b]$, $c(i) = d(i) = j$. We note the following points.

(1) The domain of φ is $I - I_b$: this is clear, for if $i \in I - I_b$ then, for all $c, d \in [b]$, $c(i) = d(i)$, so $\varphi(i)$ is well defined.

(2) The range of φ is included in $J - J_b$. For take any $i \in I - I_b$ and suppose that $\varphi(i) \in J_b$; we intend to show that this hypothesis is absurd. Since $\varphi(i) \in J_b$ there are some $i' \in I$ and $c, d \in [b]$ such that

$$\varphi(i) = c(i') \text{ and } \varphi(i) \neq d(i').$$

By definition of φ,

$$\varphi(i) = c(i) \text{ and } \varphi(i) = d(i).$$

But as both c and d are bijections, the fact that $c(i) = c(i')$ and $d(i) \neq d(i')$ implies that $i = i'$ and $i \neq i'$, which is impossible.

(3) φ is one-one. For take any i, $i' \in I - I_b$ such that $i \neq i'$. Then $\varphi(i) = b(i)$ and $\varphi(i') = b(i')$. As b is a bijection, the fact that $i \neq i'$ implies that $b(i) \neq b(i')$ and hence that $\varphi(i) \neq \varphi(i')$.

(4) φ is onto. For take any $j \in J - J_b$. Then, by the definition of J_b, for all $i \in I$ and all $c, d \in [b]$, if $j = c(i)$ then $j = d(i)$. Let i' be the individual such that $b(i') = j$. Then, for all $c \in [b]$, $c(i') = j$. Hence $\varphi(i') = j$.

Consequently, by (1)–(4), $|I - I_b| = |J - J_b|$. End of digression.]

To make the preceding idea precise, let F be any assignment function (for some given equipotent I and J). For any nonnegative integer n we define the $n'th$ *iteration* F^n of F as follows. First,

$$F^0 = F.$$

Next, suppose that F^n has been defined. Then we define, for any bijections b, c from I to J and any situation σ involving I over J,

$$b \succeq c \ (rel \ F^{n+1}(\sigma)) \text{ iff}$$
either $b \succ c \ (rel \ F^n(\sigma))$
or $b \sim c \ (rel \ F^n(\sigma))$ and $b \succeq c \ (\ rel \ F(\sigma|I_{b,n}))$,

where

$$I_{b,n} = \{i \in I : \exists c, d(b \sim c, b \sim d \ (rel \ F^n(\sigma)) \lor c(i) \neq d(i))\}.$$

(The fact that F is an assignment function guarantees that this definition is meaningful.) By the *iterative closure* F^ω of F we mean the function $\bigcap_{n < \omega} F^n$. It is easy to see that if I, and hence J, are finite, then there is some $n < \omega$ such that, for all $i > n$, $F^i = F^\omega$.

By (*Gärdenfors's*) *position-counting function*, for any fixed finite, equipotent I and J, we mean the function P such that, for all situations σ involving I over J, and all bijections b, c from I to J

$$b \succeq c \ (rel \ P(\sigma)) \text{ iff} \sum_{i \in I} rank \ (\sigma(i), b(i)) \geq \sum_{i \in I} rank \ (\sigma(i), c(i)).$$

For discussion of and intuitive considerations on Gärdenfors's position-counting function as well as its iterative closure, see [1].

5. THE CONNECTION

Sections 3 and 4 were deliberately written in such a way as to suggest a far-reaching similarity between social choice theory and assignment theory. That this similarity is not accidental will be shown in this section.

Let I and J be two fixed finite nonempty equipotent sets: I a set of individuals and J a set of items. Let B be the set of bijections from I to J. A bit loosely we may say that it is the assignment theoretician's task to find an assignment function for I over J 'with desirable properties'. It is interesting that a social choice theoretician would be perfectly able to address himself to the same problem, even though naturally he would do so in his accustomed manner: he would take his task to be to find a social choice function for I over B 'with desirable properties'.

It is tempting to ask whether an assignment theoretician and a social choice theoretician will reach 'the same result' if they have 'the same outlook'. Thus phrased this question is of course intolerably vague. But instead of trying to make it precise we shall exhibit some examples of assignment functions and social choice functions which may be said to be inspired by 'the same' intuitive considerations. As we shall see, in some cases they agree and in some cases they don't.

In the version of assignment theory that we are considering here, an individual pays attention only to what he himself will get. That means that when the social choice theoretician asks individual i to rank the bijections in B, then i will judge them by their value at i. In other words, he will prefer bijection b to bijection c if and only if he prefers item $b(i)$ to item $c(i)$. From the point of view of i, two bijections are equally desirable if and only if they assign the same item to him.

Somewhat more formally, let us say that a social choice situation s involving I over B *reduces to* an assignment situation σ involving I over J if, for all $i \in I$ and all $b, c \in B$,

$$\langle b, c \rangle \in s(i) \text{ iff } \langle b(i), c(i) \rangle \in \sigma(i).$$

For each $i \in I$, let f_i be the function from B to J such that, for all $b \in B$,

$$f_i(b) = b(i).$$

The following lemma will bring out the point made above about the connection between social choice theory and assignment theory.

LEMMA. *If s reduces to σ, then, for every $i \in I$, f_i is a homomorphism from $\langle B, s(i) \rangle$ to $\langle J, \sigma(i) \rangle$. Moreover, if $|I| = |J| = n$, then, for every $j \in J$,*

$$|\{b \in B : f_n(b) = j\}| = (n - 1)!.$$

Proof: (i) Take any $j \in J$. There are bijections b such that $b(i) = j$; let b_0 be one of them. Then $f_i(b_0) = j$. So f_i is surjective.

(ii) Take any b, $c \in B$. Then $\langle b, c \rangle \in s(i)$ iff $\langle b(i), c(i) \rangle \in \sigma(i)$ (by reduction) iff $\langle f_i(b), f_i(c) \rangle \in \sigma(i)$ (by definition of f_i). So f_i is order-preserving both ways.

(iii) If j is any item, then $f_i(b) = j$ iff $b(i) = j$. Thus the condition that $f_i(b) = j$ fixes the value of b for one argument – for each of the $n - 1$ remaining arguments anything goes. That is to say, the condition that $f_i(b) = j$ leaves $n - 1$ items in J to be assigned, in a one-to-one manner to $n - 1$ individuals in I. And there are of course $(n - 1)!$ ways in which that can be done.
Q.E.D.

COROLLARY. *If s reduces to σ, then, for all $b \in B$,*

$$rank\ (s(i),\ b) = (n - 1)!\ rank\ (\sigma(i),\ b).$$

We are now able to make certain comparisons between social choice theory and assignment theory.

THEOREM. *Any element of B is optimal according to the Borda function if and only if it is optimal according to Gärdenfors's position-counting function.*

Proof: Let s and σ be situations involving I over B and involving I over J, respectively, such that s reduces to σ. Now, a bijection $b \in B$ is optimal in s according to the Borda function if and only if

(1)   for all $c \in B$, $\displaystyle\sum_{i \in I} rank\ (s(i), b) \geqq \sum_{i \in I} rank\ (\sigma(i), c)$,

while b is optimal in σ with respect to Gärdenfors's position-counting function if and only if

(2)   for all $c \in B$, $\displaystyle\sum_{i \in I} rank\ (\sigma(i), b) \geqq \sum_{i \in I} rank\ (\sigma(i), c)$.

Since I is nonempty by assumption $(n - 1)! > 0$. Therefore, by the preceding corollary, (1) and (2) are equivalent.
Q.E.D.

In the sense made precise by the proof of the theorem, the Borda social choice function and Gärdenfors's position-counting assignment function are equivalent. However, the iterative closures of these functions are not equivalent in that sense. The remainder of the section is devoted to establishing this claim.

Consider the example given in [1], p. 337; it may be rendered as follows. Let I be a set of five individuals 1, 2, 3, 4, 5, and let J be a set of five items A, B, C, D, E. Let σ be the situation involving I over J which may be represented as follows:

$\sigma(1)$	$\sigma(2)$	$\sigma(3)$	$\sigma(4)$	$\sigma(5)$
A	A	D	D	B
B	D	B	B	D
C	C	E	E	C
D	B	A	A	A
E	E	C	C	E

(This mode of representation should be understood in such a way that, for example, $\sigma(2)$ is the smallest preference order relative to which $A \succ D$, $D \succ C$, $C \succ B$, $B \succ E$; and so on.)

As the reader easily verifies there are four bijections from I to J that are optimal in σ according to Gärdenfors's position-counting function, viz.:

$$b_1 = \{\langle 1, A\rangle, \langle 2, C\rangle, \langle 3, D\rangle, \langle 4, E\rangle, \langle 5, B\rangle\},$$
$$b_2 = \{\langle 1, A\rangle, \langle 2, C\rangle, \langle 3, E\rangle, \langle 4, D\rangle, \langle 5, B\rangle\},$$
$$b_3 = \{\langle 1, C\rangle, \langle 2, A\rangle, \langle 3, D\rangle, \langle 4, E\rangle, \langle 5, B\rangle\},$$
$$b_4 = \{\langle 1, C\rangle, \langle 2, A\rangle, \langle 3, E\rangle, \langle 4, D\rangle, \langle 5, B\rangle\}.$$

In order to apply the iteration procedure described in Section 4, one must consider the following restricted situation σ' involving $\{1, 2, 3, 4\}$ over $\{A, C, D, E\}$:

$\sigma'(1)$	$\sigma'(2)$	$\sigma'(3)$	$\sigma'(4)$
A	A	D	D
C	D	E	E
D	C	A	A
E	E	C	C

It is readily seen that in σ' only two of the four bijections considered are optimal according to the position-counting function, viz., b_3 and b_4. So b_3 and b_4 are the only bijections optimal in σ according to the first iteration of the position-counting function.

In order to continue the iteration procedure, one would have to consider the following restricted situation σ'' involving $\{3, 4\}$ over $\{D, E\}$:

$\sigma''(3)$	$\sigma''(4)$
D	D
E	E

But it is clear that again both b_3 and b_4 are optimal according to the position-counting function, and that continued iteration always leads to the same situation. Consequently – this result is in [1] –

($\sharp$) b_3 and b_4 are the only bijections that are optimal in σ according to the iterative closure of Gärdenfors's position-counting function.

Let s be any situation involving I over the set of bijections from I to J that reduces to σ. By our theorem, b_1, b_2, b_3, b_4 are precisely the bijections that are optimal in s according to the Borda function. It follows from the definition of reduction that in order to apply the iteration procedure described in Section 3, one must consider the following restricted situation s^* involving I over $\{b_1, b_2, b_3, b_4\}$:

$s^*(1)$	$s^*(2)$	$s^*(3)$	$s^*(4)$	$s^*(5)$
b_1, b_2	b_3, b_4	b_1, b_3	b_2, b_4	b_1, b_2, b_3, b_4
b_3, b_4	b_1, b_2	b_2, b_4	b_1, b_3	

(This mode of presentation should be understood in such a way that, for example, $s^*(1)$ is the smallest preference order relative to which $b_1 \sim b_2$, $b_1 \succ b_3$, and $b_3 \sim b_4$; and so on.) The situation is completely symmetric, so b_1, b_2, b_3, b_4 are all optimal in s^* according to the Borda function. And it is clear that further iteration will not change the picture. Consequently,

(*) b_1, b_2, b_3, and b_4 are the bijections that are optimal in s according to the iterative closure of the Borda function.

Thus we have established our claim that the iterative closure of the Borda function is not equivalent to the iterative closure of Gärdenfors's position-counting function, for, by (*) and ($\sharp$), b_1 and b_2 are optimal in s according to the former function but not in σ according to the latter.

6. CONCLUSIONS

It was asserted in Section 1 that 'in a clear theoretical sense assignment theory is but a special case of social choice theory'. This assertion was made precise in Section 5 by the definition of the concept of reduction. It remains to try to assess the importance of this connection between assignment theory and social choice theory.

Before proceeding to do so it may be wise to insert the following disclaimer: The connection may well lack all practical importance; it probably does. If a problem can be handled within the framework of assignment theory as conceived of here, then it is handled more simply there than the corresponding problem in social choice theory. Gärdenfors claims as much, and it is difficult to disagree with him.

The importance of the connection would rather seem to be found at the methodological level. In particular, the connection may be of interest when it comes to appraising particular assignment functions. The most striking results in social choice and assignment theory are the negative ones among which Arrow's celebrated impossibility theorem was the first, stating that functions meeting certain conditions don't exist. However, a considerable portion of the work done in these areas consists of the examination of particular functions. Often this work takes the form of a discussion of which properties the function in question has or doesn't have; some properties, it is said or understood, are desirable while others are not. Thus it is quite in keeping with tradition when Gärdenfors himself says, without further justification, that "decisiveness seems to be a desirable property" of a social choice function ([2], p. 15) or that "it seems inevitable to demand that a reasonable assignment function shall satisfy [his condition of] neutrality" while his condition of symmetry "might sometimes be regarded as too strong" ([1], p. 333). This seemingly dogmatic attitude is of course neither necessarily bad nor necessarily dogmatic. In any case it is how authors in any field of pure science seem to be writing.

This does not mean that anything goes in pure science. Consistency, for example, is one important requirement, and not only in the logical sense. In particular, if on one occasion an author favors a certain solution, then he ought to favor a similar solution on every relevantly similar occasion.

This brings us back to the topic of the paper. The Borda function and the position-counting function may be said to express – each in its field – the same idea: to let the objects of discourse be evaluated according to a certain procedure, and to let the outcome determine the group's decision. Therefore it is rather to be expected that the two functions yield the same result when they are applied to the same specific problem situation. The

Theorem of Section 5 guarantees that they do, which is encouraging.

Similarly, the iteration procedures of Sections 3 and 4 may be said to embody the same natural idea, viz., functions wanting in decisiveness may be improved by re-applying the function to whatever 'ties' earlier applications of it have produced. As we saw in Section 5, this idea may lead to different results in social choice and assignment theory. If one is encouraged by the agreement between the Borda function and the position-counting function, then one should perhaps be discouraged to a corresponding degree by the failure of their iterative closures to agree.

Some explanation is called for. Perhaps the simplest one would be to hold that the iterative procedures of Sections 3 and 4 are not really the same, even though at first we thought they were. To be sure, there is similarity; they contain the same idea of applying the favored function to a situation that is suitably restricted in a natural manner. But there is difference, too, it may be held, above all in the ways in which the restriction is effected. When in social choice theory one wants to try to order the alternatives in a tie, it is natural to omit from consideration the alternatives outside the tie; so the restricted situation will involve all the individuals over the remaining alternatives. On the other hand, when in assignment theory one wants to try to order the bijections in a tie, it is natural to omit from consideration what all the bijections in the tie agree on, i.e., certain individuals and the items assigned to them by all those bijections; so the restricted situation will involve only the remaining individuals over the remaining items.

In this manner it may be argued that the two iteration procedures don't embody quite the same idea. The argument just outlined could easily be made precise. The somewhat forbidding formalism laid down in this paper lends itself readily to that kind of purpose.

Whether the argument ought to be accepted is a different question. Yet another question is on precisely what grounds we are entitled (if we are!) to persist in regarding the Borda function and the position-counting function as expressions of the same idea. There seem to be no ready-made criteria by which to settle such questions. Indeed, one way to look at the results in Section 5 is to regard them as indicating the need for an analysis of 'the same'.

Åbo Academy

NOTE

An early version of this paper was presented to the 5th International Congress of Logic, Methodology and Philosophy of Science in London, Ontario, Canada, 1975, and has later appeared as [5]. The final version was completed at the University of Kansas while the author was on leave from Åbo Academy.

BIBLIOGRAPHY

[1] Gärdenfors, Peter, 'Assignment Problem Based on Ordinal Preferences', *Management Science* **20** (1973), 331–340.
[2] Gärdenfors, Peter, 'Positionalist Voting Functions', *Theory and Decision* **4** (1973), 1–24.
[3] Gärdenfors, Peter, 'Methods of Social Choice Theory Applied to Other Problems', working paper, mimeographed by the Mattias Fremling Society and the Department of Philosophy, University of Lund, 1973.
[4] Segerberg, Krister, Review of [1]. *Theoria* **40** (1974), 212–214.
[5] Segerberg, Krister, 'The Assignment Problem as a Problem of Social Choice', in *Filosofiska smulor tillägnade Konrad Marc-Wogau*, Ann-Mari Henschen-Dahlquist (ed.), pp. 152–157, Filosofiska Föreningen, Uppsala, 1977.

IV

PHILOSOPHICAL LOGIC

DAVID LEWIS

A PROBLEM ABOUT PERMISSION*

1. THE GAME

Consider a little language game that is played as follows.

(1) There are three players, called the *Master*, the *Slave*, and the *Kibitzer*. It would change nothing to have more than one slave, or more than one kibitzer, but let us put aside the complications that arise if a slave must serve two masters. (They say it can't be done.)

(2) There is a certain set of strings of symbols, called the set of *sentences*. A player may at any time make the move of *saying* any sentence *to* any other player within earshot.

(3) There is a certain function that assigns to any sentence $\varnothing$, at any pair $\langle t, w \rangle$ of a time t during the game and a suitable possible world w, a value 1 or 0 called the *truth value* of $\varnothing$ at t at w. (We leave off the 'at w' when w is the actual world.) $\varnothing$ is called *true* or *false* at t at w according as the truth value is 1 or 0.

(4) There is another function that assigns to any such pair $\langle t, w \rangle$ a set of worlds called the *sphere of permissibility* at t at w. Worlds in this set are said to be *permissible* at t at w.

(5) There is another function that assigns to any such pair $\langle t, w \rangle$ a set of worlds called the *sphere of accessibility* at t at w. Worlds in this set are said to be *accessible* at t at w. These worlds are the alternatives, including always w itself, that are left open by the past history of w up to t. They share that history, but they continue it in divergent ways. Spheres of accessibility are always contracting (except in trivial cases) and the contraction is irreversible: once a world has become inaccessible, it remains so forevermore. (I am not sure, but perhaps we should impose another condition: if one world is accessible at t at another, then the two worlds have exactly the same sphere of accessibility at t.)

(6) The $\langle t, w \rangle$ pairs on which the functions listed in (3)–(5) are defined include all of those such that t is a time during the game and w is accessible (at the actual world) at the time when the game begins. Let us henceforth tacitly omit from consideration all times and worlds but these.

163

E. Saarinen, R. Hilpinen, I. Niiniluoto, and M. Provence Hintikka (eds.), Essays in Honour of Jaakko Hintikka, 163–175. All Rights Reserved.

(7) There is a certain symbol ! that may be prefixed to any sentence $\varnothing$ to make a new sentence ! $\varnothing$, called an *imperative* sentence, that is true at t at w iff $\varnothing$ is true at t at every world that is both accessible and permissible at t at w.

(8) There is a certain symbol ¡ that may be prefixed to any sentence $\varnothing$ to make a new sentence ¡ $\varnothing$, called a *permissive* sentence, that is true at t at w iff $\varnothing$ is true at t at some world that is both accessible and permissible at t at w.

(9) The sphere of permissibility at any time (at any world) depends as follows on the past history of the world. When the game begins, it is the set of all worlds. Thereafter it remains unchanged except when the Master says to the Slave an imperative or permissive sentence that would be false, when said, if the sphere remained unchanged. Then the sphere adjusts itself, if possible, to make the Master's sentence true. Suppose that at t the Master says to the Slave ! $\varnothing$; and suppose that the sphere of permissibility just before t contains some worlds, accessible at t, where $\varnothing$ is false at t. Then the sphere must contract to cut those worlds out: at t, and thereafter at least until the next change, none of those worlds are permissible. If the Master changes the sphere in this way by saying ! $\varnothing$, we say that the Master *commands* that $\varnothing$. Or suppose that at t the Master says to the Slave ¡ $\varnothing$; and suppose that the sphere of permissibility just before t contains no worlds, accessible at t, where $\varnothing$ is true at t; and suppose that there do exist some such worlds outside the sphere. Then the sphere must expand to take in some of those worlds: at t, and thereafter at least until the next change, some of those worlds are permissible. If the Master changes the sphere in this way by saying ¡ $\varnothing$, we say that the Master *permits* that $\varnothing$.

(10) The Slave tries to see to it that the actual world is within the sphere of permissibility at all times. If the Slave knows, at a time t, that he acts in a certain way at t throughout the worlds that are permissible and accessible at t – for instance, if he knows that at all such worlds he begins a certain task at t – then he tries to act in that way at the actual world.

(11) Each player tries to see to it that he never says a sentence to another player unless that sentence is true at the time when he says it. The Master, when he commands or permits, is automatically truthful since the sphere adjusts to make him so; other players, and even the Master when he is not commanding or permitting, are truthful by choosing sentences to say that are true at the worlds that conform to their beliefs.

2. COMMENTS

The point of the game, as regards commanding and permitting, is to enable the Master to control the actions of the Slave. What the Slave does depends on the present sphere of permissibility, which depends in turn on the Master's previous commands and permissions. We need not ask why the Slave is willing to play his part. Perhaps he does so by habit; perhaps he is coerced; perhaps he is obligated; or perhaps he hopes that the Master's control over him will be used to his benefit as well as to the Master's. In any case, the game is played. And we may suppose it to be common knowledge that the game is played: each player expects the others to play their parts, expects the others so to expect, and so on.

In this simple example, I have tried to merge two complementary approaches to the study of imperatives. The semantic analysis of ! and ¡ given in (3)–(8) is taken, with slight changes, from Chellas [1] and [2]. The treatment of commanding and permitting as part of a social practice for enabling one person to control another is taken from Stenius [5] and Lewis [4].

If there were only commanding and no permitting, the language game could be described more simply. We could drop (4)–(10) and replace them as follows. If at any time t the imperative sentences said by the Master to the Slave before t are given by the list

(L) $!\varnothing_1$ at $t_1, \ldots, !\varnothing_n$ at t_n,

then the Slave tries at t to see to it that $\varnothing_1$ is true at $t_1, \ldots,$ and that $\varnothing_n$ is true at t_n. On this account, the only truth value that we need to associate with an imperative sentence $!\varnothing$ is the truth value of the content sentence $\varnothing$ (at the time when it was commanded). We could call *this* the truth value of the imperative $!\varnothing$, and say that the Slave tries to see to it that the Master's previous imperatives to him are made true. That was my account of imperatives in [4]. But then what do we make of permission? It is easy enough to provide for annulment of commands: the Master may at any time remove an item from the list (L), after which the Slave acts as if that command never had been given. But permissions are not, in general, annulments of particular past commands. A permission may partly undo several past commands, without fully undoing any of them. We need a device for integrating a succession of commands and permissions. A list with additions and deletions is one such device, but it is not flexible enough. The sphere of permissibility is meant to be a better device to serve the same purpose.

Commanding and permitting are not the whole of our language game. As regards all other sentence-saying, the point of the game is to enable the players to impart information to one another. Whenever truthfulness is not automatic, the hearer who expects the speaker to be truthful can infer something about the speaker's beliefs from the sentences that the speaker is willing to say; and often the hearer can go on to infer conclusions about the world, premised on confidence that the speaker's beliefs about certain topics tend to be correct. To the extent that the speaker can anticipate these inferences, he can control the hearer's beliefs by what he says. In particular, one player may wish to inform another about the present state of the sphere of permissibility – that is, about the integrated effect of the Master's commands and permissions up to now. There is nothing to keep him from doing so, given the way we have set up the language game, by using the same imperative and permissive sentences that the Master himself uses to change the sphere. One and the same sentence '! *the Slave carries rocks all day*' may be said by the Master to the Slave to reshape the sphere of permissibility and cause the Slave to carry the rocks; by the Slave to the Kibitzer to elicit sympathy; by the Master to the Kibitzer to explain why the Slave is not working on his usual chores; by the Slave or the Kibitzer to the forgetful Master to remind him what the Slave is supposed to be doing; and so on. It may even be used by the Master to the forgetful Slave as a reminder, with no further adjustment of the sphere of permissibility. Likewise '¡ *the Slave does no work tomorrow*' may be said by the Master to the Slave to grant a holiday; by the Master to the Kibitzer to point out that the Slave's lot is not so very bad after all; and so on.

While I admit to an inclination to play Old Harry with the performative/ constative fetish, I insist that I have not erased the distinction between different speech-acts that may be performed by saying an imperative sentence. The sentence may be used to command: the Master says it to the Slave, his purpose is to control the Slave's actions by changing the sphere of permissibility, and truthfulness is automatic because the sphere adjusts so that saying so makes it so. The sentence may be used to inform: either the speaker is not the Master or the hearer is not the Slave, the speaker's purpose is to impart information to one who does not yet possess it, and truthfulness is not automatic. Or the sentence may be used to remind (an intermediate case): again the Master says it to the Slave, but this time his purpose is to impart (or re-impart) information, and although truthfulness would be automatic the Master intends the sentence to be true even without any adjustment of the sphere of permissibility. Likewise for permissive sentences, except that truthfulness is never quite automatic since the Master cannot truly permit what is impossible. These are perfectly good distinctions; my point is only that they need not be part of semantics, insofar as semantics deals with truth conditions. In fact, they must not be. Only if the truth conditions are uniform from one use to another can we use the given formulation of (9).

I have no real dispute, however, with anyone who finds it intolerable to say that an imperative sentence, when used to command, has a truth value. In describing the language game I did not really use any semantic terms as primitives. I could have; but the description I actually gave is related to a description using semantic primitives as the Ramsey sentence of a term-introducing scientific theory is related to the theory itself. For instance, 'truth value' serves only as a mnemonic label for the values of the function introduced in (3) by existential quantification. If you dislike that label – or any other – feel free to substitute the euphemism of your choice.

3. PERMISSIBILITY KINEMATICS

I said that the changing sphere of permissibility integrates the effect of the Master's successive commands and permissions, but I did not say exactly how. The requirements in (9) constrain, but do not determine, the evolution of the sphere. When the Master says to the Slave an imperative or permissive sentence that would be false if the sphere remained unchanged, there will ordinarily be infinitely many alternative adjustments that would make his sentence true.

For commanding, at least, it is easy enough to say precisely how the sphere should change. Suppose that at time t (at a given world) the Master says to the Slave $!\varnothing$, and suppose that a change in the sphere of permissibility is needed to make $!\varnothing$ true at t. Let P be the old sphere just before t, and let $[\![\varnothing \text{ at } t]\!]$ be the set of all worlds where $\varnothing$ is true at t. Then the new sphere at t, and thereafter until the next change, should be the intersection $P \cap [\![\varnothing \text{ at } t]\!]$. All worlds accessible at t where $\varnothing$ is false at t must be removed from the sphere, according to (9); but it would be gratuitous to remove any further accessible worlds, since the Master has commanded that $\varnothing$ and nothing further, and it would be gratuitious to add any accessible worlds that were not permissible before, since the Master has not permitted anything but only commanded something. As for inaccessible worlds, it makes no difference which are removed or added so I have made the most convenient arbitrary stipulation.

If the sphere's evolution under the impact of commands does go by intersection in the way just proposed, then we have the proper result for the special case that there is only commanding and no permitting. Let the Master's commands before t be: $!\varnothing_1$ at $t_1, \ldots, !\varnothing_n$ at t_n. Then by successive intersections the sphere of permissibility at t is $P_0 \cap [\![\varnothing_1$ at $t_1]\!] \cap \ldots \cap [\![\varnothing_n$ at $t_n]\!]$, where the initial sphere P_0 is the set of all worlds. The Slave, according to (10), tries at t to see to it that the actual world is within the sphere of permissibility at t. That is to say that he tries at t to see to it that the actual world is in all of the sets $[\![\varnothing_1$ at $t_1]\!], \ldots, [\![\varnothing_n$ at $t_n]\!]$. And that is to say exactly what we said before about this special case: that he tries at t to see to it that $\varnothing_1$ is true at $t_1, \ldots,$ and $\varnothing_n$ is true at t_n.

One sort of commanding may seem to require special treatment: commanding the impermissible. Suppose that $[\![\varnothing$ at $t]\!]$ contains no worlds that are both accessible and permissible at t, so that $¡\varnothing$ is false at t. The Master may nevertheless wish to command at t that $\varnothing$. For instance, he may have changed his mind. Having commanded at dawn that the Slave devote his energies all day to carrying rocks, the Master may decide at noon that it would be better to have the Slave spend the afternoon on some lighter or more urgent task. If the Master simply commands at t that $\varnothing$, and if the sphere evolves by intersection, then *no* world accessible at t remains permissible; the Slave, through no fault of his own, has no way to play his part by trying to see to it that the world remains permissible. We have no idea what the Slave may do to make the best of an impermissible situation. Should we therefore say that in this case the sphere evolves not by intersection but in some more complicated way? I think not. The resources of the language game are not to blame if the Master removes all accessible worlds from the sphere of permissibility by commanding the impermissible. Rather the Master is to blame for misusing those resources. What he should have done was first to permit and then to command that $\varnothing$. He should say to the Slave, in quick succession, first $¡\varnothing$ and then $!\varnothing$; that way, he would be commanding not the impermissible but the newly permissible. We could indeed have equipped the language game with a labor-saving device: whenever $¡\varnothing$ is false, a command that $\varnothing$ is deemed to be preceded by a tacit permission that $\varnothing$, and the sphere of permissibility evolves accordingly. But this is a-frill that we can well afford to ignore, since it does not enable the Master to do anything more than he can do in the original, simpler language game.

Turning now to the evolution of the sphere under the impact of permissions, we reach the problem announced in my title. The natural parallel to evolution by intersection in the case of commands would be evolution by union, as follows: if at t the Master says to the Slave ¡ $\varnothing$, and if a change in the sphere of permissibility is needed to make ¡ $\varnothing$ true at t, and if P and $[\![\varnothing \text{ at } t]\!]$ are as before, then the new sphere at t, and thereafter until the next change, is the union $P \cup [\![\varnothing \text{ at } t]\!]$. But this sort of evolution by union, unlike evolution by intersection in the case of commands, is far from realistic. There could be a language game that did work that way – the rules are up to the players – but it would lack one salient and problematic feature of permission as we know it.

The problem is this. When the Master permits something, he does not thereby permit that thing to come about in whatever way the Slave pleases – not if the game is to be realistic. Suppose the Slave has been commanded to carry rocks every day of the week, but on Thursday the Master relents and says to the Slave '¡ *the Slave does no work tomorrow*'. That is all he says. He has thereby permitted a holiday, but not just any possible sort of holiday. He has presumably not thereby permitted a holiday that starts on Friday and goes on through Saturday, or a holiday spent guzzling in his wine cellar. *Some* of the accessible worlds where the Slave does no work on Friday have been brought into permissibility, but not all of them. The Master has not said which ones. He did not need to; somehow, that is understood.

Perhaps the incorrect principle of evolution by union in the case of permissions has some correct consequences, as follows. First, the new sphere at t should contain some world in $[\![\varnothing \text{ at } t]\!]$ that is accessible at t, if there exists some such world; that much is required by (9). Second, it should be included in $P \cup [\![\varnothing \text{ at } t]\!]$; since the Master has permitted that $\varnothing$ and nothing further, it would be gratuitous to bring worlds into permissibility where $\varnothing$ is false at t. Third, it should include all of P; since the Master has not commanded anything but only permitted something, it would be gratuitous to remove any worlds from permissibility. In short, the new sphere at t is the union of the old sphere P and some subset or other of $[\![\varnothing \text{ at } t]\!]$, where all we know yet about this subset of $[\![\varnothing \text{ at } t]\!]$ is that it must, if possible, contain some world that is accessible at t.

Let us return to our example. Hitherto the Slave has been commanded to carry rocks every day of the week, to abstain from the Master's wine, and perhaps other things besides. Now he has been permitted (on Thursday) to do no work on Friday. So the newly permissible worlds are all of the worlds that were permissible hitherto, along with some of the accessible worlds, formerly impermissible, where the Slave does no work on Friday. (If such there be; but in this case there are.) But only some, not all. The worlds brought newly into permissibility include none of those where the Slave does no work on Friday or on Saturday either; nor any of those where he does no work on Friday and drinks the Master's wine.

Why not? Various answers might be given. But though they seem sensible in this case, I do not think any of them lead to any simple and definite general principle of evolution.

* * *

Answer 1. To enlarge the sphere of permissibility so that it includes worlds where the Slave does no work on Saturday, or worlds where he drinks the Master's wine, would be a gratuitous enlargement. It would be more of an enlargement than is needed to make it permissible not to work on Friday.

I reply that the same is true of any reasonable enlargement. If the game is to be at all true to life, there will be more than one permissible way for the Slave to spend his holiday. (Even if he is required to spend the day at prayer, still he is no doubt free to choose the points in his prayers at which to take a breath.) Then more than the least possible number of worlds – more than one – must have been brought into permissibility.

* * *

Answer 2. To include worlds in the enlarged sphere of permissibility where the Slave does no work on Saturday, or where he drinks the Master's wine, would be gratuitous change, not in a quantitative but in a qualitative way. The newly permissible worlds should be selected to resemble (as closely as possible) the worlds that were permissible before.

I reply that according to my offhand judgments of similarity, that principle instructs us to select worlds where the Slave spends Friday in the gymnasium lifting weights. Among worlds where the Slave does not work on Friday, are not these the worlds most similar to the previously permissible worlds – worlds where he spends Friday carrying rocks? But surely a weight-lifting holiday is not the only sort of holiday that has been made permissible.

To be sure, the outcome depends on the relation of comparative similarity that guides the selection. Offhand judgments are no safe guide. Not every similarity relation worthy of the name gives significant weight to the obvious similarity between rock-carrying and weight-lifting. So perhaps it is true, *under the right similarity relation*, that the worlds that become permissible are those of the worlds where the Slave does no work on Friday that most resemble the previously permissible worlds. But which similarity relation is the right one for our present purpose? This is just a restatement of our original problem, and seems to me unhelpful.

* * *

Answer 3. Before the Master's permission, all worlds where the Slave did no work on Friday were impermissible; but they were not equally impermissible. Those where he also failed to work on Saturday, or where he drank the Master's wine, were more impermissible – or more remote from permissibility – than some of the others. (Whether or not they were also more dissimilar from the permissible worlds in other respects, at least they were more dissimilar in respect of their degree of permissibility.) If the Slave cannot (or will not) see to it that the actual world is within the sphere of permissibility, he may at least try for second best and keep the world as nearly permissible as he can. The relation of comparative near-permissibility determines what is second best. Perhaps it is this same relation that selects the newly permissible worlds when the Master enlarges the sphere of permissibility: the worlds that become permissible are those of the worlds where the Slave does no work on Friday that were most nearly permissible before.

I reply that this may be; and that it seems right to connect the problem of evolution under permissions with the problem of second-best courses of action for the Slave. (I am grateful to Robert Stalnaker for pointing out this connection.) Still it seems to me that again the problem has been restated rather than solved. Given the relation of comparative near-permissibility at every stage of the game, we may have a complete principle governing the evolution of the sphere of permissibility; but how does the comparative relation evolve from stage to stage? Is it so that the spheres of permissibility and accessibility at any stage suffice somehow to determine the comparative near-permissibility of worlds at that stage? If so, how?

* * *

Answer 4. Perhaps we should look outside the game to the goals it serves. It is to serve some purposes that the Master controls the Slave by commanding and permitting. The Slave either shares these purposes or at least acquiesces enough that he continues to play his part in the game.

When the Slave is permitted to do no work on Friday, some worlds remain impermissible because if they were to become permissible and the Slave were to actualize one of them, that would not serve the purposes for which the game is played. It is understood that these purposes require the Slave to work hard and to keep away from the Master's wine. Therefore worlds where the Slave does no work on Saturday, or where he drinks the Master's wine, are not readily brought into permissibility when the Master permits a holiday on Friday.

I reply that either the Slave does know what would serve the purposes in question, or he does not. If he does, then what is the point of a game of commanding and permitting? The Slave might as well simply ignore what the Master says and do whatever he judges to serve the purposes. The game is played exactly because the Slave needs guidance in serving those purposes. But if the Slave does not know what would serve the purposes, and if the evolution of the sphere of permissibility depends on what would serve the purposes, then the Slave is not in a good position to figure out how the sphere has evolved, and hence is not in a good position to figure out what is permissible. For the Slave to suffer this difficulty will itself interfere with the success of the game of commanding and permitting in serving those purposes for the sake of which it is played.

The best that might be done along these lines, I suppose, is as follows. It might be that the Slave knows just enough, and not too much, about what would serve the purposes. Since he knows enough, he is in a position to figure out how the sphere of permissibility evolves when the Master enlarges it, as by permitting a holiday on Friday. Since he does not know too much he remains in need of guidance if the purposes are to be served, and the game does not become pointless. This might be so. But I find it hard to believe that only when a delicate balance has been struck does the game I have described both retain its point and become playable.

* * *

Answer 5. At any stage, the sphere of permissibility may be specified by a list of requirements. (The list may or may not match the list of commands by which the sphere was shaped.) Each requirement on the list is satisfied at every permissible world; the worlds that are permissible are exactly those that satisfy every requirement on the list. The list might be as follows:

The Slave carries rocks all day on Sunday.

.

.

.

The Slave carries rocks all day on Friday.
The Slave carries rocks all day on Saturday.
The Slave never drinks the Master's wine.

.

.

.

Find those entries on the list that conflict with the Master's permission that the Slave do no work on Friday. There is one and only one; strike it out. The new sphere of permissibility consists of exactly those worlds that satisfy the remaining requirements.

I reply that it all depends on how you encode the sphere of permissibility by a list of requirements. If you do it the right way, as above, the technique of striking out requirements that conflict with the Master's permission will give the right answer. Unfortunately, there are also wrong ways. The same sphere could have been encoded by another list:

The Slave carries rocks every morning of the week.
The Slave carries rocks every afternoon of the week.
The Slave never drinks the Master's wine.

.

.

.

Now we cannot strike out the one and only requirement that conflicts with the Master's permission; the first two both conflict. We could strike out both of them; but that will make it permissible to do no work on Saturday. Or take this list, another that encodes the sphere:

The Slave carries rocks all day on Sunday or drinks the Master's wine.

.

.

The Slave carries rocks all day on Friday or drinks the Master's wine.
The Slave carries rocks all day on Saturday or drinks the

Master's wine.
The Slave never drinks the Master's wine.

.

.

.

Now there is no one requirement which conflicts, all by itself, with the Master's permission; but there are two that jointly conflict with it. Strike out the right one of the two, and all is well. Strike out the wrong one (or strike out both) and the results are not at all as we would wish. Strike out the requirement that the Slave never drinks the Master's wine and take the new sphere of permissibility to consist of those worlds that satisfy the remaining requirements on the list. This enlargement brings into permissibility worlds where the Slave does no work on Friday, does no work on Saturday either, and spends both days drinking the Master's wine. (It would also bring in worlds where the Slave does no work earlier in the week, except that by Thursday these worlds are inaccessible.)

So the method of listing and striking out will not work unless we choose the right one of the lists of requirements that encode the original sphere of permissibility. Which one is that? Again we have a restatement of our original problem, not a solution.

* * *

How much of a solution is it reasonable to expect? There are cases where it is really unclear which worlds have been brought into permissibility. That means that no principle can be both as definite as we might hope and clearly correct. One such case is given by Thomas Cornides in a discussion of our problem [3]. (He defends a version of Answer 5, but is well aware of the reasons why the procedure of listing and striking out will not always give a determinate answer, even if the correct list is somehow given us.) His example is as follows. First comes the command '! you play only if you do your homework.' Second comes the command '! you watch television only if you play.' And third comes the permission '¡ you watch television and you do not do your homework.' Is it now permissible to watch television, not do the homework, and not play? That is unclear; and I think it might be left unclear even if we knew all that was relevant about the players and about their reasons for playing a game of commanding and permitting. So a principle governing the evolution of permissibility cannot settle this case in a way that is clearly correct.

Princeton University

NOTES

* Thanks are due to audiences on several occasions, and especially to Robert Martin and Robert Stalnaker, for comments on a previous version of this paper.

I am told that Thomas Ballmer has developed a theory similar to that presented here. However, I have not seen any details of his work.

BIBLIOGRAPHY

[1] Chellas, Brian F., *The Logical Form of Imperatives*, Perry Lane Press, Stanford, 1969.
[2] Chellas, Brian F., 'Imperatives', *Theoria* **37** (1971), 114–129.
[3] Cornides, Thomas, 'Der Widerruf von Befehlen', *Studium Generale* **22** (1969), 1215–1263.
[4] Lewis, David, *Convention: A Philosophical Study*, Harvard University Press, 1969.
[5] Stenius, Erik, 'Mood and Language-Game', *Synthese* **17** (1967), 254–274.

VEIKKO RANTALA

POSSIBLE WORLDS AND FORMAL SEMANTICS*

Since Prof. Hintikka is one of those philosophers and logicians who are responsible for the popularity of possible worlds, and in view of his pioneering work on the methodology of the semantics of modal notions, it is perhaps understandable that my contribution to his *Festschrift* concerns possible worlds, too, no matter how amateurish it may be. But irrespective of whether this contribution has any value at all in the delicate field of possible worlds semantics, it is intended for a token of my respect to a friend and teacher of many years' standing.

Although my original purpose was to comment only briefly on an argumentation (or better, a type of argumentation) involved in an example in Plantinga (1974), I learnt soon that such a brief comment is impossible. The reason is that the philosophical possible worlds semantics, which underlies this argumentation, has been given a very peculiar form in Plantinga's book. Thus I let the example act mainly as a framework for more general remarks.

When a person who has worked mainly on extensional logic follows philosophical discussion about possible worlds, it is no wonder if he begins to panic. After several years' panic, I will now take this opportunity to try to make it clear to myself whether there could be any reasonable connection between the formal semantics of modal logic and the recent philosophical semantics of modal notions, apart from their common metaphors. Although I think that the resulting modest observations by no means are new to a philosopher who is working on modal notions, they tended to comfort me and enable me to see the value of the informal possible worlds semantics in general. I say 'in general', for unfortunately I still cannot understand some features of the kind of possible worlds semantics that is represented so ingenuously by Plantinga. Perhaps the criticism below is largely due to my philosophical narrowness, but in any case it seemed unavoidable.

E. Saarinen, R. Hilpinen, I. Niiniluoto, and M. Provence Hintikka (eds.), Essays in Honour of Jaakko Hintikka, 177–188. All Rights Reserved.

I

The example I picked from Plantinga (1974) is the following (see pp. 109–110). According to Plantinga, Socrates could have been importantly different: "He could have had many properties he lacks and lacked many he has; and this with respect to just those properties to which we look in deciding whether someone *resembles* Socrates." Among the properties Plantinga mentions are such as courage, intelligence, being the teacher of Plato, etc. Socrates shares many of his properties with Xenophon, e.g., courage, intelligence, etc. Moreover, "Socrates and Xenophon could have been such that the former was less like Socrates as he was in fact than the latter." In terms of possible worlds, this is rephrased as follows: "If we suppose that Socrates and Xenophon exist in more than one world, we could put this by saying that there is a world W distinct from α [the actual world] such that Socrates and Xenophon both exist in W, and in which Xenophon resembles Socrates as he is in α more than Socrates does." Furthermore, this can be expressed as follows:

(*)   Socrates and Xenophon could have been such that the latter
           should have resembled Socrates as he was in the actual world
           more than the former.

According to Plantinga, the sentence (*) expresses a true proposition.

The purpose of this example is to demonstrate that D. Lewis' Counterpart Theory (see Lewis, 1968) suffers from semantic inadequacies. For in Plantinga's opinion, (*) does not express a true proposition on the basis of Counterpart Theory. As he puts it, (*) is true on Counterpart Theory only if there is a world W in which Socrates and Xenophon have counterparts S_W and X_W such that X_W resembles the actual Socrates more than S_W does. But there is the following passage in Lewis (1968): "Your counterparts resemble you closely in content and context in important respects. They resemble you more closely than do the other things in their worlds."

So it might seem intuitively obvious that Plantinga is right when he says that on Counterpart Theory (*) is not true. Perhaps, on the other hand, it is not very clear intuitively that it is true on Plantinga's own system. In order to fix our intuitions about (*), we have to take a closer look both at Plantinga's system and at Counterpart Theory.

II

Let us first consider certain features of Plantinga's notions, as well as some more general aspects of possible worlds semantics which are relevant here.

Plantinga's concept of possibility is to be understood in some very broadly logical sense. Most of his discussion is based on his concept of possible world which is defined (or characterized) on pp. 44–45 of Plantinga (1974). To understand this concept we ought to understand the notions 'possible' ("in the broadly logical sense"), 'actual', 'state of affairs', and 'a state of affairs obtains' (or 'is actual'). Perhaps these notions should be considered as primitives if the following statements are regarded as definitions. Let S and S' be states of affairs. Then

– S includes S' if it is not possible that S obtains and S' fails to obtain;
– S precludes S' if it is not possible that both obtain;
– S is maximal if for every state of affairs S', S includes S' or S precludes S';
– W is a possible world if W is a maximal possible state of affairs;
– α is an actual world if α is a maximal possible state of affairs that is actual.

According to Plantinga, there are both possible and impossible states of affairs. He also argues that there is exactly one actual world, that is, exactly one possible world obtains. In so far as one can extract from the general discussion on p. 45, as well as from various examples all over the book, Plantinga seems to suppose, first, that the actual world is something which comprises at least the entire history in some sense, secondly, that there is a thing like the collection of all possible worlds (in some absolute sense, too). If he does not suppose so, his discussion is a little misleading.

Supposing that there are possible and impossible states of affairs, we nevertheless have to ask whether it follows that there really are any possible worlds in Plantinga's sense (that is, maximal possible states of affairs), i.e., whether this concept is reasonable in any sense. But even if we assume that there is a totality of all states of affairs over which we can quantify, Plantinga does not give us any evidence for possible worlds.

It is obvious, however, that there is no such thing as the collection of all states of affairs if we do not restrict somehow the concept of a state of affairs. To suppose there is will (together with the unrestricted use of states of affairs) lead to paradoxes. Since it seems (at least on the surface) that Plantinga dispenses with everything that has been done in logic and set theory in this century and since it is always possible that he does not mean that his above explication of possible worlds and related concepts should be taken seriously as definitions, it would be needless to criticize his general account unless his theorizing did not reflect a rather widespread way of dealing with possible worlds.

I shall not comment in any detail on the general logical and set-theoretical reasons for the paradoxical character of Plantinga's possible worlds because this has already been done by Dana Scott[1] (and certainly he can do it more competently). Instead, I will make some scattered, and less essential, remarks on Plantinga's possible worlds, as well as on a more fundamental confusion which seems to be involved when Plantinga speaks about 'pure semantics' and 'applied (intended) semantics'. This confusion does not seem to be very rare among philosophers dealing with modal logic.

One source of trouble is that Plantinga does not explain what 'state of affairs' and 'proposition' mean. I am specifically doubtful of his impossible states of affairs. We are immediately led to familiar paradoxes if we allow whatsoever as impossible states of affairs.

If possible, I am still more worried about *the* actual world. In the light of the above 'definitions', we could take 'actual state of affairs' and 'true proposition' to be unanalyzed primitive concepts if there were not definite examples of them. Thus Plantinga obviously supposes that we understand these concepts as well as the concept of actual world. But consider the following example of states of affairs:

(1)   Kareem Abdul-Jabbar's being more than seven feet tall.

According to Plantinga, this is an actual state of affairs. But obviously the following is also a state of affairs:

(2)   Kareem Abdul-Jabbar's being at most seven feet tall.

Presumably, (2) is not an actual state of affairs because (1) is. Would this mean that, as it were, a part of the history is not included in the actual world? It is not likely that this is Plantinga's intention, for so all-embracing his actual world seems to be, as we noticed above. Or should we think that, after all, (1) is not a proper state of affairs, but somehow elliptical? It is not quite clear to me, however, whether Plantinga would count both (1) and (2) actual states of affairs. If both are proper (actual) states of affairs (considered something like pieces of an actual course of events), is then the proposition

(3)   Kareem Abdul-Jabbar is more than seven feet tall and Kareem Abdul-Jabbar is at most seven feet tall

actually true? Does it belong to 'the book on the actual world'. This seems obvious if (1) and (2) are actual and the correspondence between states of affairs and propositions is so straightforward as is suggested by Plantinga.

No matter how trivial these considerations are, they nevertheless show that the notions 'state of affairs' and 'proposition' should be restricted or specified somehow, not only to avoid general set-theoretical and semantical paradoxes but also to avoid more trivial and straightforward inconsistencies. I doubt that there is no way to do it without rejecting Plantinga's idea of such a comprehensive actual world (possible world) altogether. He seems to suppose that our intuitions of *the* actual world are firm enough to enable us to understand his concept of possible world. Probably it is due to my deficient capacity to understand these ideas that I cannot even decide whether I am writing these objections in the actual world or in a merely possible world. Or would it be appropriate to say, e.g., that I was writing them in a merely possible world while Plantinga was writing his book?

This is not to say that the notions of actual world and possible world could not be intuitively useful but only that to use them presupposes some care. The above objections (which by no means are very original) are unavoidable if we want to evaluate Plantinga's argument about (*).

Let us next consider the following reasoning which reflects certain difficulties to swallow the formal semantics of modal logic. Plantinga makes a distinction between the pure and applied semantics of modal notions (pp. 126–128). According to him, Kripke's formal semantics "as such has no obvious connection with modal notions at all." Furthermore: ". . . it is not to the pure semantics as such that we must look for the promised insight into our modal notions – not, at least, if we take the semantics seriously rather than as a heuristic device." Instead, there is an applied semantics which can be naturally associated with Kripke's pure semantics: "In the intended applied semantics, therefore, a model structure will not be just *any* triple (G, K, R) where G is a member of K, and R is reflexive; K will be a set of possible worlds (not chessmen) – possible states of affairs of a certain kind – of which G is a member." For Plantinga, certain passages in Kripke (1963) are hints as to the intended or applied semantics.

Now, it is not quite clear to me what Plantinga means by 'applied semantics'. As every logician knows, one of the purposes of a formal system is to clarify (aspects of) some intuitive notions and ideas which as such may be vague and ambiguous and whose unrestricted use can lead to paradoxical results. Thus a purpose of formal modal logic is to clarify modal notions. The suggestive idea of possible worlds is behind its formal semantics. Hence this semantics can be considered as an enterprise to make sense of this ambiguous idea. As I understand it, the aim of this work was not only to create a new model theory but also to obtain new insights into modal notions. However, the above quotations give us the impression that this enterprise has not been successful.

Clearly it is very easy to overestimate the import of formal methods in conceptual analysis. So Plantinga's criticism of the pure semantics can be valuable. But it is also too easy to underestimate those insights which inhere in all set-theoretical constructions. For example, the following could be a consequence of these insights: 'truth', 'necessity' and the like are more or less relative concepts. That is, there are no such huge things as *the* actual world or *the* collection of all possible worlds. What we can learn, then, is precisely that although we can informally use the metaphors 'the actual world' and 'all possible worlds' as heuristical devices, they should be used carefully if we want to avoid mess. A logician who develops a formal apparatus for some purpose must have relevant insights (concerning not only the underlying formalism but also his subject matter) before he can succeed in his work. Others can obtain them from his work.

Plantinga seems to think, however, that the clue to proper understanding of modal notions is 'the applied semantics'. It is not quite clear whether he claims that his own system constitutes an applied semantics naturally associated with the pure semantics à la Kripke. Even if we leave aside the question whether it is correct to use the word 'semantics' if we deal with propositions instead of sentences or other expressions, it is not clear how such a claim would be meaningful if there were only one actual world and one collection of possible worlds. The last quotation above might suggest that Plantinga's pair (the actual world, all possible worlds) would be an informal counterpart to some of the Kripkean model structures. However, although Plantinga's concepts of actual world and possible world were meaningful, it is rather obvious that 'all possible worlds' would not form a set in any definite sense of the word. It hardly was Kripke's intention to give hints about paradoxical entities.

But if take the quotation as it is ("... *K* will be a set of possible worlds ..."), it suggests that there could be several informal counterparts (an (the?) actual world, a set of possible worlds) to Kripke's model structures. But this leads immediately to the idea noticed above, according to which modalities are context dependent, rather than absolute. Viewed this way, what could Plantinga's 'applied (intended) semantics' mean? Obviously we have here a close analogy to such troublesome notions as 'empirical interpretation' and 'intended model' as they are used by philosophers of science. We shall discuss this view a little later in connection with our specific example (*).

III

Let us first present those aspects of Counterpart Theory which are relevant to our discussion. According to Lewis (1968), Counterpart Theory is a formal theory formulated in the usual predicate calculus. Its purpose is to formalize our modal discourse as follows.

The primitive predicate symbols of Counterpart Theory are W, A (monadic), I, C (dyadic). Their intuitive meanings are the following:

$Wx:$ x is a possible world,
$Ax:$ x is actual,
$Ixy:$ x is in possible world y,
$Cxy:$ x is a counterpart of y.

These primitives are governed by eight postulates P1–P8 which we shall not repeat here. Every formula of quantified modal logic (based on first-order predicate calculus) can be translated into a formula of Counterpart Theory whose vocabulary contains the above primitives (or some of them) plus the predicate symbols occurring in the formula to be translated. Since individual constants are replaced by definite descriptions, some extra predicates may be needed. The translation of a given formula is supposed to be equivalent in some sense (intuitively?) with the given one. It is not essential here how the translation is accomplished. We will discuss certain other features of Counterpart Theory later.

We want to consider whether and how the sentence (*) could be made more precise by means of formal (alethic) modal logic, on the one hand, and by means of Counterpart Theory, on the other. (It could be handled by means of a sufficiently precise fragment of a natural language, together with a suitable semantics. However, if we are to discuss Counterpart Theory as it is defined in Lewis (1968), to which Plantinga refers, we have to translate (*) into a formal language.) As such (*) is not meaningful, not before a decision about the relevant properties.

The sentence (*) can be evaluated in the light of the usual modal logic and Counterpart Theory only on the following conditions. First, only a limited stock of properties can somehow be expressed in a formal language. Secondly, although the language in question included only a finite number of non-logical constants, the number of all 'important' properties, expressible in the language, can be infinite. Hence (*) can be translated into a formal language only if we suppose that the number of such properties is finite. Furthermore, it obviously does not affect the point of Plantinga's argumentation if we suppose that there are only, say, three important actual properties to be considered and that they do not contain modalities.

How, then, could (*) be expressed in a language of modal logic? It cannot be done faithfully at all, for it is clear that it cannot be done uniquely. For to write a formal expression presupposes not only syntactical but also some semantical commitments. However, this fact is not very essential as far as the general abstract form of (*) is considered. To fix our discussion, let us suppose that the translation is of the following kind.

Let a, b be individual constants corresponding to 'Socrates' and 'Xenophon'', respectively. (Obviously they are to be regarded as rigid designators, according to Plantinga.) Let φx, $\varphi' x$ be formulas corresponding to those (definite) actual properties of Socrates "to which we look in deciding whether someone *resembles* Socrates" such that the former is actually common in Socrates and Xenophon. Then the formula

$$\theta = (\varphi a \wedge \varphi' a \wedge \varphi b \wedge \psi b) \wedge \Diamond(\neg \varphi a \wedge \neg \varphi' a \wedge \varphi b \wedge \psi b)$$

(or something like θ) could be a possible translation of (*) into an appropriate modal language, if we suppose that the conjunctions in the parentheses are consistent.

Really a sentence like θ is not very interesting as such. It is interesting only so far as it can be related to Plantinga's argumentation.

If we abandon Plantinga's absolute concept of logical possibility, what is then left of his claim that (*) is true? If θ corresponds to (*) at all, then this claim is reduced to one of the following:

(i) θ is true in every model of Kripke's semantics or of some other formal semantics, i.e., θ is valid;

(ii) θ is satisfiable;

(iii) θ has an intended model in some sense.

It is obvious that (i) does not hold. Whether (ii) holds depends on the structure of the formulas involved. But clearly θ is satisfiable on the assumption of consistency made above.

(iii) is a very ambiguous claim, at least until the concept of intended model will be specified. Perhaps one could think that such an intended model of θ (e.g., in the Kripkean semantics) could be delineated by first taking 'a fragment of the history' in some way to act as an actual world (in the formal Kripkean sense). This would of course presuppose that we can delineate a suitable set of individuals for the domain of the actual world, as well as relevant relations (in the set-theoretical sense) for the intended extensions in the actual world of the primitive predicates. As I take it, to say that there is such a 'factual' intended model is equivalent with saying that such can be defined in some precise set-theoretical way. However, I don't know how to define one. Perhaps we can just decide to say that certain things, like Socrates and Xenophon, will be chosen for urelements when constructing the sets we need, similarly as we can choose the natural numbers for urelements in set-theoretic constructions. (Nevertheless, I can more easily imagine a set which contains a natural number than a set which contains Socrates. This poverty of imagination may be due to my insufficient experience in the semantics of natural language.) However, as far as I can see there is from the point of view of Plantinga's argumentation no essential difference between the claim (*) and the claim that Plantinga and Kripke could be such that the latter would resemble actual Plantinga more than the former. Perhaps there is even no essential difference between (*) and the corresponding claim about Sherlock Holmes and Watson (if we do not take the actual world too seriously).

These simple considerations seem to suggest that it does not make sense to think that an intended model (not even of a sentence of a natural language) should be a concrete, realistic entity in the sense that its actual world would be, e.g., something like a fragment of the actual history. At least such a concretion has not very much to do with conceptual analysis. After all, what we need are abstract set-theoretic entities which somehow structurally correspond to our intuitions in a given context.

Of course, there are lots of other problems which should be solved before we can construct set-theoretic structures which could be called intended models of θ (or perhaps of (*)). It seems to me that, generally speaking, the most difficult problems connected with the search of intended models of sentences are those which belong equally to both formal sentences and sentences of natural language. Perhaps it is to simplify matters somewhat, but I am tempted to say that those general philosophical problems which arise in possible worlds semantics, and which have been studied very extensively by Hintikka, are the same which arise when we ask how to find (general characteristics of) intended models and which are even inherent in the very notion of intended model. These problems are obviously pragmatical and epistemological, rather than purely logical. Perhaps we could say that to understand the meaning of a sentence is to know how to define or construct an intended formal model (or rather a class of such models) for the sentence. If we consider pragmatics from this point of view, we can see what is the connection of the philosophical possible worlds semantics (applied semantics, as Plantinga puts it) with the pure, formal semantics. Perhaps this view could decrease the metaphorical flavour around such concepts as 'possible world' which has sometimes been claimed to be involved in the philosophical semantics. Only after the philosophical and pragmatical work has been completed in the search of an intended model of a sentence it comes the task of formal semantics and pure logic to define exactly a set-theoretical model and to try to prove that it is a model of the sentence.

As to the present case, I cannot enter into details of how to define the structural properties of an intended model. Obviously there could be a whole class of such models (widely different from each other) which would correspond to my intuition about the abstract form of (*), even if I could fix the structural properties of the 'actual world'. (After all, it is perhaps easier to fix intuitions about the actual world than about the possible worlds or the alternativeness relation.) All told, I am not able to decide precisely whether and in what sense θ (or (*)) has an intended model, albeit it seems intuitively obvious that it has in some abstract sense of the word if we can fix the structure of those properties of Socrates and Xenophon we want to speak about. If this holds and if we interpret Plantinga's claim this way (although I think that he would not accept it), we can say that he is right in it.

Let us consider what happens to the claim that (*) is not true on Counterpart Theory. Let θ^* be the translation of θ in Counterpart Theory. Let

$$\sigma = \theta^* \wedge P1 \wedge \ldots \wedge P8.$$

Now, if we analyze (*) from the standpoint of Counterpart Theory similarly as from the standpoint of the Kripkean semantics, then to say that it is true on Counterpart Theory can mean one of the following:

(1) θ^* is true in every model of P1 $\wedge \ldots \wedge$ P8, that is, P1 $\wedge \ldots \wedge$ P8 $\rightarrow \theta^*$ is logically true (true in every model for our language);

(2) θ^* is satisfiable in a model of P1 $\wedge \ldots \wedge$ P8, that is, σ is satisfiable;

(3) σ has an intended model.

Obviously enough, (1) does not hold. Although I do not want to try any proof for (2), it seems rather clear that it holds. Thus (*) would be true on Counterpart Theory in an analogous sense as on the Kripkean semantics. However, Plantinga says that it is not true on Counterpart Theory since the truth conditions assigned to it by Counterpart Theory are not fulfilled. But what are those truth conditions in Lewis (1968) to which Plantinga refers? Counterpart Theory is a formal theory formulated in the usual first-order predicate calculus. Thus the truth conditions are the familiar truth conditions of predicate calculus. Lewis (1968) does not state any special truth conditions.

But the point of Plantinga's criticism can be directed to (3), as the quotation from Lewis (1968) we presented in Section I shows. There seems to be a great disparity between Lewis' formal Counterpart Theory and his informal descriptions of the counterpart relation. That is, there are no hints in the postulates P1–P8 which would show that the formal counterpart relation is a similarity relation of some kind. I think that it is not appropriate to call such descriptions truth conditions as far as they do not contain any definite criteria of truth.

What, then, has these descriptions of similarity to do with the postulates governing the counterpart relation? Obviously they can be considered as methodological devices in the search of intended interpretations for formal Counterpart Theory. According to Lewis, however, Counterpart Theory serves to rule out the problem of individuation occurring in modal contexts. But if the understanding of the meaning of a modal expression presupposes that an intended interpretation can be found for Counterpart Theory, then the problem of individuation has only been replaced by the problem of similarity. (In our analysis of (*) by means of the Kripkean semantics the problem of transworld identity presumably is not more difficult than the problem of similarity which arises if we rely on Counterpart Theory.) To think that a syntactical relation symbol (together with those rather loose postulates which do not even specify an equivalence relation) would rule out the problem of individuation since this symbol is said to denote similarity would be analogous to thinking that if Hintikka's world lines were raised into the syntax, this would solve the methodological problem of transworld identity. Although I do not believe that Lewis thinks so (for he says that the counterpart relation is problematic in the way all relations of similarity are), Counterpart Theory would have profited from some systematic methodological principles concerning intended interpretations of the counterpart relation (despite of the references he gives to other authors who have studied similarity).

Now, whether Plantinga is right (from the standpoint of our analysis) in his claim that (*) is false on Counterpart Theory depends on the intended meaning of the counterpart relation in the case at hand. Certainly he may be right if all the 'important' properties of Socrates, i.e., "those properties to which we look in deciding whether someone *resembles* Socrates", can be packed into a finite number of first-order formulas. (Of course, we have to ask: How about the other individuals involved?) If not, then the claim has no precise meaning, for (*) cannot be dealt with by means of Counterpart Theory, in so far as Counterpart Theory is not applied to stronger modal logics. Then a sentence of the form θ^* (or any first-order sentence) is not a faithful translation of (*). Even in this case, however, he is right in some loose sense which corresponds to Lewis' own intuitive idea of the correspondence relation.

The Academy of Finland

NOTES

* I am indebted to Dr. Esa Saarinen for his criticism.
[1] When I was writing this paper, Dr. Saarinen drew my attention to an unpublished paper by Scott, 'Is There Life on Possible Worlds?', where he deals with Plantinga's and Lewis' systems. I have tried to minimize overlapping. Unfortunately, it was not possible to avoid it altogether.

BIBLIOGRAPHY

Kripke, S., 'Semantical Considerations on Modal Logic', | Modal | and | Many-Valued Logics, Acta Philosophica Fennica 16 (1963), 83–94.
Lewis, D., 'Counterpart Theory and Quantified Modal Logic', Journal of Philosophy 65 (1968), 113–126.
Plantinga, A., The Nature of Necessity, Oxford University Press, Oxford, 1974.

ESA SAARINEN

CONTINUITY AND SIMILARITY IN CROSS-IDENTIFICATION*

One of the central problems of quantified intensional logic is the problem of cross-identification. Briefly put, this problem arises when we try to explicate the conditions under which an individual figuring in a possible world is the same as an individual figuring in another possible world.

In this paper we shall discuss some aspects of this problem. We shall try to analyse the status of continuity principles in cross-identification, emphasized recently by Hintikka, and relate this discussion to Hintikka's earlier point about the duality in cross-identification methods, and to some well-known proposals of David Lewis and Alvin Plantinga. We shall also discuss Kripke's widespread view according to which the problem of cross-identification is a pseudo-problem. We shall attempt to show that this viewpoint is theoretically unsatisfactory.

1. TRACING INDIVIDUALS BACK TO 'COMMON GROUND'

Hintikka has recently argued that "one of the main vehicles, perhaps the main vehicle, of cross-identification is just continuity in space and time". (Hintikka, 1975, p. 127.) This sounds surprisingly simple. It is relatively uncontroversial that the most important vehicle of *re*-identification is continuity in space and time. Thus if Hintikka's suggestion is correct, it follows that cross-identification does not present us any new unsurmountable difficulties, over and above those we would face anyway when studying the philosophical foundations of re-identification.

In order to see Hintikka's reasons for stressing the import of continuity principles in cross-identification, let us quote him at some length here:

Let's consider a typical situation we face in cross-identifying individuals, let's say a situation in which we are talking of what someone, say John, believes. The 'possible worlds' involved here are all the worlds compatible with everything John believes. We

189

E. Saarinen, R. Hilpinen, I. Niiniluoto, and M. Provence Hintikka (eds.), Essays in Honour of Jaakko Hintikka, 189–215. All Rights Reserved.
Copyright © 1979 by D. Reidel Publishing Company, Dordrecht, Holland.

may think each of them represented by one of Richard Jeffrey's 'complete novels'. They all usually have a recognizable part in common, viz. that specified by John's definite beliefs about the world. Usually that part is rather extensive. We may think of it as printed in a way different from the rest of the novel, say in red ink.

Cross-identification is now like taking a character mentioned in one novel and a character mentioned in another one and asking whether they are one and the same individual. One way of answering this question is implicit in what has already been said. We follow each individual back and forth in his respective world, trying in both instances to reach the red-letter part (common ground) of the two novels. This we can try to do as soon as our individuals are continuous in space and time. When we come to the red-letter part, we can decide the cross-identification question simply by seeing whether the individuals coincide there.

If you pause and think for a moment how we actually cross-identify between (say) my knowledge-worlds, you will soon see that this is just what we in fact do. What does it mean for a person to be identical with my friend Bill in a possible future course of events compatible with everything I know? It means for him to be a spatio-temporal sequel in that world to the career I know Bill has so far enjoyed. (Hintikka, 1975, pp. 127–8.)

Thus, according to Hintikka, the way we typically cross-identify between (say) b's belief-alternatives (the possible worlds compatible with everything b believes) is that we trace individuals back and forth in space and time and try to reach the common ground of the possible worlds. This common ground, in turn, is precisely that part of the world which is specified by b's beliefs.

There is surely something very tempting in this simple model of cross-identification advocated here by Hintikka. Yet we wish to argue that the scheme, as it is expressed by Hintikka in the quotation just given and elsewhere, is drastically inadequate when applied to propositional attitude contexts.

This is a negative result. On the positive side, we shall argue that Hintikka's model is however applicable in a number of other intensional contexts, e.g., in most contexts involving counterfactuals.

Three notions are crucial for Hintikka's schema for cross-identification described above: continuity of individuals in space and time, common ground of possible worlds, and 'coincidence' of individuals at the common ground. We shall take the first notion in our discussion as uncontroversial.

Let us start our analysis by reflecting a situation where things *do* work according to the simple model put forward by Hintikka. For this purpose, let us make a distinction between (possible) *states of affairs* and (possible) *courses of events*. We shall not try to analyse these notions much further here as their intuitive contents seems to us rather clear. One important aspect of the two notions is yet worth stressing. The way we understand the terms, a course of events is a class of linearly ordered states of affairs. Intuitively speaking, the essential difference here is that a state of affairs, unlike a course of events, does not have any future or history. According to our intuitions, it makes sense to say that one and the same state of affairs is a temporal slice in several different courses of events. This assumption will be made throughout in the discussion that follows.[1]

Assume now we are considering possible future courses of events for certain states of affairs (stage of a possible world). (Call that states of affairs or stage s.) The relevant possible courses of events can be naturally represented in an obvious way by means of a tree-structure, the root of which is the states of affairs or stage s. In this situation Hintikka's model will work as follows. Pick out individuals i and j from two different possible courses of events, say from h_1 and h_2. Follow now i in the possible course of events h_1 back to the initial states of affairs or stage s of h_1. Do the same with j in h_2. i and j are cross-identified if they coincide in s.

To put the same somewhat differently, what we do here is that we *re-identify* i with an inhabitant of the states of affairs s in the course of events h_1; then *re-identify* j with an inhabitant of s in the course of events h_2. After making these two re-identifications in the usual way, we just see whether the two inhabitants of s mentioned are one and the same. When this is the case, the common states of affairs or stage of the relevant possible courses of events will serve as a Hintikkian 'common ground'.

In this case it is also clear what it means for two spatio-temporally continuous individuals to 'coincide' at the common ground.

While Hintikka's model of cross-identification works straightforwardly in the case discussed, the assumptions we have made so far are severe. Indeed, most of the most interesting applications in the field of propositional attitudes seem to fall outside the scope of our discussion so far. This is the case because in most cases we are simply not cross-identifying between possible courses of events which have at least one stage (states of affairs) in common.

Take for instance a case, mentioned by Hintikka in the quotation above, where we are cross-identifying between all future courses of events which are compatible with everything I know. The assumption that all these courses of events have at least one stage in common now means that I know *everything*, each and every aspect of the world at a certain moment. But this is a totally unrealistic assumption.

2. 'SMALL WORLDS' AS COMMON GROUND

Something can be said in order to rectify the situation, however. To see the point, assume that we are considering the situation at a certain lecture room at a certain moment of time. It clearly makes perfectly good sense to reflect various alternative courses of events which initiate from that state of affairs in the classroom. What we only keep fixed is the situation in the lecture room; what happens at the same time outside that room is left completely unspecified. As a consequence, the various possible courses of events will have no *full* stage in common. (There will not be any time slice that each of the possible courses of events would have in common.) What they only have in common is a *part* of a stage (part of a time slice).

Another way to put the present point is to talk of 'small worlds' as Hintikka calls them (Hintikka, 1975, p. 195). In other words, we are not discussing the whole of a possible state of affairs but only a (spatially restricted) part of such. In still other words, we deal with possible states of affairs which are *locally* identical even though they are not *globally* so.

It is readily seen that a 'small world' (or a part of a time slice of a course of events) such as the states of affairs at the lecture room at a given moment of time in the above example can serve as a Hintikkian 'common ground' in the model of cross-identification being discussed here. When we are given two individuals i and j of two different alternative courses of events h_1 and h_2, we can cross-identify i and j provided we can spatio-temporally continuously follow i in h_1 and j in h_2 back to the same individual figuring at the states of affairs at the lecture room (at the given moment).

Again, all we need to do for cross-identification is re-identification in the standard way. No dubious principles enter the picture. Yet, even though the situation is now more satisfactory than before, there are again only a limited number of cases in which we could actually use the present model of cross-identification in the theory of propositional attitudes.

Consider again a situation where we are dealing with all future courses of events which are compatible with everything I know. For the above model to be applicable, we would have to assume that there is some stage of the world such that my knowledge specifies *completely* at least some (spatially restricted) part of that stage. In other words, I would have to know everything concerning some part of the world (at some moment of time). I would have to know some spatially restricted part of the world in complete detail. But again these assumptions are readily seen to be quite unrealistic.

Even apart from these considerations, notice that the above model could say nothing of the cross-identification of those individuals which cannot be re-identified with any inhabitants of the common parts. This fact alone would mark a restriction for the applicability of the model.

3. POSSIBLE HISTORIES DIVERGENCING BACKWARDS IN TIME

After having just pointed out some restrictions of the range of applicability of the present model of cross-identification, it is worth mentioning some other, less obvious cases where the model does work.

We have argued that cross-identification is uncontroversial as long as we deal with possible courses of events which have a stage in common, and our individuals are continuous in space and time. It is worth noting what is the range of possibilities that we allow here. It is customary to allow possible courses of events to diverge in the future, that is, to allow two possible courses of events h_1 and h_2 to differ from a moment t onwards even though they would coincide up to t. This kind of structures will be needed when we discuss what we mean when we say, for instance, that it is true of Aristotle that he might never have gone into pedagogy, etc.

Our model allows also the much less obvious possibility of two courses of events h_1 and h_2 to coincide from a moment t onwards but *differ* before t. In other words, we do not allow just the possibility, observed often enough, that possible courses of events may diverge forwards in time, but we allow them to diverge also backwards in time. If we call s the (momentary) state of affairs of a course of events h at moment t, then the possibility of backwards divergency will allow us to cope with several possible courses of events which could lead to the state of affairs s. It should be clear enough that such possibilities should not be excluded. For instance, when I am writing this paper, there is an issue of *Philosophical Studies* on the table. Clearly we can reflect various courses of events which all lead to that particular state of affairs: maybe I picked the issue from the bookshelf only a moment ago, maybe somebody else did it yesterday; maybe it was in my office before it was moved here, maybe in somebody else's or perhaps even in the departmental library. Reflecting such examples, it will not take us long to appreciate the point that several different courses of events can amount to one and the same state of affairs. Thus it is not in vain that we allow the possibility that the courses of events which our model of cross-identification can handle include also ones that divergence as we move backwards in time.

(Notice how naturally we could cross-identify the various issues of *Philosophical Studies* across the different courses of events all amounting to the present state of affairs (on my table). An issue of the journal figuring in a course of events h_1 would be cross-identified with one figuring in another course of events h_2 if the two could be re-identified inside h_1 and h_2, respectively, with the issue there actually is now on the table, i.e., with one and the same individual of the present state of affairs.)

4. ON THE ALLEGED NECESSITY OF ORIGIN

What has just been stated must be protected against one possible source of misunderstanding. For it has been argued by several scholars that an individual's origin is its necessary property. This point, the best known advocate of which is Saul Kripke (1972), has been made recently by J. L. Mackie as follows:

Queen Elizabeth – this very woman – might never have become a queen; but she could not have been born of different parents. It is indeed possible that she was not born of those of whom we now believe to be her parents; but given that she was born of them, she was necessarily born of them; anyone not born of them, though she might have done all the things that Elizabeth has done since infancy, would not have been this woman. (Mackie, 1974, p. 551.)

The same point has been made in a somewhat more sophisticated form by Colin McGinn:

Now I think it helpful to distinguish three relations between entities of different kinds in which the origin of a person may be said to consist: first, the relation between the fertilized egg – the *zygote* – and the person it is destined to become; second, the relation between the egg and sperm – the *gametes* – and the zygote (and hence person) they fuse to produce; third, the relation between the gametes and the parents of the resulting person. Our task, then, is to give some account of the *rigidity* of these relations; i.e., to explain why it is that when entities stand in these relations they necessarily do, why is it that in any world which they exist these entities are related as they are in the actual world. (McGinn, 1976, p. 131.)

At the face of it, these positions of Kripke, Mackie and McGinn are strictly opposed to the one we are here advocating. For we have argued that cross-identification is possible even in case we deal with backwards diverging possible courses of events. Once such a possibility is allowed, then there is no security of the similarity of origin of two cross-identified individuals.

There is thus an obvious disagreement. However, it seems to me this disagreement carries over on the surface only.

It is one of the basic ideas of possible worlds semantics for intensional concepts that these concepts involve (semantically) a simultaneous consideration of several possible worlds. Furthermore, and this is critical for the issue being discussed, the class of possible worlds that we have to consider varies as a function of intensional concepts. In particular, the class will be different for different possibility concepts (practical, physical, logical possibility, etc.) and for different propositional attitudes.

Even though I will not argue for this point in more detail here, it seems clear that Kripke *et al.* have had in mind one particular intensional concept when they have put forward the thesis of the *de re* necessity of origin. When they argue that it is not *possible* for an individual who is figuring in a possible world *w* and who do not have the same origin as Queen Elizabeth, to be cross-identified with the actual Elizabeth, they have in mind one particular concept of possibility. It is not possibility in the sense of (say) 'logically possible' that they have in mind but rather something like a concept of possibility which we *ordinarily use* in our everyday language and everyday considerations. Mackie, who discusses backwards diverging histories, has indeed made the point clearly enough as follows:

> The truth is rather that we don't normally consider that sort of possibility [backwards divergencies]. The counter-factual possibilities – might-have-beens – that interest us are nearly always forward divergencies from actual history. Still we can concede this much to Kripke, that *among the possibilities that we ordinarily consider* it is the origin of an individual person or thing that is necessary for that very individual, whereas all its subsequent history is contingent. (Mackie, 1974, p. 556. Italics in the original.)

Thus the Kripkean claim is that origin of an individual comes out as a *de re* truth across all those possible worlds which our ordinary counterfactual (or possibility) statements introduce us.

The present point can also be made by observing that there lurks an ambiguity when we say, e.g.

(1)   Queen Elizabeth could not have been born of different parents she actually did.

For (together with a standard possible worlds semantics for modalities) this statement can mean either one of the following two statements:

(i)   There is no possible world where Queen Elizabeth is born of different parents she actually is; or

(ii)   There is no possible world which the modality of (1) introduces
              us to consider where Queen Elizabeth is born of different
              parents she actually is.

In case the concept of possibility used in (1) is not thought to be the logical
possibility, there is a difference between (i) and (ii). (ii) makes a statement
of only a subclass of all (logically) possible worlds, whereas (i) makes a
statement about *all* of them.

From these considerations it follows that Kripke's point cannot be
generalized (at least without further arguments) to cases where we consider
larger classes of possible worlds than what our normal counterfactual
statements introduce us. Thus, it may very well be the case that the origin
of an individual were not a *de re* truth across all possible worlds *all*
intensional concepts (e.g., propositional attitudes) force us to consider.

It will not take us long to observe that in our conceptual scheme we
often have to cope with backwards diverging histories, and with individuals
which have to be cross-identified even though they do not have an identical
(or even similar) origin. For consider what typically happens in propositional attitude contexts. Consider the actual world as it is now and an
individual (say Nixon) figuring there. It makes perfectly good sense to
reflect the various courses of events which are compatible with everything
we know or believe (say) and which lead to the present state of affairs
and to Nixon being precisely as he actually is now.[2] We are not likely to
know or believe practically anything specific concerning the circumstances
of Nixon's origin, much less of the relevant McGinnian zygotes or gamets.
Thus there will be courses of events, compatible with everything we know
or believe, where Nixon's origin varies considerably enough from one
course of events to another. Thus it will not be a *de re* necessity that Nixon
had the origin he actually had across the possible courses of events we are
considering here. Yet there will be no problems in cross-identifying Nixon
across the relevant courses of events: all these courses of events have a
stage in common (viz. the actual world as it is now) and all we have to do
is to follow Nixon back and forth in space and time inside the boundaries
of the various courses of events. (Here we of course assume that we know
that persons are continuous in space and time.)

In sum, our observations above concerning cross-identification are
consistent with the view, defended by Kripke, Mackie and others, of the
necessity of origin. But this is the case only if we carefully keep in mind
that the point of Kripke *et al.* holds only of *certain* intensional concepts,

not all. There are contexts where two members i and j of two different possible worlds u and v are (and intuitively should be) cross-identified even if i and j do not have the same origin.

Such cross-identification between u and v do not constitute a counter-example to the thesis of Kripke *et al.* for the simple reason that u and v could not both be among the class of possible worlds the consideration of which is presupposed by the notion of possibility Kripke *et al.* is interested in.

One should always carefully distinguish problems of defining the class of possible worlds which are relevant in different intensional contexts, and problems of cross-identification. It seems to me that this fundamental point has often been overlooked in the literature. Thus we find, e.g., McGinn claiming (in the passage quoted above) that *any* individual figuring in *any* possible world who do not come from the same fetuses and zygotes as Nixon actually comes cannot be cross-identified with the actual Nixon. Were this the case, we or indeed poor Nixon himself could not have practically no attitudes *de re* about Nixon for there are always bound to be courses of events, compatible with our and Nixon's attitudes, in which Nixon's fetuses and zygotes are not the same. Our attitudes simply do not specify those entities up to anything like identity.

5. SIMILARITY ENTERING THE PICTURE

It will not take us long to observe where the reason for our difficulties lies. The main problem seems to be that we are requiring that the 'common ground', which serves as the 'base' of our cross-identification considerations, have to be *the same, completely* similar. Whether that 'base' is a state of affairs (a stage in a course of events) or only a part of such, we are in both cases making this same strong requirement. The requirement excludes the present model from working in most propositional attitude contexts for the simple reason that our propositional attitudes practically never specify any part of the world completely.

The next obvious step is to try to enlarge the applicability of the present model of cross-identification by considering stages of possible courses of events (or parts of such stages) which are not *completely* similar. This line of thought immediately raises the following question: how similar should stages of courses of events (or parts of such) be for the above model to work?

In order to appreciate this problem, let us consider the following simple example. Assume we are dealing with a person's, say a's, beliefs. His beliefs specify some spatially restricted parts of the world better than others. For instance, his beliefs will specify his own home and his own body in more detail than what they specify the home or body of the Prime Minister of China (or some other equally remote person). Yet, it is safe to assume, a's beliefs do not specify his own home, his own body or any part of the world completely, thus making the simple Hintikkian model inapplicable. What marks a difference between the different parts of the world of which a's beliefs say something is *in what detail* a's beliefs specify them. This is a matter of degree.

Reflect a situation where we are trying to cross-identify a himself across a's doxastic alternatives. If a is like most of us, such cross-identification should presumably be possible. We surely do have *de re* beliefs of ourselves. Yet a's beliefs do not, we may safely assume, specify *all* of his properties. For instance, he is not likely to believe anything definite about the number of cells in him, or the number of hairs in his hair. Such features of him are always to remain unspecified by his beliefs. Accordingly, a's properties in a's different doxastic alternatives will vary from one world to another. This can hardly be interpreted as a demonstration of the impossibility of the cross-identification of a across a's doxastic alternatives.

Yet cross-identification on the basis of the Hintikkian model discussed above is impossible here. This is the case for the reason that there is no spatio-temporally restricted part of the various courses of events compatible with everything a believes which would include a. (One must be careful with the order of quantifiers here; for all possible courses of events compatible with everything a believes there is a spatio-temporally restricted part which includes a. But those spatio-temporally restricted parts are not *the same* – even though they resemble each other very much indeed.)

The only rationale for us to cross-identify a across his doxastic alternatives seems to be that a's beliefs specify him in *sufficient* detail. But this seems tantamount to saying that we cross-identify a across the relevant alternative worlds on the basis of *similarity* considerations.

This comes close enough to the suggestions put forward by David Lewis (1968). Hintikka has however criticized Lewis as follows:

David Lewis has suggested that the world lines joining to each other 'counterparts' (his term) in different worlds are based on the similarity of the counterparts in question, the similarity being something like a weighted average of many different kinds of similarity considerations. This is a misleading way, I want to argue. . . . It is misleading because by far the most important vehicle of trans world comparisons is given to us by various continuity principles. . . . This leaves [cross-identification] principles largely at the mercy of the laws of nature which serve to guarantee the continuity of our 'natural' individuals (e.g. physical objects) in space and time. (Hintikka, 1975, p. 209.)

In this criticism of Lewis' position, we find Hintikka again stressing continuity principles as the basis of cross-identification. Above, we have argued that continuity principles can work as the vehicles of cross-identification only in case we deal with courses of events which have at least one stage (temporal slice) in common, or a (spatially restricted) part of such. However, this is practically never the case when we deal with possible worlds compatible with someone's propositional attitudes. For instance, in the above case we tried to cross-identify *a* across his doxastic alternatives. Presumably *a* believes he is continuous in space and time. By continuity principles, we could accordingly follow his *prima facie* counterparts back and forth. We could draw *stage lines*, lines that bind together the same individual at different moments of time inside a course of events. But in the situation considered continuity does not yield us any means of binding several such stage lines to a full-fledged *world line*, a line that picks out one and the same individual from different courses of events. (Or if we did wish to use David Kaplan's well-known phrase, maybe we could say that continuity gives a *domestic* TWA ticket but not an oversea one.)

I conclude that Hintikka's criticism of the similarity approach, when applied to the theory of propositional attitudes, is ill-founded. In propositional attitude contexts our main vehicle of cross-identification is not provided by various continuity principles but by similarity considerations.

6. HINTIKKA'S DEFINITION OF 'COMMON GROUND' TOO INCLUSIVE

It is in fact relatively easy to see what has led Hintikka to overstate the import of continuity principles as a vehicle of cross-identity in propositional attitude contexts. For we have already noticed that according to Hintikka, when we are discussing (say) what *b* beliefs, a 'common ground' is just any part of the world that is specified by *b*'s definite beliefs about the world. Now if this kind of definition of 'common ground' were acceptable in the model of cross-identification being discussed here, it is of course trivial that the model would be applicable also in propositional attitude contexts.

Yet Hintikka's definition of a 'common ground' is not acceptable. We shall demonstrate this by means of an example later in this section. Meanwhile, let us quote the general reason for our conception. It seems to us that one's (say) definite beliefs about the world do not define any 'common ground' for doxastic alternatives for the reason that one's beliefs do not (normally) specify any (spatio-temporally restricted) part of the world. What one's beliefs do is that they specify the world *partly* – which is not to say that they did specify any *part* of it.

In order to appreciate the point, let us consider a concrete example.

Assume a police officer does not know who broke into the Mayor's house at a certain time. However, assume he has certain unspecific beliefs concerning the burglar (whoever he is). Thus we are considering various courses of events, each compatible with everything the police officer believes. In each of these courses of events a person broke into the Mayor's house at a certain time; but who that person is varies from one case to another. Consider now in particular two possible courses of events which we can assume both to be compatible with everything the police officer believes: First, a course of events h_1 in which a fellow Joe Smith breaks into the Mayor's house at moment t. After that Joe does in h_1 various things into which we need not enter here. Second, we consider a course of events h_2 in which another chap Bob Brown breaks into the Mayor's house at moment t after that going into business the nature of which is not of interest here. We may assume that we can follow Joe Smith in h_1 and Bob Brown in h_2 back and forth in space and time. (We can, e.g., assume that the police officer believes all persons to be continuous in space and time.)

h_1 and h_2 have *something* in common; in both of them at moment t somebody breaks into the Mayor's house. More generally, h_1 and h_2 have that much in common as is fixed by the police officer's definite beliefs about the world. In Hintikka's terminology h_1 and h_2 thus have a 'common ground'. This 'common ground' is the partially described state of affairs which is defined by the statement that at the moment t somebody broke into the Mayor's house.

Hintikka's model of cross-identification says that normally all we have to do in cross-identification is to follow individuals back and forth in space and time and see if they coincide at the common ground. In the above example, we could follow Joe Smith back and forth in the course of events h_1 and similarly Bob Brown in the course of events h_2. We can even trace them back to the 'common ground', i.e., to that 'part' of the world which is specified by the police officer's beliefs. This 'part' of the world is described by saying that in h_1 and h_2 at t the following holds:

(2)   Somebody breaks into the Mayor's house.

Do Joe Smith and Bob Brown 'coincide' at this 'common ground'? It seems that they do for (2) describes the 'common ground' and the intended values of the existential quantifier in h_1 is Smith and in h_2 Brown. Yet of course Smith and Brown should not be cross-identified. On the other hand, if we claim that Brown and Smith do *not* 'coincide' at the common ground of h_1 and h_2, we ought to give some kind of systematical arguments for our conception. No such arguments seem to be forthcoming.

Thus, if we follow Hintikka in defining a common ground between (say) doxastic alternatives of a's to be just any old 'part' of the world specified by all of a's beliefs, the whole Hintikkian scheme of cross-identification discussed here loses its credibility. It is no use in tracing individuals back to common ground if we have no methods of cross-identifying *between* the common grounds. No such methods are provided by Hintikka, according to whom cross-identifying between common grounds amounts 'simply' to seeing whether the individuals 'coincide there'.

If we take the Lewisian similarity approach, the situation changes. Now Bob Brown of h_1 and Joe Smith of h_2 could not be cross-identified because they are not similar enough in their relevant properties.

The difference between the present case – where a common ground of (say) a's doxastic alternatives is specified by a's beliefs about the world, however incomplete – and those discussed above – where a common ground is either a whole temporal slice or a stage of a course of events, or a part of such – is striking. In the latter case Hintikka's simple model of cross-identification, based essentially on uncontroversial *re*-identification and continuity principles, works neatly but has a restricted range of applicability in the theory of propositional attitudes. In the former case one would have a large range of applications but there the model does not work in the straightforward way presupposed by Hintikka.

7. HINTIKKA'S DESCRIPTIVE VS. PERSPECTIVAL CROSS-IDENTIFICATION AS TURNING ON LEWISIAN SIMILARITY CONSIDERATIONS

A moment's reflection in fact shows that our present conclusions concord well with Hintikka's earlier points concerning the nature of cross-identification. The important earlier point Hintikka has made on cross-

identification is the *duality* in cross-identification methods to which he has
called our attention. Hintikka has argued that we can establish trans world
identities in two essentially different ways, depending on whether we rely
our considerations on descriptive (physical) criteria or on perspectival
(acquaintance) criteria. (See Hintikka, 1969 and Chapter III of Hintikka,
1975.)

The duality in cross-identification methods advocated by Hintikka
seems to me a valid observation, and several scholars have put forward
further evidence for it. (See Howell, 1972; Thomason, 1973; Saarinen,
1978a and 1978b; Smith, 1978.) The following question readily suggests
itself. How does Hintikka's point about the duality in cross-identification
methods relate to his new proposal to the effect that our main vehicle
of cross-identification is provided by continuity principles? This
question seems to have been neglected so far, both by Hintikka and
others. It is about this problem that I now wish to put forward a
suggestion.

The reader will have to be content with the sketchiest of presentations,
partly because we cannot go into the details of Hintikka's two cross-
identification methods here. If we first look at cross-identification
between alternative worlds, not between the actual world and alternatives,
it seems to us that both of Hintikka's cross-identification methods turn
essentially on certain kinds of similarity considerations. In the descriptive
(physical) case, we base our similarity considerations on relevant descriptive (physical) properties. In the perspectival case the attitude
studied fixes a system of co-ordinates (a 'perspective') on the basis of
which, or to the relation of which, we then make the similarity considerations.

In case we deal with the actual world and its alternatives, the situation
is more or less the same except that now we have to take into account also
the relevant causal elements. (See Hintikka, 1976.)

In brief, Hintikka's earlier points about cross-identity do not strike us
totally unlike the proposals of David Lewis. In particular, in cases where
Hintikka's new continuity model is not applicable (as is typically the case,
e.g., in propositional attitude contexts) one must establish the relevant
trans world identities by means of descriptive or perspectival cross-
identification methods. Neither one of these methods works in the way
Hintikka's 'continuity' model does, but both of them turn on considerations very much like those advocated by David Lewis.

8. CONTINUITY PRINCIPLES REPLACING SIMILARITY IN IMPERSONAL MODAL CONTEXTS

What has just been argued about the role of similarity considerations in cross-identification should not be taken to imply that in each and every intensional context similarity were the main vehicle of cross-identification. In fact, it seems to us that in *impersonal* intensional contexts, unlike in propositional attitude contexts, the continuity principles stressed by Hintikka in his recent writings are operative.

An example will illustrate the situation. Consider

(3)   It is possible that in five years my brother will resemble me (as I am now) more than I will then.

Examples like this are often presented as counterexamples to Lewis's similarity-theory. (See, e.g., Plantinga, 1974.) For assume (3) is true. Then reflect the world as it is in five years from now (call that states of affairs v). If cross-identity is solely based on similarity, then, since my brother in v resembles me now more than I do in v it follows that my brother in v is cross-identified with me. This is absurd.

Examples like (3) show clearly enough that similarity cannot be *the* vehicle of cross-identification. It is here that Hintikka's emphasis of continuity principles has its rationale. The correct analysis of (3) is not to take two temporal slices (two states of affairs without any history or future) and ask which individuals figuring in those slices (states of affairs) resemble most which of the other one. Rather – and this concords well with one's intuitions – we ought to consider different futures for the actual world as it is now (the actual states of affairs now). In (3) we claim that in one of these possible courses of events a certain state of affairs will actualize in five years time, viz. a state of affairs in which my brother looks more like I look now than I will. Now how is the cross-identification carried out here? It is of course carried out by *relying on continuity principles*: we follow individuals back and forth in space and time in that course of events we are envisaging. One such life span (or stage line) will connect myself, as I am now, with a person figuring in the relevant future state of affairs (in that course of events). Similarly for my brother. What (3) then says is that the person picked out from the relevant future states of affairs by my own stage line resembles less myself now than the person picked out in five years' time by the stage line of my brother's.

This explication of the meaning of (3) goes well together with our intuitions. The principles on which cross-identification here is based – continuity in space and time – are uncontroversial and yield intuitively correct results.

Our observations so far suggest a number of more general points. One of them is that we ought to rely our cross-identification considerations on continuity principles whenever we can. In particular, whenever continuity principles are applicable we should base our considerations on them rather than on similarity principles. This yields a partial vindication of Hintikka's point, quoted in the beginning of this paper.

While we have found continuity principles not to be applicable in most propositional attitude contexts, examples like (3) readily suggest an important class of sentences where the situation is different. For when dealing with impersonal modalities, we are often tacitly considering possible *courses of events*, rather than states of affairs, and this is a prerequisite for the applicability of continuity principles.

What was just mentioned could be put in a stronger form as follows. When discussing the nature of cross-identification, theoreticians have usually been occupied by the problem of establishing trans world identities between two given states of affairs. This way to reflect the problem easily leads one astray, however. Take for instance (3). One might be tempted to say that what we have to do there is to establish world lines between two possible states of affairs, viz. the actual states of affairs (now) and an alternative future states of affairs. This way to reflect the situation can be misleading, however, because it easily leads to disregarding continuity principles. In particular, it easily leads us to disregarding the point – which is crucial here – that the two possible states of affairs between which we are cross-identifying in (3), are stages in one and the same course of events. And to disregard this point is of drastic consequence because that will in effect make it impossible to use continuity principles as the vehicle of cross-identification.

Assume somebody presents us two possible states of affairs and asks which individuals, if any, figuring in one of them is identical with a certain individual figuring in the other one. This is how the problem of trans world identity or cross-identification is typically presented. How is one to answer the question? The point we wish to put forward here is that essentially different kinds of answers to the question can be given, depending whether we are taking the two states of affairs to be stages in (different stages of) a possible course of events, or not. When we are dealing with courses of events continuity principles are likely to play an essential role, which is not the case when we deal with states of affairs not connected by any course of events. The striking point here is that our actual semantical practice often presupposes that we are dealing with possible courses of events, even when this is not explicitly reflected in the surface forms of the sentences we are discussing. In other words, even in cases where we *prima facie* are not dealing with courses of events we as a matter of fact do so.

One example which illustrates the situation is (3). Another type of sentences in the same vein is provided by those involving *counterfactuals*. It seems to me that most problems of cross-identification which arise in counterfactuals tacitly presuppose that we are dealing with courses of events rather than with distinct and unconnected states of affairs. Thus the main vehicle of cross-identification in such cases is continuity. Take for instance Kripke's (1971) example

(4)   If Nixon had only given a sufficient bribe to senator X, he would have gotten Carlswell through.

Most of the discussion of cross-identification in the connection of counterfactuals tries to solve the problem as to how to cross-identify between a given states of affairs (the actual world) and a *counterfactual situation*. This is what, e.g., Kripke does when he discusses (4). I will return to Kripke's specific proposals on this topic in a moment. Meanwhile, it suffices to point out how this view to look at (4) overlooks one important aspect of the situation at the outset. For a counterfactual situation is a temporal slice, it is a state of affairs with no history and no future. Thus if we characterize the cross-identity problems inherent in (4) by saying that they amount to comparisons between the actual states of affairs (the actual world) and a counterfactual situation, thus disregarding the course of events in which the two states of affairs are stages, we readily exclude the possibility of using continuity principles as the criteria of cross-identity.

And yet the most plausible and intuitively natural way to look at (4) involves cross-identifying in terms of continuity principles. Concentrate yourself for instance on cross-identifying Nixon across the worlds relevant for (4). The semantics for (4) that I wish to propose analyses it somewhat as follows. First go backwards in time until you come to a moment during Nixon's *regime*. Actual world at that moment (that actual historical state of affairs) has several possible futures. In other words, there are several different future courses of events which all have the actual world at that time as their initial stage. These courses of events include the actual one where the Watergate scandal takes place in due course of time, where Nixon does not get Carlswell through, etc. They also include unrealized courses of events in some of which Nixon gives a sufficient bribe to Senator X, and in some of which Nixon also gets Carlswell through.

Look now at all the courses of events in which Nixon gives a sufficient bribe to Senator X *and* in which Nixon is spatio-temporally continuous. We shall return to the rationale behind adopting the latter assumption in a moment. As for the former, we wish to point out that whatever the truth conditions of counterfactuals in general and (4) in particular turn out to be in the last analysis, this much seems plausible to us: All the counterfactual situations or possible worlds that require our attention in (4) are *included* in the class just mentioned.

How can we cross-identify Nixon throughout these possible worlds? The answer is obvious now that we are dealing with possible courses of events. We of course follow individuals back and forth in space and time (in the various courses of events) and see whether they can be traced back to Nixon as he appears at the initial stage. This is all we need for cross-identifying in this case; Hintikka's simple model is applicable because we are dealing with a class of possible courses of events which have a stage – this time the initial stage – in common, and the individual (Nixon) whom we wish to cross-identify figures at that stage.

The differences between the present analysis of (4) and a Lewisian one are not hard to come by and appreciate as speaking *pro* our analysis. For surely it is possible that after having settled in the White House Mr. Nixon had put on weight enormously, lost suddenly all his hair, gone for a nose specialist, etc., in brief, it is possible that he would have been subject to drastic changes as regards his physical appearance. So drastic, in fact, that when the relevant bribing of Senator X would be due to take place nobody would believe, looking (say) at his photographs now and then, that he was the same man that entered so cheerfully the White House. Such incidents affecting Nixon's physical appearance, so apt to lead similarity considerations astray, will cause no problems for a theorist basing cross-identification on continuity principles.

The following two assumptions underly our analysis. We have assumed that only those possible courses of events are relevant for the truth conditions of (4) in which, first (at least) Nixon is spatio-temporally continuous. Second, we assumed that there is a moment t (during Nixon's lifetime) such that all the relevant courses of events are possible future courses of events to the actual world as it is at that moment t (actual states of affairs at the moment t).

In the particular case of (4) both these assumptions seem harmless. Yet this may be peculiar to our example. A moment's reflection will in fact show that there are counterfactual sentences at least *prima facie* referring to Nixon in which we could not use the above model. A case in point is

(5)   If Nixon had been born one year later than he actually was born, he would look younger now.

It will readily be observed that our second assumption above excludes a similar analysis of (5) as what was proposed to (4).

These observations do not heavily speak against our theory, however. For one thing, we have suggested above that when we cannot, for one reason or another, use continuity principles as the basis of our trans-world comparisons, then we ought to look for other principles (similarity in particular) for a possible vehicle of cross-identity. And indeed it seems to me that something like this will be suitable for (5). Furthermore, it seems doubtful how important notorious sentences like (5) are in the first place. It seems our intuitions are far from being certain here and it can even be claimed that (5) does not make intuitively any clear sense if interpreted *de re*. (One can of course read it without difficulty *de dicto* but that will involve no trans world comparisons and is not of interest here.)

9. PLANTINGA ON TRANS WORLD IDENTITY

We may enhance the credibility of our theory by relating it to some recent authorative discussions of trans world identity. It is interesting to see that, e.g., such an eminent scholar as Alvin Plantinga does not as much as *mention* continuity principles when he discusses cross-identification in his important *The Nature of Necessity*. Yet most of Plantinga's examples are most naturally analysed in terms of the model relying on continuity principles set forth above.

Plantinga for instance discusses the problem whether "Socrates could have been an alligator" (p. 65 onwards). Plantinga argues on this problem as follows:

... shall we say that any mind-alligator-body composite an alligator, or must the mind be a special, relatively dull sort? If the first alternative is correct, then I think Socrates could have been an alligator; for I think he could have had an alligator body during part of his career. We have no difficulty in understanding Kafka's story about the man who wakes up one morning to discover that he has the body of a beetle, and in fact the state of affairs depicted there is entirely possible. In the same way it is possible that I should awaken one morning and discover . . . that my body had been exchanged for an alligator body. Socrates, therefore, could have had an alligator body; if this is sufficient for his having been an alligator, then Socrates could have been an alligator. (Plantinga, 1974, p. 65.)

What are the principles on the basis of which we cross-identify Socrates with an alligator or Kafka's Gregor Samsa with a beetle? Anything like a satisfactory answer to this question is not provided by Plantinga. And yet the answer is obvious. For by far the most natural way to look at, say, Kafka's penetrating metamorphosis is to say that poor Mr. Samsa and the beetle are one and the same because they are spatio-temporally connected in the relevant course of events. Thus it is again continuity principles that we find operative here. Also, notice how well the same model goes together with Plantinga's intuitions of the Socrates-alligator case. ". . . it is possible that I should awaken one morning and discover . . . that my body had been exchanged for an alligator body. Socrates, therefore, could have had an alligator body." What is being described by Plantinga is a possible *course of events* in which my body (or Socrates' body) is transformed from a human body to that of an alligator's. Problems of cross-identity are therefore not insuperable because my identity, or Socrates' identity, is established on the firm foundation of continuity principles. Inside the relevant course of events we can follow myself (or Socrates) back and forth in space and time. This we can do irrespective of the bodily transformations which the corporal outfit is subject to.

The reason why we can cross-identify Socrates with an alligator is that we can envisage a course of events which leads from a state of affairs where Socrates is his snubnosed himself to another state of affairs where he is an alligator (i.e., where he at least looks like an alligator). Given *just* these two possible states of affairs without the connecting course of events at the background it would be however impossible to draw the intended world line. For depending on the course of events as parts of which we reflect the two states of affairs we in fact get different trans world identities.

Assume that in the states of affairs with alligators from which we are trying to locate Socrates there are two alligators. Which one is Socrates?

This question is unanswerable as such. But the situation changes drastically once we stipulate that the state of affairs we are considering is a part of a course of events, at some stage of which Socrates can be identified without controversy. Once such a course of events is put forward we can establish which alligator, if either, Socrates is in the relevant state of affairs. This can be done by applying continuity principles in a straightforward manner.

It is of interest to note how close Plantinga has come to our analysis without making the right conclusions. For recall what Kafka's story inspired Plantinga to proclaim "At least he [Socrates] could have had an alligator body during *part* of his career." (Italics added.) Thus Plantinga accepts the intuition, shared by us, that in those courses of events in which Socrates is an alligator, he is thus and so only part of the time. This is the case for good reasons, if we are right. For the model of cross-identification based on continuity is applicable to Socrates only if the relevant course of events have at least one stage (or a part of a stage) in which Socrates figures in common with the actual history. Since Socrates was not actually an alligator, it follows that our model can locate Socrates only from courses of events in which he was not all of the time an alligator.

(There is not any moment t during Socrates' lifetime such that there would be a possible future course of events for the actual states of affairs at t in which Socrates would be an alligator *all* of his lifetime.)

These observations show how well the model of cross-identity based on continuity does justice to Plantinga's intentions and intuitions. The model is based on uncontroversial principles which readily explains why situations where it is applicable are intuitively so simple to grasp. When other kinds of principles have to be used our intuitions also tend to fluctuate and lose their sharpness.

For instance, could Socrates have been *all* of his lifetime an alligator? Our intuitions here are vague, and not unpredictably in view of the above points. For as already mentioned now the model proposed does not work. (It may perhaps be added, however, that if we accept the possibility laid down, we seem to rely on the relevant trans world comparisons on some kind of similarity considerations. The exact nature of these considerations – as so often is the case with similarity considerations – is likely to remain rather unclear, however.)

10. KRIPKE ON TRANS WORLD IDENTITY: PREFABRICATED INDIVIDUALS

We promised above that we would return later to the specific proposals

Kripke has made as regards cross-identification. Time has now come to take up this promise. Kripke's position is wide-spread and perhaps the most often advocated one. It is therefore of interest to see how our proposals relate to those of Kripke's.

Briefly put, Kripke's idea is that there is no problem of trans world identity: the apparent problem arises out of a wrong way to view possible worlds. Trans world identities, so the theory proceeds, are trivial because they are taken for granted, individuals as it were individuate themselves. (See Kripke, 1971; 1972.)

Kripke himself has argued for his position as follows:

[It is claimed that] we must describe counterfactual situations purely qualitatively and then ask the question, "Given that the situation contains people or things with such and such qualities, which of these people is (or is a counterpart of) Nixon, which is Carlswell and so on. This seems to me wrong. Who is to prevent us from saying "Nixon might have gotten Carlswell through had he done certain things"? We are speaking of *Nixon* and asking what, in certain counterfactual situations, would have been true of *him*. We can say that if Nixon had done such and such, he would have lost the election to Humphrey. Those I am opposing would argue, "Yes, but how do you find out if the man you are talking about is in fact Nixon?" It would indeed be very hard to find out, if you were looking at the whole situation through a telescope, but that is not what we are doing here. (Kripke, 1971, pp. 147–148.)

By way of evaluating Kripke's proposals, let us first concentrate on his point, made in the very beginning of our quotation, to the effect that it is *wrong* to describe counterfactual situations purely qualitatively and then proceed to asking who's who. We could not agree more. Indeed it is wrong to view the situation thus because that will amount to focusing attention to counterfactual but possible *situations*. We have argued above that given a counterfactual situation we typically cannot decide which actual individual (if any) figures in that situation. This we cannot do before it is stipulated in which *course of events* the counterfactual situation is assumed to be a stage (or part of such). But once such a possible course of events is put forward, we will face no difficulties in finding out who is Nixon and who is Carlswell in the counterfactual situation. This will be done by using continuity principles, in the way described above.

Notice also how right Kripke is in pointing out how difficult it would be to cross-identify between two possible states of affairs, if we were looking at those states of affairs through a telescope. Indeed this is the case, as readily follows from our discussion. That kind of telescoping could not establish the relevant world lines for the simple reason that no matter how much we would telescope a state of affairs, we could not unequivocally decide in which course of events it is a stage (or part of such).

In sum, we find ourselves in agreement with two of the points Kripke makes in the passage quoted above. Yet there is a disagreement about the basic point. For according to Kripke, the facts put forward show that there is no *need* for cross-identification. A counterfactual statement like

(4)   If Nixon had given a sufficient bribe to Senator *X*, he would have got Carlswell through

does not present us any kinds of problems of cross-identifying Nixon across the counterfactual situations. (4) is *about* Nixon, "we are speaking of *Nixon* and asking what, in certain counterfactual situations, would have been true of *him*."

Kripke's view about cross-identification, widespread among scholars, is in our view inadequate. We shall offer arguments in favour of our viewpoint below.

First, let us consider what Kripke's point would amount to in the simplest possible case involving *one* kind of cross-identification, viz. in a straightforward temporal context. Consider for instance

(6)   Once upon a time Nixon was an idealistic teenager.

Here we are speaking of *Nixon*; we are asking what is true of *him* in a counterfactual situation – viz. in a state of affairs which once has obtained in the actual history. (Notice that we can indeed say that the situation or state of affairs in which Nixon is considered (6) is *counterfactual*: surely Nixon is not *now* an idealistic teenager.) Does this show that it is a pseudo-question (or at least a trivial one) as to *who* in the relevant counterfactual situation is Nixon? Does it follow that it would be idiotic to ask how do we know that a certain individual in that counterfactual situation is Nixon? Of course it does not. That (6) is *about* Nixon – i.e., that 'Nixon' has a larger scope in (6) than the past tense operator – is secured by certain conceptual principles whose mechanism is non-trivial and a subject of much philosophical and semantical controversy.

In sum, it seems to me that Kripkeans at least owe us an explanation as to why Nixon's identity in the counterfactual situations relevant for (6) is taken for granted, trivial, but why this is not the case in the counterfactual situation(s) relevant for (4).

Notice how naturally things work for a non-Kripkean. (6) and (4) are both of Nixon, they are *de re*. In both cases it is relevant, and meaningful, to ask how do we know that the relevant person in the relevant possible states of affairs is indeed Nixon. The only difference is that in the former case this answer is more easy to answer than in the latter. But this is what we could have expected; for surely simple tenses are a more straight-forward phenomenon, a phenomenon more amenable to analysis, than counterfactuals are.

Another inadequacy in the Kripkean model is revealed by the following observation. On Kripke's view, individuals are primary to possible worlds. It is implicit in the approach that, e.g., the counterfactual states of affairs relevant for (4) are constructed 'around' Nixon and Carlswell. We pick out the actual Nixon and the actual Carlswell; then we construct the relevant worlds around them.

This scheme looks nice and an explanation for the irrelevance of all trans world identity comparisons is readily in the offing.

But what does it mean that a possible world is constructed 'around an (actual or one-time actual) individual'? It can scarcely mean that we would pick up such an individual, then undress him from all his properties and then dress him up again to all those properties we want him to carry in the new possible world. Indeed, could individuals be viewed in the first place completely naked, totally without properties? If they can, what is the criterion of identity and non-identity (in the *intra*-world sense) among them? Is this question meaningful at all? For if *a* and *b* are bare individuals which are identical or which are not identical with each other then it appears we in both cases readily get a property for both of them, some-thing that should be excluded by their definition. But if we do not have any criterion of identity, one finds it hard to see how we could call those entities 'individuals' in any reasonable sense, however inclusive.

On the other hand, it appears we *should* undress the individuals, that is, it appears we should start the construction of a possible world from a bare, naked individual. For if we did not do this but did require that our individuals carry over to new possible worlds some of the properties they actually have, we would immediately run into difficulties. Who is to decide which properties we drop and which we leave? Consider, e.g., Nixon in the states of affairs relevant for (4). It would seem that in order to 'construct' these states of affairs around Nixon, we are allowed to drop such actual properties of Nixon's as 'the first U.S. president to resign from his office' or 'an individual who lives in California'. It seems natural to think that we can drop such properties, and it is equally natural to think that we cannot drop a property like 'to be (for at least part of his lifetime) a human being'.

But what are the criteria which guide our judgment here? It seems hard to avoid any other conclusion but this: We can drop some properties but not others because the former are 'accidental' to that individual whereas the latter are not. In other words, we can drop those properties which do not touch upon our individual's *identity*.

This is precisely the problem of cross-identification. To answer the question as to which properties we can drop and which not when constructing possible worlds 'around' individuals is to answer the question of trans world identity. Thus the same question is there to be answered for the Kripkeans, too, and we have gained nothing by starting from 'prefabricated' individuals.

A look at the simple temporal case is instructive here. Reflect again (6). Would it be natural to say that in (6) the relevant counterfactual situation is somehow 'constructed' around the actual Nixon? Of course not. What we do in a sentence like (6) is that we pick out a counterfactual state of affairs which was actualized at an earlier moment in the history of the actual world. We then say something of an individual figuring in that state of affairs. If somebody asks how do we know that *that* feller and nobody else is Nixon, we just refer to continuity principles which connect that particular teenager with the present actual Nixon in the actual course of events. But the crucial point is that we have to refer to such principles: It is not trivial as to how to identify Nixon from the relevant counterfactual circumstances. We are speaking about *Nixon* all right but what that means is a non-trivial philosophico-semantical problem. This is the vital observation which has gone unnoticed by Kripke.

Academy of Finland

NOTES

* I am indebted to Ilkka Niiniluoto for helpful comments.
[1] Throughout this paper, we shall understand states of affairs and courses of events realistically in the sense that we shall assume that they are not relativized to a given language or conceptual framework. As we understand the terms, a possible state of affairs or a possible course of events is independent of any conceptual framework; their properties need not be characterizable by any language we may be employing or even in principle. The methodological point I am here making says that all possible state of affairs and courses of events modalities semantically introduce to us are of this kind. The possible state of affairs modalities introduce us to consider are real-life alternatives

to the actual state of affairs, not those aspects of such real-life alternative states of affairs which are characterizable in a given language or conceptual framework. (Similarly for courses of events.)

The criticism which I will level against Hintikka's model of cross-identification as applied to propositional attitude contexts presupposes that we take the relevant states of affairs or courses of events realistically in this sense. Hintikka's model of cross-identification (based on continuity principles) could perhaps be applied with success to propositional attitude contexts too, if we did not make this assumption.

Why should we then understand states of affairs or courses of events realistically? I am not going to argue for the point in detail here, partly because of lack of space, partly because the point has been argued for me forcefully by Hintikka himself. In his insightful paper on Carnap's semantics, we read:

> "Montague reports that, according to Carnap's verbal suggestions, too, 'possible worlds [are] identified with models'. In other words, possible worlds are not thought of by Carnap as the real-life situations in which a speaker might possibly find himself, but as any old configurations – perhaps even linguistic – exemplifying the appropriate structures . . . it is this apparently small point that precludes Carnap from some of the most promising uses of possible worlds semantics." (Hintikka, 1973, p. 378.)

Later in the same article we find Hintikka criticizing Carnap's doxastic logic in the same vein:

> Carnap was apparently prevented from analysing the concept of belief in this way [in the way it is done in modern possible-worlds semantics] by the very same peculiarity which made us say above that he never reached full-fledged possible-worlds semantics, viz. by his failure to interpret his models as genuine possible worlds, i.e., real-life alternatives to our actual world. (Hintikka, 1973, p. 380.)

The key idea in the criticism of Carnap levelled by Hintikka is thus the way we ought to look philosophically at the situation. Technically, the basic idea of possible worlds semantics is to define that a sentence 'Op' (where 'O' is a one-place intensional operator) is true in a model M iff 'p' is true in all (alternative) models N. This technical insight Carnap came close to making, but the really crucial step which Hintikka finds lacking in Carnap is in the way one ought to look at those alternative models N. Those alternative models should be taken as real-life alternative courses of events or states of affairs in which the speaker could conceivably find himself. They are not, in particular, state descriptions or any other kind of linguistic entities, but extra-linguistic entities in the same sense as the actual world (actual states of affairs or course of events) is extra-linguistic.

[2] One ought to be cautious when reflecting those properties Nixon now actually has. Call s the present state of affairs at the present moment t. There are various properties one is inclined to say Nixon has at s but which, if stated of Nixon, will lead to a contradiction with what we have argued above about backwards diverging possible histories. For instance, assume we ask how old Nixon is at s. The obvious candidate for the age is Nixon's *actual* age n at t. (We assumed that s was the actual state of affairs at moment t.) But saying that Nixon's age at s is n readily excludes such backwards

diverging histories which lead to s and where Nixon's birth takes place earlier (or later) than n years ago.

The solution to the puzzle is not hard to come by. All we have to do is to assume that only those properties of Nixon's at s are well defined which do not involve explicit or implicit reference to earlier or future states of affairs. This is a step which is natural enough for we have assumed that s (like all states of affairs in our sense) is a state of affairs without any future or history. This assumption comes close to saying that s (like all states of affairs) can be a stage in different courses of events, as respect to both the history and the future.

There is no paradox in our saying that one cannot assert how old Nixon is at the present actual state of affairs s. What gives the air of paradox is that one is tempted to understand the present actual state of affairs to involve also the whole of the actual history (of the actual world). This is not how we are employing the notions, as has already been pointed out. For us, the present actual state of affairs s could be the output of several different courses of events, only one of which is the actual course of events. Properties of Nixon's at s are not well-defined as soon as they refer to earlier states of affairs. It is only when we stipulate the course of events as a stage of which we are considering s that these properties of Nixon's become well-defined.

BIBLIOGRAPHY

Hintikka,Jaakko, *Models for Modalities*, D. Reidel Publ. Co., Dordrecht, Holland, 1969.

Hintikka, Jaakko: (1973), 'Carnap's Semantics in Retrospect', *Synthese* **25**, 372–397. Reprinted in Hintikka (1975).

Hintikka, Jaakko, *Intentions of Intentionality and Other New Models for Modalities*, D. Reidel Publ. Co., Dordrecht, Holland, 1975.

Hintikka, Jaakko, 'Information, Causality, and the Logic of Perception', *Ajatus* **36** (1976), 76–94. Also in Hintikka (1975).

Howell, Robert, 'Seeing As', *Synthese* **23** (1972), 400–422.

Kripke, Saul, 'Identity and Necessity', in M. K. Munitz (ed.), *Identity and Individuation*, New York University Press, New York, 1971.

Kripke, Saul, 'Naming and Necessity', in D. Davidson and G. Harman (eds.), *Semantics of Natural Language*, D. Reidel Publ. Co., Dordrecht, Holland, 1972.

Lewis, David, 'Counterpart Theory and Quantified Modal Logic', *Journal of Philosophy* **65** (1968), 113–126.

Mackie, J. L., '*De* What *Re* Is *De Re* Modality?', *Journal of Philosophy* **71** (1974), 551–560.

McGinn, Colin, 'On the Necessity of Origin', *Journal of Philosophy* **73** (1976), 127–134.

Plantinga, Alvin, *The Nature of Necessity*, Oxford University Press, Oxford, 1974.

Saarinen, Esa, 'Intentional Identity Interpreted', *Linguistics and Philosophy* **2** (1978), 151–223a.

Saarinen, Esa, 'Propositional Attitudes, Anaphora, and Backwards-Looking Operators' (to appear), 1978b.

Smith, David W., 'The Case of the Exploding Perception', *Synthese* (to appear), 1978.

Thomason, Richmond H., 'Perception and Individuation', in M. K. Munitz (ed.), *Logic and Ontology*, New York University Press, New York, 1972.

V

EPISTEMOLOGY

ISAAC LEVI

SERIOUS POSSIBILITY*

1. CREDAL STATES

Strict Bayesians emphasize the importance of the evaluation of truth value bearing hypotheses with respect to subjective or credal probability for both practical deliberation and scientific inquiry. According to strict Bayesians, agent X's *credal state* $B_{X, t}$ for hypotheses expressible in language L containing neither modal nor epistemic operators or predicates is representable by a function $Q(h; e)$ from sentences in L to real numbers satisfying the following conditions.

(1) $Q(h; e)$ takes all $h \in L$ in the first argument place and all and only $e \in L$ such that the truth of e is a serious possibility according to X at t. The value of the function is nonnegative, real and finite for every such pair $(h; e)$.

(2)   If the falsity of $e \equiv e'$ and $h \equiv h'$ is not a serious possibility according to X at t, $Q(h; e) = Q(h'; e')$.

(3)   If the falsity of $e \supset h$ is not a serious possibility according to X at t, $Q(h; e) = 1$.

(4)   If the falsity of $e \supset -(h \& f)$ is not a serious possibility according to X at t, $Q(h \vee f; e) = Q(h; e) + Q(f; e)$.

(5) $Q(h \& f; e) = Q(h; f \& e)Q(f; e)$.

Agent X need not be a person but could be a group or institution. But whether X is a person or group, we cannot expect X to be capable of adopting a credal state for hypotheses expressible in L satisfying these conditions on all occasions. Emotional or sociological disturbances, failure of personal or group memory and lack of computational facility are among the factors which will normally prevent X from satisfying the requirements just prescribed. In real life, X will normally fail to be ideally situated.

Nonetheless, we may prescribe that X conform to the prescriptions for an ideally situated agent insofar as he is able. In this sense, we may suppose

219

E. Saarinen, R. Hilpinen, I. Niiniluoto, and M. Provence Hintikka (eds.), *Essays in Honour of Jaakko Hintikka*, 219–236. *All Rights Reserved.*
Copyright © 1979 by D. Reidel Publishing Company, Dordrecht, Holland.

that less than ideally situated X should be committed to a credal state representable by a Q-function satisfying the conditions just specified even though he may excusably fail to live up to his commitments. It is such a commitment which I shall take to be X's credal state.

Even so, it seems to me that the demands imposed by strict Bayesians on ideally situated agents are excessive. I shall not argue the point here, but it seems to me that ideally situated agents are not only permitted to have credal states representable as sets of more than one Q-function satisfying the conditions (1)–(5) but that under suitable circumstances they are obliged to adopt such credal states. But whether $B_{X, t}$ is representable by a single Q-function or by a nonempty set of such functions (satisfying a convexity condition), the conditions (1)–(5) presuppose that X is committed to evaluating all truth value bearing hypotheses expressible in L with respect to serious possibility in a manner which meets the following conditions:

(6)   The sentences in L are partitioned into those belonging to the *corpus* $K_{X, t}$ and those which do not where $h \in K_{X, t}$ if and only if the falsity of h is not a serious possibility according to X at t.

(7) $K_{X, t}$ is deductively closed in L.

(8) $K_{X, t}$ contains the *urcorpus UK* for L
          (a) *UK* is deductively closed
          (b) *UK* contains all truths of first order logic set theory and mathematics expressible in L
          (c) *UK* contains all other incorrigible hypotheses expressible in L.

If $e \; \varepsilon \; K_{X, t}$, $Q(h; e) = Q(h)$. Hence, $Q(e; e) = Q(e) = 1$ and $Q(-e) = 0$. On the other hand, if $Q(-e) = 0$, it does not follow that $e \in K_{X, t}$ and, hence, it does not follow that the falsity of e is not a serious possibility according to X at t.

X's corpus $K_{X, t}$ at time t is his *standard for serious possibility at t*. All and only hypotheses consistent with $K_{X, t}$ are serious possibilities according to X at t.

2. CONFIRMATIONAL COMMITMENTS

At t, X is committed to a corpus $K_{X, t}$ and credal state $B_{X, t}$ for hypotheses expressible in L which together satisfy conditions (1)–(8). But there are many pairs (K, B) other than $(K_{X,\,t}, B_{X,\,t})$ which meet these requirements. Any such pair is a *potential corpus cum credal state.*

I shall assume that at time t, ideally situated and rational X adopts a rule stipulating for each potential corpus K what the credal state should be in the eventuality that X adopts K as his standard for serious possibility. Less than ideally situated but rational X is committed to some such rule. Such a rule is representable as a function $C(K) = B$. Let $C_{X,\,t}$ be the function to which X is committed at t (his *confirmational commitment at t*). (K, B) is *accessible* to X at t if and only if $C_{X,\,t}(K) = B$. Clearly $(K_{X,\,t}, B_{X,\,t})$ should be accessible to X at t. That is to say, $C_{X,\,t}(K_{X,\,t}) = B_{X,\,t}$.

Let $K \subseteq K'$ where K' is obtained from K by adding e consistent with K and forming the deductive closure. If X should change from K to K', he has *expanded* his corpus. If he shifts from K' to K, he has *contracted* his corpus.

Let both (K, B) and (K', B') satisfy (1)–(8).

B' is the *conditionalization* of B with respect to K and K' if and only if for every $Q \in B$ there is a $Q' \in B'$ and for every $Q' \in B'$ there is a $Q \in B$ such that for all h in L and all f in L consistent with K', $Q'(h;f) = Q(h; f \,\&\, e)$.

The principle of *confirmational conditionalization* stipulates that X's confirmational commitment be a function C such that $C(K')$ must be the conditionalization of $C(K)$ with respect to K and K'.

Setting aside minor technicalities, all consistent potential corpora are expansions of UK. Given (UK, B_{uk}) accessible according to X at t for the urcorpus UK, the confirmational commitment is uniquely determined for all other potential corpora.

Confirmational conditionalization does not by itself regulate revisions of credal state over time. It is a condition on a rule to which X is committed at t for revising his credal states with respect to revisions in his evaluations of hypotheses with respect to serious possibility. If X lives up to his commitments as a rational agent and has no good reasons for revising his confirmational commitment over time, revisions in his credal state will manifest his endorsement of confirmational conditionalization.

Thus, if X expands his corpus by shifting from K to K' while keeping his confirmational commitment constant, the shift from B to B' will be a *temporal credal conditionalization*. Should X shift from K' to K, the corresponding shift from B' to B will be an *inverse temporal credal conditionalization*. If, however, X revises his confirmational commitment in the interim, the shift in credal state need not be either a temporal credal conditionalization or its inverse.

Many authors have characterized legitimate changes in credal state as conforming to temporal credal conditionalization. Doing so presupposes that confirmational commitments should never be revised and that all revisions of corpus are expansions. In my opinion, the untenability of both assumptions is obvious. The obviousness of these absurdities has been masked, so it seems, by a failure on the part of those who have discussed conditionalization to distinguish carefully between confirmational conditionalization, temporal credal conditionalization and inverse temporal credal conditionalization.

3. CONSISTENCY AND POSSIBILITY

h (in L) is seriously possible according to X at t if and only if h is consistent with X's corpus or standard for serious possibility $K_{X, t}$ at t. This suggests that we can also consider a notion of serious possibility relative to potential corpus K. h is a serious possibility relative to K if and only if h is consistent with K.

In this case, the relativity of serious possibility to X's cognitive state at t is replaced by a relativity to a potential corpus K. Observe, however, that the importance of serious possibility relative to K depends on the fact that K is a corpus which an agent is capable of adopting as his standard for serious possibility and the agent may wish to calculate how he would evaluate hypotheses expressible in L with respect to serious possibility were he to adopt K as his corpus.

Thus, we must not automatically suppose that every time one can construct a notion of A-consistency – i.e., consistency with some set of sentences – we have a useful notion of A-possibility – i.e., serious possibility relative to A. The set A may not qualify as a potential corpus and even if it does, it may not be a potential corpus which agents are likely to endorse. In that case, there is no useful purpose in having a notion of A-possibility additional to the notion of A-consistency.

To illustrate, consider the notion of logical possibility. Some authors construe this to be equivalent to logical consistency – i.e., consistency with the truths of first order logic – perhaps, including identify.

However, if L is sufficiently rich, the set of such logical truths is not a potential corpus. A rational agent should be committed to a corpus containing these logical truths but much more as well. The weakest potential corpus UK should be substantially stronger. Consequently, there is little point in considering how serious possibility should be evaluated relative to the set of first order logical truths. Of course, the notion of logical consistency is of first rate importance. But there is no need to have another term 'logical possibility'.

We can make 'logical possibility' do work for us if we equate it with consistency with the weakest potential corpus UK; for we are often interested in how serious possibility should be evaluated in case UK is adopted as the standard for serious possibility. Indeed, there are many authors who think rational men should always endorse UK as the standard for serious possibility insofar as they are able. But even if we do not follow such an extreme view, we may be interested in how X evaluates hypotheses with respect to serious possibility in a state of modal ignorance.

One sort of evaluation of hypotheses with respect to serious possibility often invoked is relative to a corpus which is a specified transformation of the currently adopted corpus. Thus, if X is prepared to affirm 'if h were the case, g would be the case', he is evaluating the falsity of g as not being a serious possibility relative to a corpus obtained from $K_{X,\,t}$ by first removing $-h$ in a manner which yields a minimal loss of informational value and then adding h. That is to say, $-g$ is inconsistent with the corpus obtained from $K_{X,\,t}$ in this fashion.

On other occasions, serious possibility is evaluated relative to a corpus described in a manner which makes no reference to its relation to $K_{X,\,t}$. This is true of logical possibility construed as consistency with UK.

In the philosophical literature, we often hear talk of physical possibility, psychological possibility and the like where we mean consistency with the laws of physics, psychology and the like.

Now the deductive closure of UK and the true laws of physics expressible in L does, indeed, qualify as a potential corpus in the sense covered by conditions (6)–(8). However, it is far from clear that there is any interesting basis for evaluating hypotheses with respect to serious possibility relative to such a corpus.

Not only is it highly unlikely that any agent will ever adopt a corpus containing all the true laws of physics, it is even more unlikely that such an agent would ever endorse a corpus containing only the true laws of physics (in addition to items in UK). And it is obscure to me why one should wish to evaluate serious possibility relative to such a corpus.

To be sure, there is an interesting notion of physical possibility. If I were asked to run a four minute mile at a track meet, I might declare that it is physically impossible for me to do so. I am not thereby claiming that 'Levi will run a four minute mile at the meet' is inconsistent with the true laws of physics. Indeed, I think it is consistent with the true laws of physics. The claim is that I lack the physical ability to run the race in four minutes.

If I know that I lack this ability, I shall, indeed, regard the truth of 'Levi will run a four minute mile at the meet' as not a serious possibility. But even if I did believe that I had the ability, I might appraise the truth of the sentence as not being a serious possibility. I might know, for example, that I will not enter the race at all. Or, alternatively, I might know that I did run the race but failed to run a four minute mile even though I was capable of doing so.

In this discussion, I focus on various types of possibility parasitic on the notion of serious possibility and not on notions of ability, constraint and disposition. There are, indeed, useful notions of physical possibility, psychological possibility, economic possibility and technological possibility; but they concern abilities of various kinds and have little to do with consistency with the laws of physics, psychology, economics and (heaven help us!) technology.

4. KNOWLEDGE AND BELIEF

Consider any propositional attitude or activity such as knowing, believing, desiring or choosing. Given such an attitude A, let $A_{X, t}$ be the set of hypotheses expressible in L such that X A's these hypotheses true at time t.

As before, we have a notion of A-consistency readily available to us. Should we also introduce a notion of A-possibility?

If the attitude A is that of evaluating the falsity of hypotheses as not being serious possibilities, then $A_{X, t}$ is the corpus $K_{X, t}$. On the other hand, the set of hypotheses desired true does not serve as X's standard for serious possibility and, indeed, does not qualify as a potential corpus.

What about the set of hypotheses X knows at t to be true or the set of hypotheses X believes at t to be true?

Presystematically, knowledge and belief are highly ambiguous notions. Any proposal for explicating them ought to recognize the need to select from this abundance of riches and should keep in mind the purpose for which a selection is made.

The concern here is with how knowledge and belief ought to function as relevant factors in practical deliberation and scientific inquiry. The thesis advanced here is that, in at least one important sense, partial belief is explicated in terms of credal probability and that our full beliefs when taken together with their deductive consequences constitute our standard serious possibility.

On this view, 'X fully believes that h' does not imply that X has a maximal degree of introspectible conviction that h or that X is disposed to a maximally intense affirmative response to interrogation as to the truth of h. With due apologies to Hume and Quine, the relevance of conceptions of belief such as these to deliberation and inquiry remain obscure.

Hintikka distinguishes between what X knows at t and what X fully believes at t. Corresponding to the distinction between knowledge and belief, Hintikka recognizes two notions of possibility: possibility for all that X knows and compatibility with everything X believes.

I do not quarrel with the introduction of a body of knowledge and a body of belief and speaking of consistency relative to the one set and relative to the other. Nor does it really matter that Hintikka talks of possibility in the one case and compatibility in the other.

What I am concerned about is this: Should X's body of knowledge serve as his standard for serious possibility? Should his body of full beliefs? Or, perhaps, neither should.

X should be urged to avoid a double standard for serious possibility. Hence, if neither X's corpus of knowledge nor body of beliefs has the exclusive right to be X's standard, neither corpus should serve as the standard at all. In that event, it becomes unclear what relevance either knowledge or belief have for deliberation and inquiry and what point there is in speaking of epistemic or doxastic possibility.

These considerations present us with a dilemma. On the one hand, there is the ancient philosophical tradition of distinguishing between knowledge and belief to uphold. On the other hand, if we identify either X's body of knowledge or X's body of belief as his standard for serious possibility, we remain with the problem of determining what the relevance of the corpus which has not been identified as X's standard for serious possibility is as far as deliberation and inquiry are concerned.

To avoid the dilemma, I propose that we distinguish between X's point of view at t and the point of view of a third party Y who contemplates X's propositional attitudes.

From X's point of view at t, all items in his standard for serious possibility $K_{X, t}$ are infallibly true. That is to say, if $h \in K_{X, t}$, the falsity of h is not a serious possibility according to X at t.

I contend that under these conditions X fully believes that h is true and, moreover, is committed to the view that whatever he believes true at t is true.

I also think it appropriate to say that from X's point of view at t, X knows all items in $K_{X, t}$ to be true so that $K_{X, t}$ qualifies as X's corpus of knowledge at t. I also call $K_{X, t}$ X's corpus of evidence at t. If $h \in K_{X, t}$, X accepts h as evidence at t.

Terminology is unimportant. What is crucial is that, from X's point of view, there is no interesting distinction between what X knows and what X believes. The philosophical tradition which suggests otherwise is based on an illicit transfer of a distinction which has some value from Y's point of view at t' when appraising X's cognitive state at t to the case where X at t contemplates his own cognitive state at t.

While assessing X's corpus $K_{X, t}$, Y is committed to his own corpus $K_{Y, t'}$ expressible in L and to a corpus $K_{Y, t'}^1$ expressible in the metalanguage $L_1 \supseteq L$. In general, $K_{X, t} \neq K_{Y, t'}$. $K_{Y, t'}$ will contain items inconsistent with $K_{X, t}$. Y is committed (in L_1) to assuming that such beliefs belonging to X's corpus are false beliefs and not knowledge.

Of course, Y will also identify items in X's corpus expressible in L which belong to his corpus expressible in L. $K_{X, t} \cap K_{Y, t'}$ should be deductively closed and contain UK. From Y's point of view, all items in L which Y knows to be false but which belong to $K_{X, t}$ should be removed from X's standard for serious possibility at t. Should that standard be restricted to $K_{X, t} \cap K_{Y, t'}$?

$K_{X, t} \cap K_{Y, t'}$ should, from Y's point of view, be contained in X's standard for serious possibility at t but need not coincide with it.

First, Y might identify h as an element of $K_{X, t}$ even though neither $h \in K_{Y, t'}$ nor $h \notin K_{Y, t'}$. From Y's point of view, it is a serious possibility that X knows that h. However, Y does not know that X knows that h.

Second, Y might assume, in his metacorpus $K_{Y, t'}^1$, that there is a true sentence in $K_{X, t}$ which is not in $K_{Y, t'}$. In this case, not only is it, from Y's point of view at t', a serious possibility that X knows something Y does not know but Y knows this to be the case.

Thus, Y may usefully distinguish between what X knows and what X believes without saddling either himself or X with a double standard for serious possibility. However, X at t cannot distinguish in the same way between what he knows and what he fully believes at t. His predicament is similar to what Y's would be were Y in perfect agreement with X.

Perhaps, someone who is prepared to concede that from X's point of view his corpus of beliefs and corpus of true beliefs are indistinguishable will, nonetheless, insist on a distinction between X's corpus of true beliefs and his corpus of knowledge.

This objection rests on a confusion of knowing with coming to know.

At t, X may contemplate adding new information to his current corpus $K_{X,\,t}$. One appropriate potential expansion strategy might be to add g to $K_{X,\,t}$ and form the deductive closure $K_{X,\,t}^{g}$. From X's point of view prior to expansion, it is a serious possibility that g is true and also that it is false. Thus, implementing the strategy might lead to importation of error or it may fail to do so. If error is imported, X would come to believe that g falsely. It would not be an instance of coming to know the items in $K_{X,\,t}^{g} - K_{X,\,t}$ even when they are true.

Suppose, however, that g is true so that implementation of the expansion strategy is error free. Would X then come to know all items in $K_{X,\,t}^{g} - K_{X,\,t}$? Not unless he was justified in choosing that expansion strategy over the alternatives available to him.

Thus, from X's point of view in the context of inquiry concerned with evaluating expansion stragleties, the distinction between coming to believe without error and coming to know is an important and intelligible one.

Suppose, however, X has implemented an expansion strategy and has shifted to the new corpus $K_{X,\,t}^{g} = K_{X,\,t'}$. From X's vantage point prior to expansion when his corpus was $K_{X,\,t}$, X may not have been justified in implementing this strategy. Hence, from that vantage point, the shift is not an instance of coming to know. However, once X has implemented the strategy, he has a new standard for serious possibility. All items in $K_{X,\,t'}$ are then judged to be true, infallibly true and certainly true. There are no invidious distinctions between them due to the way they gained admittance into his corpus. From X's new point of view at t' all items in $K_{X,\,t'}$ qualify as objects of his knowledge regardless of pedigree. He draws no distinction between what he truly believes at t' and what he knows at t'.

The serious point lurking behind the verbal byplay is that questions of justification of belief arise in the context of revising a standard for serious possibility either by adding new items to a corpus (expansion) or removing items (contraction). Unless we are considering a context where alternative strategies for revision are contemplated, there is no point in talking about justification of belief or of drawing distinctions between true belief and knowledge.

5. MODAL REALISM

'The truth of h in L is seriously possible according to X at t' bears a truth value. It describes X as having a propositional attitude – namely as appraising the truth of h to be a serious possibility.

Furthermore, rational X evaluates only truth value bearing hypotheses with respect to serious possibility. Presumably, therefore, h has a truth value.

What about X's evaluation of the truth of h as a serious possibility? Does it have a truth value? Clearly some propositional attitudes have truth values and others do not. Believings have truth values. Desirings do not.

One can, to be sure, claim that desirings have truth values after all. One might claim that X desires that h be true if and only if X believes that the truth of h is desirable. In this way, 'the truth of h is desirable' is taken itself to be a truth value bearing hypothesis The truth of that hypothesis is, given that X believes it true, a necessary and sufficient condition for the truth of X's belief and, hence, for the truth of X's desiring that h be true.

Of course, those who adopt this strategy are confronted with the task of offering some account of desirability of a proposition's being true where 'It is desirable that h is true' is not equivalent to 'X desires that h be true' or 'It is desirable that h be true according to X'. If there are good reasons for pursuing the strategy, the task must be faced. I do not know of such good reasons and regard the maneuver as an illustration of an incursion into gratuitous metaphysics.

Thus, in my opinion, believings do (at least sometimes) bear truth values while desirings do not. What about evaluations of hypotheses with respect to serious possibility? I contend that they are more like desirings than like believings.

Recall that we are concerned here not with X's beliefs about what his evaluations of hypotheses with respect to serious possibility are or with his beliefs about what the evaluations of other agents are but with his evaluations. The claim is that these evaluations lack truth values.

If they do have truth values, they presumably are to be construed as cases of believing truth value bearing hypotheses of some kind to be true. We are thus supposed to construe 'It is seriously possible that h according to X at t' as equivalent to 'X believes at t that it is objectively possible that h' where 'It is objectively possible that h' is not itself a description of X's evaluations of hypotheses with respect to serious possibility or a description of anyone else's.

Anyone who endorses such a modal realist view of appraisals of hypotheses with respect to serious possibility faces the task of explicating the notion of the objective possibility of truth value bearing hypotheses. As in the case of desiring, so here I find the realist maneuver unnecessary and the incursion into metaphysics gratuitous. Moreover, there is yet another complication.

Let L^m be an extension of the language L obtained by adding a new modal operator '$\Diamond$' for objective possibility. If evaluations of hypotheses expressible in L with respect to serious possibility have truth values in the sense just indicated, there should be some interpretation of the modal operator such that the following condition obtains:

(I)   It is seriously possible that h according to X at t (where $h \in L$) if and only if X fully believes at t that $\Diamond h$ – i.e., if and only if '$\Diamond h$' is in X's corpus at t expressible in L^m.

It is not seriously possible that h according to X at t if and only if X fully believes at t that $-\Diamond h$.

Let h be a sentence in L. Either h is in X's corpus at t expressible in L, $-h$ is in that corpus or neither h nor $-h$ is in the corpus. In the first and the third case, '$\Diamond h$' is in X's corpus at t expressible in L^m. In the second case, '$-\Diamond h$' is in X's corpus at t expressible in L^m. These results are secured by condition (I).

Thus, X's corpus expressible in L^m at t must contain either '$\Diamond h$' or '$-\Diamond h$'. X cannot suspend judgment between these rival alternatives. In my opinion, this is unacceptable. X should not be compelled to make up his mind in all cases concerning the truth values of hypotheses purportedly describing objective conditions.

If someone persists in endorsing *de dicto* realistically construed possibility statements in L^m, the following weakening of (I) might be adopted.

(II)   If the truth of h is a serious possibility at t according to X, then X fully believes that $\Diamond h$ at t.
        If X fully believes that $-\Diamond h$ at t, the truth of h is not a serious possibility according to X at t.

(II) fails to yield necessary and sufficient conditions for the truth of X's evaluations of hypotheses expressible in L with respect to serious possibility.

We might try to obtain such truth conditions by substituting some other truth value bearing propositional attitude for full belief in the condition (I). Let that propositional attitude be accepting hypotheses in some sense where X may accept h, accept $-h$ (reject h) or do neither. Not only would we have to determine what significance is to be given to the notion of acceptance thus introduced but we would still face the predicament that X should either accept $\Diamond h$ or reject it.

Thus, if the only motive for introducing truth value bearing modal statements is in order to specify truth conditions for appraisals of truth value bearing hypotheses with respect to serious possibility, the motive is insufficient. Evaluations of hypotheses with respect to serious possibility lack truth values.

None of this should be surprising. X's credal state – i.e., his system of appraisals of truth value bearing hypotheses in L with respect to credal probability lacks truth value in just the same way.

Suppose to the contrary that such appraisal has truth value. That is to say, if X assigns h degree of credence r, he fully believes that h is objectively probable (in some sense) to degree r.

Consider now a situation where X suspends judgment as to whether the degree of objective probability that h is 0.4 or 0.6. Let y be his degree of credence that the objective probability is 0.4 and $1 - y$ that the objective probability is 0.6. X's degree of credence that h is, under these circumstances, equal to $0.4y + (1 - y)0.6$. As long as y is positive and less than 1, X's degree of credence that h must be different from 0.4 and from 0.6. But this means that X must fully believe that the degree of credence is different from 0.4 and from 0.6 counter to the assumption that he is in suspense between these two rivals.

I have been arguing that categorical evaluations of hypotheses expressible in L with respect to serious possibility lack truth values. But if the system of evaluations adopted by X at t lacks a truth value, so do alternative systems. Consequently, conditional appraisals of hypotheses with respect to serious possibility also lack truth values. This suggests that open and counterfactual conditionals construed as expressions of such hypothetical appraisals are neither true nor false.

Jaakko Hintikka has insisted that possible worlds alternative to the actual one are to be described by means of counterfactual conditionals. David Lewis proceeds in the reverse fashion and specifies a semantics for subjunctive conditionals in terms of possible world semantics. Both views imply that subjunctive conditionals have truth values and, more-over, that the truth conditions make no reference to the subjective states of the utterers (except, of course, insofar as such subjective states are described in antecedents or consequents of such conditionals.)

On my view, counterfactuals have truth values only insofar as they are construed as descriptions of the agent's conditional evaluations with respect to serious possibility. The appraisals themselves lack truth values and, as a consequence, so do the conditionals construed as expressions of such appraisals.

Hintikka must have in mind some alternative conception of how subjunctive conditionals are to be construed than the construal I favor here. Neither he nor Lewis nor any other modal realist has, to my knowledge, established an alternative construal which bears the relevance to deliberation and inquiry which evaluations with respect to serious possibility both categorical and hypothetical have.

I cannot prove conclusively that realistically construed notions of *de dicto* modality both conditional and categorical are *verdoppelte Metaphysik*. But the onus is on those who deny this to explain why the introduction of such conceptions is not gratuitous insofar as we are concerned with questions pertaining to epistemology, scientific inquiry and practical deliberation.

6. TRUTH

Quine contends that we judge truth as earnestly and as seriously as can be relative to our evolving doctrine. I construe him as claiming that we 'judge' truth on the assumption that all items in our corpus are true. With this much I agree.

But, for Quine, to judge that h is true seems to amount to little more than declaring that h is true as earnestly and as seriously as can be. My contention is that in seeking to improve our evolving doctrine, we seek to avoid error where error and truth is judged relative to the very same evolving doctrine.

I do not believe that Quine intended to make this sort of claim. Indeed, there is ample evidence to indicate that avoidance of error is never a desideratum for him in inquiry and deliberation.

If agent X is to have the sort of aims I urge upon him, it seems clear that he is at least committed to the view that all items in his corpus $K_{x,\,t}$ expressible in L are true in L. But this assumption is most conveniently construed as expressible in X's corpus $K_{X,\,t}^1$ at t expressible in a meta-language L_1 rich enough to specify truth conditions for sentences in L. I follow the practice of letting L_1 contain L and, hence, letting $K^1_{X,\,t}$ contain $K_{X,\,t}$.

7. KNOWING THAT ONE KNOWS AND THAT ONE DOES NOT KNOW

Suppose $h \in K_{X,\,t}$. Should X be committed to including this information in his corpus $K^1_{X,\,t}$?

Remember that an affirmative answer does not entail insistence that rational X be consciously or explicitly are of the contents of his corpus expressible in L. But we may insist that insofar as limitations of computational facility, memory, emotional instability and the like do not prevent him, he should know which items expressible in L belong to $K_{X,\,t}$.

Suppose X is offered a gamble on the truth of the hypothesis that the integer in the billionth place in the decimal expansion of pi is 9. X is committed either the assuming the truth of that hypothesis or assuming the truth of its negation. That is so because either the hypothesis or its negation is entailed by the urcorpus and, hence, by X's corpus.

Hence, X is committed to assigning the hypothesis the degree of credence 0 or the degree of credence 1.

Yet, X may not be in a position to carry out the calculations required to identify his commitments. He may not, therefore, be able to evaluate the gamble in accordance with the dictates of the credal state to which he is committed.

Decision theories of the sort for which accounts of credal states and appraisals of serious possibility like mine are designed to provide an underpinning cannot pretend to offer wise counsel for such predicaments. Some second best criterion for choice may have to be invoked.

Nonetheless, it does not seem sensible to jettison an account of rational decision making designed for ideally situated agents because predicaments such as this can and will arise.

To the contrary, it seems preferable to obtain a clear view of the demands imposed on ideally situated agents so that we can help ordinary agents approximate the behavior of ideally situated agents.

In the example just cited, X does not know what is in his corpus. Yet, it seems clear that he should know if he is able.

If, for the moment, we overlook the fact that Hintikka treats 'X knows that' as an operator whereas I treat it as a metalinguistic predicate, there appears to be substantial agreement between us concerning the propriety of endorsing the requirement that if X knows that h, he should know that he knows that h.

Suppose, however, that $h \notin K_{X, t}$. Should X know this? Hintikka answers in the negative. I disagree.

If $h \notin K_{X, t}$, then we still need to consider X's credal state for h and $-h$. Suppose that both h and $-h$ are not members of $K_{X, t}$. That is to say, both are serious possibilities. Suppose further that $B_{X, t}$ contains exactly one numerical assignment for h – say, x. $Q(-h) = 1 - x$. If X were confronted with a decision where the payoffs for rival options depended on the truth values of h and $-h$ but did not know the value of x, he would be suffering from disabilities similar to those facing the agent offered a gamble on a hypothesis about the integral value in the billionth place in the decimal expansion of pi. An ideally situated agent would not be in such a predicament and we are considering requirements imposed on such ideally situated agents.

Now if x is different from 0 or 1, X is committed to the view that neither h nor $-h$ belongs in $K_{X, t}$. It is but a small step from this observation to the conclusion that if $h \in K_{X, t}$, X is committed to knowing this.

Following this approach does not incur commitment to the claim that an effective decision procedure exists for membership in X's corpus in L. An ideally situated agent should be able to know what is and what is not in his corpus; but this does not imply that membership in his corpus expressible in L must be effectively decidable.

Hintikka represents X's cognitive state in a language L^k containing iterable epistemic operators. He adopts as an epistemic logic a variant of $S4$. If I were to represent X's cognitive state in such a language L^k, the underlying epistemic logic I would employ would be a variant of $S5$.

Someone might wonder why I objected previously to evaluations of hypotheses with respect to serious possibility being construed as beliefs as to the truth of statements of objective possibility on the grounds that X would be obliged to be opinionated about objective modality while I am prepared to insist that X should be opinionated as to what is and what is not in his corpus expressible in L.

I have already explained that in requiring that X be opinionated as to what he knows, I am concerned only with X's commitments. Insofar as X fails or is incapable of living up to these commitments, he will be incapable of using his standard for serious possibility effectively in deliberation and inquiry.

Suppose, however, that X is ideally situated in this regard and is capable of living up to the requirement that he be opinionated concerning what he knows. He should then be in a position to use his standard for serious possibility for hypotheses expressible in L in deliberation and inquiry. This is so whether X's evaluation of h as a serious possibility is equated with X's assuming that it is objectively possible that h or X's evaluation is not regarded as bearing a truth value at all.

Thus, the requirement that X be opinionated concerning the contents of his corpus expressible in L is motivated by the fact that failure to satisfy this condition weakens the effectiveness of X's corpus expressible in L as a standard for serious possibility useful in deliberation and inquiry. Considerations such as this serve as an excuse for requiring ideally situated X to be opinionated in this way.

No such excuse exists for being so opinionated concerning hypotheses about objective possibility – assuming such hypotheses to be intelligible. That is to say, there is no excuse for our obligating ideally situated X to be opinionated in this way. In my opinion, we should always allow X the option of suspending judgment between rival hypotheses – unless there is some countervailing consideration.

8. QUANTIFYING IN

I have not considered the appeal which Hintikka and others who adopt similar approaches make to the virtues of appealing to possible worlds semantics in dealing with quantification into epistemic contexts.

Keep in mind, however, that I am concerned with corpora of knowledge as standards for serious possibility and with the revision of such standards.

My thesis is that only truth value bearing hypotheses are evaluated by agents with respect to serious possibility. This thesis is companion to the thesis that only truth value bearing hypotheses are evaluated by agents with respect to credal probability. I know of no problem in decision theory or scientific inquiry where evaluations of either sort are required to be *de re*.

If this thesis is correct, X's standard for serious possibility or corpus of knowledge is representable by a set of sentences containing no modal operators. A change in X's knowledge is representable by a change in the set of sentences known to be true.

No doubt it will be objected that this view is a dogmatic and cavalier dismissal of linguistic evidence to the effect that we do quantify into epistemic contexts – evidence which even Quine acknowledges.

The kernel of truth in this charge is located in the same place as the kernel of truth in the charge that I ignore the distinction between knowledge and belief.

Recall the important differences noted between X's point of view when assessing his corpus and X's point of view when assessing X's corpus. From X's point of view, there is no difference between what he knows and what he fully believes (at present). From Y's point of view, only a subset of X's corpus constitutes knowledge.

We have also just noted that an ideally situated X is committed to an identification of each h in L as belonging to $K_{X,\,t}$ or its complement.

But even an ideally situated Y need not know of every sentence in L whether it is in X's corpus or not. On the other hand, Y may have information that X's corpus contains some sentence in L satisfying conditions C without Y having in his corpus a sentence 'X's corpus contains h and h satisfies C' where 'h' is a standard designator for a sentence in L.

Thus, Y may assume that X knows who killed Cock Robin. That is to say, Y knows that X's corpus contains a true sentence which qualifies as a potential answer to the question 'Who killed Cock Robin?'

In my view, this sort of story can be told in all cases where it is alleged that X's evaluations with respect to serious possibility are *de re*.

9. CONCLUSION

Hintikka's work on epistemic and doxastic logic has pioneered in efforts to understand relations between modality and cognitive attitudes. By focusing attention on the role of evaluations of hypotheses with respect to serious possibility and credal probability in deliberation and inquiry, the conclusions I have reached tend to conflict rather strongly with Hintikka's ideas.

Hintikka and those who share his views may, perhaps, be able to respond to the difficulties I have raised from my quasi Bayesian viewpoint. Hintikka himself is after all very much interested in probability judgment. It would be worthwhile to understand his approach to integrating judgments of possibility and judgments of probability. My rather desultory comments will have served their purpose if they provoke Hintikka to some discussion of these matters.

Columbia University

236   ISAAC LEVI

NOTE

* I wish to thank Charles Parsons for having saved me from some errors. In spite of his gentle benevolence, however, I have resisted his efforts to save me from others.

DAVID WIGGINS

ON KNOWING, KNOWING THAT ONE KNOWS
AND CONSCIOUSNESS

It is an honour to receive a formal invitation to honour Jaako Hintikka, a logician, philosopher and scholar of preternatural energy and erudition. I cannot however match a private or unofficial tribute which I paid to him in 1967–1968, when the sense that a new ideal of order and system had overtaken the work I was doing caused me to abandon a manuscript I had been working on for some years about self-deception and the opacity of consciousness. The ideal in question was that represented by Hintikka's *Knowledge and Belief: An Introduction to the Logic of the Two Notions* (Cornell University Press, Ithaca, New York, 1962) – a book with which I disagreed, but which I found I could not meet on its own terms.

Anyone who now takes up these subjects has to reckon better than I could then, not only with *Knowledge and Belief*, but also with a mass of further work which that book has since provoked. And the tiro will regret as keenly as a philosopher as disaccustomed to these problems and as unread in the recent literature as I now am that there is at the moment no complete account of Hintikka's own present opinions on these matters, nor any consolidated review by him of the developments of the last fourteen years. Let the lacuna be my excuse for wheeling out in Hintikka's honour one rusty and antiquated but just conceivably salvageable fragment of the work which ten years ago I decided he had somehow, without refuting it, made obsolete. And let it be my excuse also for not now embarking on the seemingly indefinite task of refurbishing these thoughts with references to the new authorities and new work that Hintikka has inspired.

In his paper 'Is There Only One Correct System of Modal Logic?'[1] John Lemmon used an argument of the following sort in order to disprove the implication 'if a man knows that p then he knows that he knows p' (which I shall represent $Kp \supset KKp$): If someone asks me at time t whether I know the answer to a question then one thing I can reply is 'I don't know whether I know the answer or not, wait a moment'. I can then try very hard to recall relevant and related incidents, stir up my memory,

237

E. Saarinen, R. Hilpinen, I. Niiniluoto, and M. Provence Hintikka (eds.), Essays in Honour of Jaakko Hintikka, 237–248. All Rights Reserved.

and after some struggle say 'Yes, the answer to the question is p'. Don't I then do what I was unsure whether I could do, namely recall the answer? Surely I find that I *do* know the answer, having not previously known whether I knew the answer or not.

Lemmon backed this argument with a Rylean account of knowledge (in terms of *having learned and not forgotten*) which is vulnerable to objection. But this is inessential, and it could have been backed by other accounts with stronger claims to represent what we mean in English by 'know'. We should reach the same conclusion as Lemmon did, for instance, if we held, as Peter Unger once held, that for x to know that p is for it to be no accident that x is right to believe that p, and if we substituted this account of knowledge for Lemmon's Rylean account. Again the argument could equally well have been backed by another account of knowledge, to which I subscribed at the time of writing about self-deception.[2] This account was not quite of the same mould as the account at Aquinas' *Summa Theologica* II.I Qu. II A.3, or the better known modern causal theories (e.g., Alvin Goldman's first account, given in *Journal of Philosophy* **64** (1967)), though for certain cases it came down to something somewhat similar. I shall temporarily postpone all further consideration of this conception, however, because my first concern here is not with the account of knowledge itself, but with a curious objection to Lemmon's counter-example to $Kp \supset KKp$ which escaped attention.

If I can remember something, while temporarily being unable to produce it, then perhaps the production of the utterance 'I don't know the answer to that question', where the question is the question to which p is the answer, cannot of itself guarantee the truth of what is said by the utterance 'I don't know the answer'. One must look further, the objection says; and see that when I asserted 'I don't know whether I know the answer', I was simply *wrong*, as it turned out. The objector will hold that the sequel and my later recall of the answer to the question show that I knew that I knew, but had temporarily forgotten that I knew that I knew.

The objection is both interesting and wrongheaded. We want to be able to point to a man asleep, and say in the normal sense of 'know' which is unspoiled by theory that he knows the product of 2 and 2. And we surely want to say, in that same normal sense, that he *knows that he knows* that product. But we do also want to be able to say of a man like Lemmon's subject, who is awake, knows the answer to some question and eventually recalls it but says 'I don't know if I know the answer to that question', that he can be right about this, and not know whether he knows that thing which it later proves he does in fact know. And it will be a great pity if we cannot achieve a statement of the difference between these two cases which leaves room for us to arrive at an account of the mind's awareness of itself and its surroundings in terms of nothing more mysterious than propositional knowledge and propositional knowledge of propositional knowledge.[3] At first blush the two examples, taken in concert, may discourage us from holding onto what would otherwise seem plausible, that there is some conceptual connexion between consciousness and knowledge of one's own mental states (such as knowledge itself).

That there must be something wrong with the objection to Lemmon we can see in a *prima facie* sort of way if we consider the plausibility of the principle

(1)   (Find out q at t_1) $\supset$ Not(Kq immediately before t_1).

Taking 'Kp immediately before t_1' as a substitution instance of q, we get

(2)   (Find out at t_1 (Kp immediately before t_1)) $\supset$ (Not(Kp immediately before t_1 (Kp immediately before t_1))).

Principle (1) reinforces Lemmon's objection to any general implication $Kp \supset KKp$. But perhaps someone's confidence in (1) will only last till the moment when he applies it and arrives at (2). He may object that, even if (1) were correct, the man is only really investigating the answer to the original question (compare Plato *Charmides*), not the answer to the question whether he knows the answer to the question. But here I should retort on Lemmon's behalf that what the man does is to try to recall; and that there is a difference between finding out whether one knows the answer to a question, whose answer is p, by trying to recall the answer, and finding out whether p, which is a mad way of describing a successful attempt to recall p.

Lemmon's argument against $Kp \supset KKp$ depends on remembering's having a special relation to knowing, being in fact a species of it. It will clarify this status to consider now an opposite objection to his argument. With Lemmon, we maintained that before the moment of recall the man knows but does not know that he knows. The first objection maintained that the man both knows and knows that he knows before the recall but has forgotten that he does. The other and opposite objection, holds that at the moment of recall the man comes both to know and to know that he knows, being before the recall in a state of lacking, however temporarily, both knowledge and knowledge of knowledge.

This second objection may be put as follows.[4] Of the man who says at *t* 'Wait a moment, I don't know whether I know the answer to that question or not' it may be said that at *t* and thereafter, *until he recalls it*, he *forgets* that *p*. But then, on the strength of this forgetting, it may be maintained that, at the moment when the man wonders what the answer to the question is and finds himself unsure whether he knows the answer, he is in a state of not even knowing that *p*. He knows it only when he recalls it; in which case the proper answer to the question whether he knows at *t* the answer to the question posed at *t* is not *I don't know* but *No*. No doubt he *had* known that *p* some time before he was asked the question. But this knowledge had since lapsed. It is said that it is only later, and by the effort of reconstruction or recall, that the man came again to know it and came again, at the same moment, to know that he knew it.

This is a different strategy for restoring $Kp \supset KKp$ and responding to Lemmon's counterexample. But the objection runs directly counter to the principle that a man can only recall what he remembers, and can only count as remembering if immediately before recalling he did know. When the man tries to recall and succeeds in doing so he discovers something. No truthful account can omit this. And, as before, I should insist that no plausible account can conflate the discovery that he *knows or remembers* that *p* with the discovery that *p*.

Moving forward to a less familiar point, I would also remark that the new objection overlooks a complexity in the notion of forgetting. Just as 'remember' is a verb with a capacity sense or use (one says of a sleeping man that he remembers – or retains the capacity to recall – the two-times table) and also an exercise of capacity use or sense ('suddenly I remembered that my plane left in ten minutes'), so we should expect a similar complexity in 'forget'. We should expect to find one use or sense denoting the state of having failed to retain, i.e., lost a capacity, and another sense or use denoting the failure to exercise a capacity on an occasion suitable for its exercise. (I hedge here between sense and use because I suspect but am not certain that the difference in the two uses could be explained by a general theory of verb-aspects which was by no means special to forgetting and remembering.) Examples of the former would be 'I have now forgotten my sixth birthday' (loss of capacity), and of the latter 'I forget his name for the moment' or 'he forgot that he needed to lock the back door' or 'he forgot that his wife would buy the potatoes' (failure to exercise a capacity and/or temporary difficulty in doing so). One important point here is that knowledge persists through forgetting (failure to exercise) but not through having forgotten (loss of capacity). Another is that there is no similar bifurcation to be found within the dispositional 'know' itself.[5] It follows that it is an adequate precaution against confusion (more than adequate, in fact, if what I say at the end of this paper is right) to stipulate that in the implication $Kp \supset KKp$ the constant symbol 'K' represents the dispositional long-term mental state of knowing, a relational state which, once entered, can persist through sleep and inattention but can be lost by confusion or brain damage or having *finally* forgotten.

I believe that these rebuttals of two mutually opposed objections to Lemmon will suffice to create a strong *prima facie* case for taking his counterexample to $Kp \supset KKp$ seriously; and likewise for insisting (pending a positive argument concerning some particular value of p in favour of the actual coincidence of knowing that p and knowing that one knows it) that the distinctness of the propositional objects of Kp and KKp will generally suffice for the *prima facie* distinctness of the states themselves. What we need now is not more defence of this contention, or further elaboration of such counterexamples as Lemmon's, but better understanding. But this is impossible without some definition, characterization or further elucidation of knowledge.

It will promote clarity, and improve the chances of our working back to the central sense of 'know' from a central use of it, to focus on the third person case (as in 'he knows where his father is/who his friends are/ why his bicycle won't brake/that his train leaves at 3.15 p.m.).[6] Suppose that there is something about a person's total state which cannot be explained otherwise than by his being in a certain doxastic state ψ with respect to p (*believing* that p, *supposing* that p, *thinking* that p, *strongly suspecting* [?] that p – I do not know how to give the principle which generates this list, which the conservative will restrict to the single term *believing* that p). Then what would be required of his state of ψ-ing that p in order for him to count as knowing that p? One suggestion, which has already been mentioned, is that the man knows if and only if it is no accident that he is right with respect to whether p. A second and cognate proposal is that he knows that p if and only if the truth of p figures essentially in the best full explanation of why, *given* that he is in some doxastic state ψ with respect to whether p, he ψs that p rather than ψ-ing that not-p. There is a plausible and ancient idea (a version of the principle of sufficient reason in fact) that, against a certain background held constant, a phenomenon confirms a theory at t just to the extent that the theory is the best explanation discoverable at t of the phenomenon.[7] Analogy will

suggest that, if there is anything right about the second suggestion which
has just been mentioned, then there must equally be something right about
a third proposal to which I used to be drawn[2] –: that x knows that p if
and only if there is a doxastic state ψ of x such that, *relative to x's being
in that state ψ*, it must be that p.[8]

I have less confidence than I used to have in any of these three proposals. I have also come to doubt, for reasons not immediately relevant,
that they are completely equivalent. If they should prove to be discrepant,
maybe the second will best deserve preservation. In the interim, however,
it is worth bringing the third to the same state of relative clarity as the
first two proposals. Let us say that x knows that p if and only if there is a
doxastic state ψ such that

(i) x ψs that p

(ii)   there is a sentential value for q such that x ψs that p because
           q[9], where

(ii/a)   a premiss to the effect that q and a premiss to the effect that
           x ψs that p constitute all or part of some sound (but not
           necessarily deductive) argument to the truth of p, and

(ii/b)   this argument cannot dispense with the proposition that
           x ψs that p, or dispense with any other of its premisses.

Thus a man may be said to know that there is a tree in front of him
because (i) he believes there is, (ii) he believes it because he is in that visual
state which would go with seeing it, and (ii/a and b) there is a sound
argument from this *particular* (normal, sane, undeluded, visually un-
handicapped) man's being in that visual state, from the particular circumstances of his being in that state, and from his believing that he sees a tree,
to the conclusion that there is a tree in front of him. More contentiously,
but for the purposes of the definition equally straightforwardly, knowledge
may be ascribed to the man in J. L. Watling's example in his paper
'Inference from the Known to the Unknown', who can tell reliably over
and over again, even though he does not know how he tells or feel any
confidence in the matter, the position of someone else who stands on the
other side of a thick high wall.[10] In convincing ourselves that he is reliable
in this respect we convince ourselves that he satisfies just the sort of
requirement which each of the three definitions attempts to codify.[11]
Note that we can satisfy ourselves of this without *identifying* at all the
'mechanism' by which the man gets to know.

I have put down all three proposals, not so much with the purpose of taking out insurance against objections to any one of them, as with the humbler purpose of providing some however vague general illustration of what it would be to frame an elucidation of knowing independently of the issue of $Kp \supset KKp$ but in conscious reaction against absurd and unrealistic definitions which have created paradox[12] and have seemed to many critics to make a joke of philosophical epistemology. The three proposals do however show how there *might* be a relatively straightforward explanation of why $Kp \supset KKp$ should have been expected to fail as a general principle. I hope that they will also prove to embody a conception of knowledge consistent with the statement of a non-arbitrary distinction between the case of the sleeping man's knowledge of knowledge and the case of the waking man's ignorance of knowledge.

First, if knowledge is at all what these three accounts represent it as being, then it is always a disposition or state, and a relational one. Second, it is a state trained upon something which can be either some condition of the external world or *another* state of the knower himself; states of knowledge being individuated by reference to their propositional objects. Finally, if a man knows that he knows that p then the reflexivity which this involves may be analysed in terms of the K state of knowing p, plus the in principle distinguishable KK state whose propositional object is that he knows that p. All three proposals vindicate the distinctness of this first order state from the second order state for which it is as a propositional object.

Dispositions are not the property of the behaviourist but, for the sake of argument, let us make to him the reasonable and minimal concession that for every belief (or doxastic state) there is some circumstance, however unlikely or hopelessly remote from actual, such that something or other about x in those circumstances can only be explained by reference to his having the belief (or whatever) that p. But the nature of the manifestation required in any particular case depends, not only on the circumstances, but also upon the *object* of the state.[13] So we can only make a fair comparison between the waking man and the sleeping man who is superior to the waking man in knowledge of knowledge if we arrange the comparison to be in respect of one and the same proposition. But as soon as we do this any problem there might seem to have been about the contrast disappears. The sleeping man knows that he knows, say, the distance of the moon from the earth; and what shows this is that, if and when placed in the same circumstances as the waking man (who was genuinely unsure whether he knew this distance and had to think very hard to discover whether he did by recalling the distance, or by recalling that he could recall this distance), the sleeping man would perform in a way far more impressive than the waking man does. Just to the extent that, when suitably circumstanced, he does better than Lemmon's man, the at present sleeping man knows that he knows, whereas the waking man does not.

Exempli gratia, let us now apply the third definition of knowing to our main problem. Suppose that a man knows that d is the distance between the earth and the moon. Then, by the third proposal for the elucidation of 'know', we shall expect that there is some answer q to the question "why does he believe d is the distance between the earth and the moon"? Then q and the man's belief that d is that distance will yield a sound argument to the conclusion that d is indeed the distance. But this can hold without the man's even believing or supposing that he *knows* this distance. Without his belief that he knows we cannot even get to the point of asking whether there is a sound argument from his *believing* that he knows the distance to his actually knowing that he knows the distance. But then, when we see the matter in this way, there is no plausibility at all in $Kp \supset KKp$.

Here it could be asked whether the weaker implication ((Kp & BKp) $\supset$ KKp) is necessarily true. Intuitively however it seems that not even this implication should hold. To avoid complications which are irrelevant let us shift to a more homely example. Suppose x has looked in an out of date timetable (the cover is lost, say) and read there that his train goes at 3.15 p.m. Independently of this, he may also have been told by someone that it goes at 3.15 p.m. The role of this independent testimony is that, *if* it had been discrepant, x would have mounted a third inquiry. Since it cohered, he let matters rest. He believes the train goes at 3.15, then, and he believes he knows it does. *Question:* does it follow that he knows that he knows that 3.15 p.m. is the time of the train? *Answer:* (a) whether or not he knows that he knows, x does know when the train goes. For suppose that his informant was perfectly reliable, did himself know, and was well placed to contaminate anyone else with the timetable-knowledge which he had and to protect them from error. That anyone believes a train runs at 3.15 p.m. and believes it at least in part because he has been told by this particular man *does* give a sound argument to 3.15 p.m.'s being the time of the train. So, by the definition, x does know when the train goes. (b) We can perfectly well suppose, however, that the reason why our subject believes he knows that the train runs at 3.15 is *not* that this informant told him. He believes that he knows but the reason why he *believes* he knows is that the timetable said 3.15 p.m. At best, he thinks, the informant simply confirmed that time. Then it is only an accident that x is right to *suppose that he knows* when the train leaves. So we cannot conclude from the fact that he knows and believes that he does, that he knows that he knows.

Again the formalities of definition bear out this finding. There is no sound argument for 'he knows when the train leaves' (still less for 'there is a sound argument from his belief it leaves at 3.15 p.m. to its leaving at 3.15 p.m.') to be contrived from the fact that, because that timetable says it does, he believes that he knows that it leaves at 3.15. We might try to contrive such an argument by using the other conjunct of the antecedent of the would be implication, viz. Kp, and by throwing into the putative argument extra premises, as is permitted by my elucidation of 'know'. But the only extra premiss which will do the trick is 'Kp' itself (or something which it entails, viz. its own analysis). But this would render the premiss 'BKp' redundant, in flagrant violation of clause (ii/b) of the definition of 'know'.

There is here a very clear contrast between this way of arguing questions of knowledge and Hintikka's way:

If the rule (A.PKK*) ... is to be applicable to what the person referred to by a is said to know we have to assume ... that the persuasion which may be directed against him may be based ... on everything he would undertake to defend if he said in so many words, 'I know' ... among these commitments is the very fact *that* he (= he*) knows. This assumption is satisfied if by 'a knows' we mean that he has enough information to say, correctly, 'I know'. It is not satisfied if 'Knowing that' merely means being aware of the fact in question (p. 33, *Knowledge and Belief*).

One habituated to working within Hintikka's system will have long wanted to protest that, right from the start, my willingness to accommodate Watling's hesitant wall guessing example as a case of knowledge shows that, apart from all other questions of method, there is an utter difference between his aims and mine which derives from our interests in quite different concepts of knowledge.

Hintikka made amply clear that he was proposing rules not for any and every sense of the word "know" but only for the verb "in its most typical sense" (p. 17). And it may be that there is no alternative but to admit several senses (even several propositional senses) of the word 'know'. It *may* then be that I was simply aiming at a different sense from Hintikka. In that case the proposed reconciliation will be correct.[14] I would only say, in conclusion of my paper, that I do not think anyone is yet in a position to make this claim. Senses are not to be multiplied beyond necessity. My own methodological preference would be for an exploration of the possibility that one propositional sense of 'know' will suffice for all purposes. If it will, then it is likely that the central or unifying sense of know is closer to 'being aware of' than Hintikka was disposed to imagine – a weak or flat sense, that is, and a sense around which there can gather (especially in the first person case which has impressed and confused philosophers so much) the phenomena which, at the time when I was trying to write about self-deception, H. P. Grice was ordering and beginning to explain in his theory of 'implicature' and 'maxims of conversation'. Such a theory of knowledge, belief and consciousness could not aspire to the orderliness or elegance of Hintikka's. But it would aspire to build up from the modest notion of a dispositional state of knowledge, via the various utterly disparate *objects* of knowledge, towards the verisimilitude and amplitude required of a good description of self-deception and self-ignorance, and self-awareness and self-understanding.

That there is hope of this let me illustrate by another reference to our ordinary kind of example. It may be wondered how, if you use 'know' in such a way that a sleeping or unconscious man knows (in this sense) that he knows (in this same allegedly central sense) that $2 \times 2 = 4$, the required connexion can be made with the phenomenon of consciousness. But the sleeping man does not know even in this flat sense that we are talking about him, or that the sun is shining or that it is getting colder or that he is lying prone. There is nothing about him by reference to which we can ascribe these particular pieces of knowledge or belief to him. To count as having these *particular* beliefs he would have to manifest a quality of response which a sleeping man cannot manifest. What *consciousness* would require, however, is a state in which a man can come by indefinitely much propositional knowledge (and propositional knowledge of propositional knowledge) of indefinitely many facts like these facts about how things are here and now with himself and with his immediate and bodily environs.

Bedford College, London

NOTES

[1] *Proceedings of the Aristotelian Society*, Supplementary Volume **33** (1959), p. 38.

[2] Cp. 'Freedom, Knowledge, Belief and Causality' in *Knowledge and Necessity* (ed. by G. N. A. Vesey), Macmillan, London 1970.

[3] I have heard it claimed that we have to abandon the idea of any simple affinity between awareness or consciousness in the full sense and the propositional kind of knowledge which figures in Lemmon's argument. I think that this conclusion is related to something I have heard people say about its not being propositional knowledge which figures in knowing what one is doing, or in knowledge of oneself, or acting knowingly, or even in knowing the position of one's limbs, and knowing the state of one's intentions etc. The idea is presumably that 'know' and 'aware' are, in one flat sense of each, as inter-definable as one would expect; but that the *know* which is in question in self-awareness, consciousness, etc. is a special kind of *know*. I think it is supposed that it is related to the *know* in knowing *how to*, and perhaps a founding member of a class of verbs which Gilbert Ryle used to call *adverbial verbs*. This is a position I should expend more effort in rendering fully intelligible only when satisfied that the simple approach mentioned in the text was unworkable. See also the concluding paragraphs of this essay.

[4] This objection Lemmon did address. See his later article 'If I Know, Do I Know That I Know?' in *Epistemology: New Essays in the Theory of Knowledge* (ed. by Aurum Stroll), Harper & Row, N.Y. 1967. See especially p. 64. Hintikka has responded in *Synthese* **21** (1970) to this article of Lemmon's.

[5] There is however a parallel for 'know' with a *distinct* bifurcation within 'remember', viz. *start* to have the capacity to recall *versus* have the capacity. 'Then (at t) he knew, i.e., realized, that he was going to fall' means that the state of knowledge that he was going to fall began then at t. In the absence of an argument for $Kp \supset KKp$ which is not shown up as invalid by Lemmon's counterexample, there is no reason to suppose that even at the first moment of knowing a man must know that he knows. Here I claim more for Lemmon's counterexample than it seems he did. See p. 80, *op. cit.* previous note. (The discrepancy arises from Lemmon's conceding more to the good evidence account of knowledge than I do.)

[6] For the relations of these apparently distinct constructions see Lemmon, *op. cit.* 'If I Know do I Know that I Know?', p. 59.

[7] Cp. Gilbert Harman's account of the role of the inference back to best explanation in 'The Inference to the Best Explanation', *Philosophical Review* **64** (1965).

[8] And if this is right, we can see the hazard of the first person case. The claim that, *relative to his own state*, it *must* be that p is a claim whose strength may combine with the exigency of the maxim or precept 'not to make assertions which one cannot back fully' in such a way as to discourage x himself from asserting 'I know that p', and discourage him even when he does in fact know that p. He needs more than knowledge that p to *assert* that he knows that p. Compare H. P. Grice's work on logic and conversation.

[9] i.e., the explanation of x's ψ-ing that p is q. The 'because' need not correspond to the content of any thought by x. For a fuller account of the nature and ambiguity of such a *because*, see 'Freedom, Knowledge, Belief and Causality', *op. cit.*, p. 142. Appearances perhaps to the contrary, the word 'because' in the present paper always plays the

indicative role which is there isolated. If it also plays the subjunctive or opaque role in some cases, then that is due to its playing both at once, with a sort of dual occurrence now familiar in connection with phenomena of opacity.

[10] *Proceedings of the Aristotelian Society* **55** (1954–1955), 83.

[11] Note that for soundness in an argument all premisses must be *true*; and that soundness is *defeasible*. The criteria of soundness in an argument (a) take in considerations which are domestic to a subject matter; (b) do not in general preserve the monotonicity property: if p gives a sound argument for q it does not follow that p *and* r would give one. The question which always confronts one when presented with an argument of the kind I am talking about is: Is there a true sentence which could be added to the premisses and which would disrupt the plausibility of the conclusion?

The following kind of counterexample has been urged against the account. A small boy who has very indulgent parents has only to believe that something will be given to him or done for him and it is given or done. One can argue soundly from his belief that he will receive a rocking horse to his future receipt of one. Does he know that he will get a rocking horse? But I reply *yes*; that is what real security comes to! If his parents were really omnipotent he would be omniscient. If not all such counterexamples can be treated in this way, however, then I think then the second proposal will still be available.

[12] See 'Freedom, Knowledge, Belief and Causality', *op. cit.*, p. 138.

[13] It has sometimes been claimed in epistemology that the standards of 'justification' required of a man if he is to have knowledge vary with subject matter. That may or may not be so (though by my definition, which says nothing directly of evidence, the only way this could come about is by subject matter dependent variation in standards of sound argument from ψp to p). But, whatever the affinity, this is *not* the point I am making in this sentence.

[14] Esa Saarinen points out to me that *Models for Modalities* (D. Reidel, Dordrecht, 1969) is particularly relevant here. There Hintikka says (p. 8) that he doubts whether *any* cases can be produced from natural language where Kp and KKp are synonymous. Throughout the discussion in the first essay in *Models for Modalities* he is making the point that only for some 'basic meaning' of 'know' – a sense not necessarily exhibited in the surface of natural language – is the KK-thesis valid. The disagreement between us must then focus on the question whether the subject is best organized by a theory of knowing which builds up from such a basic sense of 'know', rather than from the flat dispositional sense which I champion for this role.

ILKKA NIINILUOTO

.

KNOWING THAT ONE SEES

In his essay 'On the Logic of Perception' (1969) Jaakko Hintikka argued that the logic of perceptual terms ('perceive', 'see', 'hear', etc.) is a branch of his more general theory of propositional attitudes. According to this thesis, perception – just like knowledge, belief, and memory – is a propositional attitude in the sense that it involves a relation between a person (the 'percipient') and a proposition. Moreover, the syntax and the semantics of expressions of the form '*a* perceives that *p*' can be analyzed by means of the possible-worlds semantics essentially in a similar fashion as Hintikka has treated epistemic logic in his classical *Knowledge and Belief* (1962) and in a series of later papers.[1]

Hintikka has discussed and clarified a number of technical questions which are related to his 'logic of perception': the failure of existential generalization and of the substitutivity of identity in perceptual contexts, the distinction between physical and perceptual cross-identification, the corresponding distinction between two sorts of quantifiers, and the reduction of object-perception ('*a* sees *b*') to the propositional perceiving that-construction.[2] He has also applied his account to several classical problems within the philosophy of perception: sense datum theories, the role of causality in perception, and the intentionality of perception.[3]

This paper is an examination of some selected aspects of Hintikka's logic of perception. It consists of questions, comments, and suggestions concerning the formalism of this logic, on its basic laws, on its expressive power, and on its ability to give a reasonable account of various perceptual situations. As the title suggests, special attention is given to problems related to the interplay between epistemic and perceptual operators.

I

Before presenting the basic formalism for the logic of perception, some clarifying comments on the nature of our study are in order. The aim of

249

E. Saarinen, R. Hilpinen, I. Niiniluoto, and M. Provence Hintikka (eds.), Essays in Honour of Jaakko Hintikka, 249–282. All Rights Reserved.

the *logic* of perception is not to replace the *philosophy* of perception – no more than the philosophy of perception tries to replace the *psychology* and the *physiology* of perception. Every attempt towards a comprehensive theory of perception has to face a number of philosophical as well as factual questions. Formalization is here not an aim for its own sake, but rather one eventually hopes to achieve some clarity in philosophical matters by means of it.

An explicit account of the syntax and the semantics of perceptual statements provides us a tool for making precise several different views about the nature of perception. This tool itself may be philosophically non-neutral in its specific features, and the decision to use it may already involve commitments in important philosophical matters.[4] On the other hand, the crucial virtue that one expects from a formalism is flexibility or expressive power. It is interesting to formulate and to discuss different notions of perception, as characterized by different sets of postulates concerning the formal behaviour of perceptual terms, and to try to see what commitments these different sets of assumptions implicitly contain. The situation here, in the case of perception, is precisely the same as in epistemic logic.

The epistemological viewpoint is not the only one from which a logic of perception is interesting. It is relevant to general linguistics, too, since it gives an explicit semantics for a fragment of English containing perceptual terms. One need not claim nor presuppose that there exists a unique, most natural way of using perceptual terms in natural language which would be the best one for all purposes. The overall aim is rather to give a systematic account, in precise terms, of the range of the different interpretations that perceptual statements in fact have.

In this paper, I shall restrict my attention exclusively to visual perception, i.e., seeing, without attempting to guess what features of seeing are common to other forms of perception as well. The principles which are discussed and in some cases tentatively adopted in this paper should be understood in the weak and hypothetical sense indicated above – as principles with a realistic (psychological or linguistic) interpretation they are of course controversial and highly idealized.

II

Let $PL(O_1, \ldots, O_n)$ be a language which is obtained from ordinary propositional logic PL (with propositional letters p, q, ..., and connectives $\sim$, $\vee$, $\wedge$, $\supset$, $\equiv$) by adding n unary intensional operators O_1, ..., O_n. Thus, if A is a sentence in $PL(O_1, \ldots, O_n)$, then so is O_iA for any $i = 1, \ldots, n$.[5]

The semantics for language $PL(O_1, \ldots, O_n)$ is given by specifying a *frame* $\langle W, \mathcal{O}_1, \ldots, \mathcal{O}_n \rangle$, where W is a non-empty class of *possible worlds* and each $\mathcal{O}_i (i = 1, \ldots, n)$ is a binary relation in W, the O_i-*alternativeness relation*. The truth-conditions for sentences containing intensional operators are determined by the following rule: for each world w in W and for each sentence A

(1) O_iA is *true* at w iff A is true at w' for all w' such that $w\mathcal{O}_iw'$.

It follows from (1) that O_iA is false at w if and only if A is false at some w' in W for which $w\mathcal{O}_iw'$. A sentence A is *valid* in W if A is true at every w in W.

Three sorts of intensional operators will be considered in this paper: K_x, B_x, and S_x, where 'x' may be replaced by an individual constant 'a', 'b', ... which refers to some person. Here K_a and B_a are the familiar epistemic operators and S_a is the operator for seeing. We thus have the readings:

$$K_ap = \text{'}a \text{ knows that } p\text{'}$$
$$B_ap = \text{'}a \text{ believes that } p\text{'}$$
$$S_ap = \text{'}a \text{ sees that } p\text{'}.$$

In this paper, the possible worlds w in W are usually thought to be alternative states of affairs which concern only some 'small' part of the universe.[6] The intuitive interpretation of the alternativeness relations is the following: $w\mathcal{K}_aw'$ (resp. $w\mathcal{B}_aw'$, $w\mathcal{S}_aw'$) if and only if w' is compatible with what a knows (resp. believes, sees) in w. The truth-conditions for sentences of the form K_aA, B_aA, and S_aA are determined by rule (1). In particular,

(2) S_aA is true at w iff A is true at w' for all w' in W such that $w\mathcal{S}_aw'$.

In less formal terms, the basic conditions for the *S-logic* (i.e., the logic of operators S_a) are the following:

(3) a sees that p iff in all states of affairs compatible with what a sees it is the case that p

(4) a does not see that p iff in some possible state of affairs
           compatible with what a sees it is not the case that p.

(Cf. Hintikka, 1969, p. 155.)

From these conditions it follows that the S-logic is at least as strong as a
normal system of modal logic (cf. Segerberg, 1971, p. 12), i.e., the operator
S_a satisfies the principles

(S1) $S_a(A \supset B) \supset (S_a A \supset S_a B)$
(S2) $S_a(A \equiv B) \supset (S_a A \equiv S_a B)$
(S3) $S_a T$, if T is a propositional tautology
(S4) $S_a(A \wedge B) \equiv (S_a A \wedge S_a B)$
(S5) $S_a A \supset S_a(A \vee B)$.

(The names of these principles should not be confused with the names of
the Lewis systems of modal logic.) Note that the principle

(S6) $S_a(A \vee B) \supset (S_a A \vee S_a B)$

is not valid, however. On the other hand, it seems reasonable to assume
that no person can see an explicit contradiction:

(S7) $\sim S_a(A \wedge \sim A)$.

According to S4, principle S7 is equivalent to

(S8) $S_a A \supset \sim S_a \sim A$.

S8 says that if a sees that p then he does not see that non-p. S8 is equivalent
to the assumption that each world w in W has at least one S_a-alternative
w' in W (see Hughes and Creswell, 1968, p. 301). Together with S1 and
S3, principle S6 entails that $\sim S_a p$ for any p which entails a contradiction.

It is interesting to observe that the operation of obligation in deontic
logic too satisfies principles corresponding to S1–S5, S7–S8. Moreover,
the principle of S-logic which would correspond to Prior's debated rule
for obligation, viz.

$$S_a A \supset ((A \supset S_a B) \supset S_a B)$$

is by no means obvious, and we shall not assume its validity.[7] Principles
K1–K5, K7–K8, B1–B5, B7–B8, which are obtained from S1–S5, S7–S8
by replacing S_a by K_a or B_a, respectively, are valid principles of epistemic
logic.[8] Moreover, the notion of knowledge satisfies a success condition

(K9) $K_a A \supset A$,

while the corresponding condition B9 for belief fails.[9]

Many epistemologists have taken the phrases 'a perceives that p' or 'a sees that p' to mean something like 'a comes to know that p through visual means'. In this strong sense, propositional *seeing* is a species of *knowledge*, and hence it should satisfy a success condition

(S9) $S_a A \supset A$.

(Cf., for example, Warnock, 1965, pp. 61–62; Chisholm, 1965, p. 473; Dretske, 1969, pp. 78–79; Armstrong, 1973, p. 27.) An attempt to build a logic of perception upon this basis has recently been made by Romane Clark (see Clark, 1976). Further support for S9 can be given by the observation that seeing that-locutions have in many cases a success grammar in ordinary language: if I see that my house is burning, then my house is burning (cf. Unger, 1972; Thomason, 1973; Armstrong, 1976). Principle S9 – which is equivalent to the assumption that $\mathscr{S}_a$ is reflexive – is also a convenient condition from the technical viewpoint: it would make the S-logic at least as strong as the familiar system S4 of modal logic. Richmond Thomason has indeed argued that "there is really no choice but to make" the assumption S9 in the logic of perception (Thomason, 1973, p. 263).

In ordinary discourse, perceptual statements are sometimes used in a weaker sense which can best be expressed – in Ayer's terms – in the 'language of appearance'. Thus, 'a sees that p' may mean something like 'it appears to a that p', 'it looks to a that p' or 'a seems to see that p'. Hintikka does not accept the success condition S9, since he interprets seeing that-statements precisely in this weaker sense (Hintikka, 1975, p. 68).

We shall follow Hintikka in the weak interpretation of S_a and in the denial of the validity of S9. At this point, this is an important question of strategy in the logic of perception, and the reasons for our choice will become evident when we proceed. We want to be able to give an account of various perceptual situations (such as veridical seeing, visual illusions and hallucinations) without introducing other perceptual operators than S_a, but it seems *prima facie* very difficult to give such a formal treatment if condition S9 is accepted. For example, it seems that Thomason fails to reconstruct seeing as-statements in his approach just for the reason that he assumes S9 (cf. Thomason, 1973, p. 284).

So far, the principles governing K_a, B_a, and S_a have been discussed only separately. We shall now inquire whether there are between these operators any valid interconnections which are simple in the sense that they do not yet contain iterations of operations.

Perhaps the most uncontroversial principle of this sort is the claim that knowledge entails belief:

(C1) $K_a A \supset B_a A$,

where of course the K- and B-operators have the same index 'a'. (See, however, Williams, 1970.) The converse of C1 is not valid even in the strengthened form:

(C2) $B_a A \wedge A \supset K_a A$.

Mere true belief does not guarantee knowledge, since it may be based upon inconclusive or even wrong evidence. In the so-called classical definition of knowledge C2 is replaced by

(C3) $B_a A \wedge A \wedge E_a A \supset K_a A$,

where $E_a A$ means that a has *adequate evidence* for A (cf. Hilpinen, 1970). The question of giving a reasonable analysis of $E_a A$ is presently a source of a lively debate among the epistemologists. One tentative suggestion might be

(C4) $B_a A \wedge A \wedge S_a A \supset K_a A$.

As we have already rejected rule S8 in Section II, it is clear that $S_a A$ alone is not sufficient for $K_a A$, i.e., that $S_a A \supset K_a A$ is not valid. If it appears to me that p, I may be mistaken in claiming that p. However, if I seem to see that p, if p is in fact true, and if I further have reason to believe that p, does this show that I know that p? The answer is probably negative, since there is reason to think that C4 has to be qualified by adding some material conditions of adequacy for the premise $S_a A$.[10]

Let us next ask whether

(C5) $K_a A \supset S_a A$

is valid. *Prima facie*, C5 seems very unnatural: I know that $2 + 2 = 4$, but is it reasonable to say that I therefore *see* that $2 + 2 = 4$? To answer this question, note first that the possible-worlds approach which was accepted in Section II commits us to the principle S3 which says that $S_a p$ holds for any tautological proposition p. Moreover, it follows that $S_a p$ holds for any proposition p which is a logical consequence of propositions $q_1, \ldots, q_n$ such that $S_a q_1, \ldots, S_a q_n$. In other words, the percipient a is assumed to be logically omniscient with respect to his knowledge, belief, and perception.[11] Secondly, it is important to make a distinction between *immediate* and *mediate* (propositional) *perception*. Whenever person a sees that p without being visually aware of the state of affairs described by p, a sees mediately that p; otherwise his seeing that p is immediate.[12] To use Berkeley's example, by seeing immediately that a bar of iron is red one can see mediately that it is hot (Armstrong, 1961, pp. 19–20). Similarly, one can mediately see that there is cold outside by looking at the thermometer. Mediate perception thus typically involves inference, i.e., it is always to some extent based upon the knowledge of the percipient. This suggests that the distinction between immediate and mediate perception is a matter of degree, and that situations where we 'see' tautologies or where we 'see' something which we already know are limiting cases of mediate perception.

Does the existence of mediate perception (including its limiting cases) make it reasonable to adopt C5? It is well known to psychologists and to philosophers that our perceptions and perceptual judgments are very strongly influenced or preconditioned by our knowledge or our convictions. For example, it has been argued that scientific observations are always 'theory-laden' in the sense that they depend upon the theoretical assumptions that the scientists take for granted.[13] Hintikka has referred, in a footnote, to a related fact:

... background knowledge (and background beliefs) is used to weed out many of the possible courses of events apparently compatible with one's perceptions within a given specious present *before* we cross-identify between the states of affairs compatible with one's perceptions. Thus there probably is no such thing in the last analysis as cross-identification by purely perceptual means, but only by means of perceptual information *cum* collateral non-perceptual information. (Hintikka, 1975, p. 75.)

Considerations of this sort may suggest that the relation between background knowledge and perception should be understood in the following strong sense: if a state of affairs w' is not compatible with what I know in world w, then I don't take w' as a serious perceptual alternative to w. In other words, possible worlds which would contradict my knowledge are not accepted as S-alternatives to w. This would amount to assuming that

(5)   If $w\mathscr{S}_a w'$, then $w\mathscr{K}_a w'$,

which immediately entails C5. We could thus conclude that whenever a knows that p then a (immediately *or* mediately) sees that p.

This defense of C5 – even if there is some degree of truth in it – is not adequate, however, since there are clear counter-examples to C5. It is possible that I see that $\sim p$ even if I know quite well that p, i.e., there are situations w where $K_a p \wedge S_a \sim p$ is true. If C5 were valid, then also $S_a p \wedge S_a \sim p$ would be true at w, but this is impossible in view of S4 and S7. For example, in the case of the familiar pair of Müller-Lyer lines, I *know* that the lines have the same length, but I cannot help the fact that they nevertheless *appear* unequal to me. Such cases of *conscious illusions* (cf. Hintikka, 1975, pp. 198–199) provide counter-examples to thesis C5. At the same time, they refute the principle $B_a A \supset S_a A$.

A simple modification of C5 which seems to take care of the presented counter-examples is the following:

(C6) $K_a A \wedge \sim S_a \sim A \supset S_a A.$

C6 says that a sees (immediately or mediately) that p if he knows that p and does not see that non-p. In other words, knowing that p entails seeing that p unless one (immediately) sees otherwise. This means in fact that conscious illusions are the only possible counter-examples to C5: if our immediate perception is not sufficiently definite to make $\sim p$ true at *all* S-alternatives, then our knowledge that p has the force of weeding out all those S-alternatives where $\sim p$ is true, so that p will be true at all the remaining S-alternatives.[14] Note further that S8 and C6 together entail

(C7) $K_a A \supset (S_a A \equiv \sim S_a \sim A),$

while S1, S4, and C6 entail

(C8) $K_a(A \supset B) \wedge (S_a A \supset \sim S_a \sim B) \supset (S_a A \supset S_a B).$

A principle recognized by many philosophers is the rule that 'seeing is believing':

(C9) $S_a A \supset B_a A.$

For example, Howell explicitly accepts C9 for the weak sense of seeing that.[15] Yet conscious illusions refute C9, too, by showing that $S_a p \wedge B_a \sim p$ is possible – I see that the Müller-Lyer lines are unequal but I believe that they are equal. Thus, if C9 were valid, then $B_a p \wedge B_a \sim p$ could be true, which contradicts B4 and B7. Still, seeing is in many cases a source of belief (and of knowledge), so that in normal circumstances we have at least a disposition to believe what we perceive (cf. Hamlyn, 1970, p. 184). Following Armstrong, one may suggest that "cases where we disbelieve our perceptions" only occur "where we have independent information that runs counter to the 'evidence of the senses'" (Armstrong, 1961, p. 85). The following principle may then be proposed in analogy with C6:

(C10) $S_a A \wedge \sim B_a \sim A \supset B_a A.$

If I see that p and if I don't have reason to believe otherwise, then I adopt the belief that p. The same point is expressed by G. Vesey as follows: "What an object looks like to a person is what he would judge that object to be if he had no reason to judge otherwise" (Vesey, 1965, p. 83). Note further that the principle

(C11) $S_a A \supset (B_a A \equiv \sim B_a \sim A)$

is a consequence of C10.

IV

Iteration of intensional operators creates many interesting problems. In *Knowledge and Belief*, Hintikka accepted for knowledge the *KK-thesis*

(K10) $K_a A \supset K_a K_a A,$

which together with K9 entails

(K11) $K_a A \equiv K_a K_a A.$

Moreover, knowledge satisfies a principle of transmission:

(K12) $K_a K_b A \supset K_a A.$

Of the corresponding conditions for belief, only B10 is valid.[16]

In order that formulae $K_a K_b p$ and $K_a B_b p$ are meaningful, one has to assume that $K_b p$ and $B_b p$ express states of affairs which can act as objects of a's knowledge. In the case of epistemic operators, such an assumption seems reasonable: one may, for example, identify b's knowledge and b's beliefs with certain structured dispositional states of b's mind (cf. Armstrong, 1973, pp. 3–21). In a similar way, formulae $K_a S_b p$ and $B_a S_b p$ seem relatively unproblematic: there is a complex state of the world which corresponds to the event of b's seeing that p, and this state of affairs can be the object of a's knowledge and a's beliefs. However, the interpretation of formulae S_aK_bp and S_aB_bp raises a number of difficult questions. If b's belief that p is a dispositional state, then presumably a can immediately see at most some of the possible manifestations of this state in b's action. Therefore, a cannot immediately see that b believes or knows something. This claim holds also in the case where a and b are the same person: even if a could 'perceive' his mind and its state by some sort of introspection, this would not be an instance of immediate visual perception. These remarks do not show, however, that a person a could not *mediately* see that K_bp. We shall therefore assume that formulae of the form S_aK_bp, S_aB_bp, S_aK_ap, and S_aB_ap are meaningful in the mediate sense of seeing – and in this sense only.

The situation is slightly different in the case of formulae S_aS_bp and S_aS_ap. Such formulae may of course be meaningful in the mediate sense, but it is also possible that a sees b and thereby is able to immediately see that b sees that p. Further, a may in some situations see himself – through a mirror, perhaps – and see that he sees something.

As formula S_aK_bp is meaningful precisely in those cases where seeing is understood in the mediate sense, the simplest assumption we can make is the following: S_aK_bp is true if and only if a knows that K_bp, i.e.,

(I1) $K_aK_bA \equiv S_aK_bA.$[17]

In particular, I1 entails

(I2) $K_aK_aA \equiv S_aK_aA.$

If the KK-thesis K11 is assumed, then I2 gives

(I3) $K_aA \equiv S_aK_aA.$

I3 may be called the *SK-thesis*: it says that *seeing that one knows* is (because it is always mediate) equivalent simply to *knowing*. In the same way, one may assume

(I4) $K_aB_bA \equiv S_aB_bA.$

In particular,

(I5) $K_aB_aA \equiv S_aB_aA.$

Hence, by K9 and I1,

(I6) $S_aB_aA \supset B_aA$

(I7) $S_a B_a A \supset B_a B_a A,$

where the converse of I6 is not valid if $B_a A \supset K_a B_a A$ is rejected.

What then is the relation between seeing and knowing that one sees? In one direction, the answer is clear by K9:

(I8) $K_a S_a A \supset S_a A.$

The converse of I8 is problematic, however. Those philosophers who accept the success condition S9 for seeing will reject it, since they have to treat visual illusions as mistakes in our beliefs that we are perceiving some state of affairs (cf. Armstrong, 1961, p. 98; Dretske, 1969, p. 79). For us the situation is different: if a is having a visual illusion that p, then it appears to a that p and hence $S_a p$ is *true*. We can indeed defend the principle

(I9) $S_a A \supset K_a S_a A$

by noting that if it appears to me that p then I may be in doubt concerning the truth of p (cf. the failure of S9), but in normal circumstances I should not doubt the fact that I seem to see that p. This line of thought has been accepted by many philosophers,[18] even if it has sometimes been challenged.[19] If I cannot (as a matter of fact) be mistaken in thinking that it looks to me that p, then $B_a S_a p$ should imply $S_a p$. By combining this observation with I8 and I9, we obtain

(I10) $S_a A \equiv B_a S_a A \equiv K_a S_a A.$

I10 expresses the *KS-thesis* (and the *BS-thesis*) according to which *knowing that one sees* (resp. *believing that one sees*) and *seeing* are equivalent.

Three possible objections to the KS-thesis should be answered immediately. First, unlike the SK-thesis, the KS-thesis is not based upon the KK-thesis. More generally, we have not assumed that knowledge entails logical incorrigibility or indubitability. Therefore, the KS-thesis should be consistent with those approaches where knowledge presupposes some sort of *de facto* incorrigibility and the KK-thesis is rejected (cf. Armstrong, 1973, pp. 160, 197, 212–213).

Secondly, one might suggest that the existence of so-called 'subliminal perception'[20] conflicts with I10. However, if propositional seeing is not always conscious, then epistemic attitudes need not be conscious either. Thus, if I unconsciously see that p, then I10 says only that I unconsciously believe that I see that p.

Thirdly, the KS-thesis concerns only *propositional* perception. There-fore, it is not as such inconsistent with Dretske's claim that there exists cases of non-propositional perception which do not entail any belief-sentences (Dretske, 1969, pp. 6–12).[21] What is perhaps more important, it does not follow from I10 that if I am looking at some person whom I am not aquainted with – Peter Strawson, say – then I know that the man in front of me is Peter Strawson. The question of how strong commitments the KS-thesis has relative to various perceptual situations is discussed in Section VII below.

Let us finally consider the properties of iterated S-operators. In the first place, the failure of S9 for immediate perception shows that the principles $S_aS_bA \supset S_bA$ and $S_aS_aA \supset S_aA$ are not valid. Hence, the principle S11 corresponding to K11 is not valid, either. As seeing is in some cases transmissible – for example, if b immediately sees c and if a immediately sees that b sees c, then a immediately sees c – the principle

(S12) $S_aS_bA \supset S_aA$

has some plausibility. Nevertheless, S12 is not acceptable. The reason for this is the fact, discussed above in Section III, that propositional per-ception is preconditioned by the convictions (and the conceptual system) of the percipient. Thus, a may be able to use information about b's epistemic attitudes to infer that, in some situation, it appears to b that p, even if a himself sees that something else is the case. For example, I might know that my daughter mistakenly thinks that all bald men are old. When we both see a young man who has lost his hair, then I can (mediately) see that she (mediately) sees that the man in front of us is old, even if I see that he is young.

As the rules $S_aA \supset S_aS_bA$ and $S_aA \supset S_bS_aA$ obviously may fail to be true, there remains for consideration only the *SS-principle*

(S10) $S_aA \supset S_aS_aA$.

Thomason (1973) has adopted S10 without any supporting arguments.[22] Note first that C6, S1, S3, and I3 give the result

(I11) $K_aA \wedge S_a \sim S_a \sim A \supset S_aS_aA$.

Further, a slightly weaker principle than S10 follows from C6 and I10:

(S13) $S_aA \wedge \sim S_a \sim S_aA \supset S_aS_aA$.

If we could argue for the principle

(S14) $S_a A \supset \sim S_a \sim S_a A,$

then S10 would follow from S13 and S14. This shows that all the counter-examples to the SS-thesis would be of the following kind: $S_a p$ and $S_a \sim S_a p$ are true at the same time, i.e., $S_a(p \wedge \sim S_a p)$ is true. In the case of know-ledge and belief, such situations cannot arise, so that the principles K14 and B14 corresponding to S14 are valid (cf. Hintikka, 1962, pp. 68, 83). It does not seem possible to create such situations in the case of perception, either. Note that here $S_a \sim S_a p$ could not be true for the reason that $K_a \sim S_a p$, since that would imply a contradiction. The conjunction $S_a p \wedge S_a \sim S_a p$ would therefore describe a situation, where it appears to a that p and a is having a mistaken visual impression that he does not see that p. If such situations do not exist, as I am inclined to think, then S14 and hence S10 are acceptable.[23]

V

The most important technical feature of Hintikka's logic of perception is his treatment of quantification. Hintikka's theory is based upon a distinction between two different methods of identifying individuals appearing in different possible worlds. One of them relies on the physical properties of individuals, such as continuity in space and time or similarity with respect to descriptive properties. The other relies on the similar role that the individuals play when viewed from a given perspective. The former method is called *physical* or *descriptive cross-identification*, the latter *perspectival* or *demonstrative* cross-identification. In particular, *perceptual* cross-identification is a species of the perspectival method where the role of individuals is viewed from the perspective of a particular percipient. Two individuals existing in a's S-alternatives w and w', respectively, will thus be perceptually identified if they have the same role in a's visual field.[24]

Let W be the class of possible worlds, and for each w in W let $\mathrm{Dm}(w)$ be the class of individuals which exist in w.[25] If $\varnothing \neq V \subseteq W$, then a function f defined in V such that $f(w)$ is an element of $\mathrm{Dm}(w)$ for each w in V is a *world-line* on V. A world-line thus picks one individual from each relevant possible world. If f is a world-line on V, then f is a *physical world-line* just in case $f(w)$ and $f(w')$ for all w and w' in V are correlated by means of physical cross-identification. Similarly, f is a *perceptual world-line* on V relative to a world w in W and to a person a in $\mathrm{Dm}(w)$ if (i) for all w' and w'' in V, $w' \neq w$, $w'' \neq w$, the individuals $f(w')$ and $f(w'')$ are correlated by means of perceptual cross-identification from the perspective of the percipient a existing in w, and (ii) if w belongs to V, then $f(w)$ is linked to the individuals $f(w')$, $w' \in V$, $w' \neq w$, through a causal connection (cf. Hintikka, 1975, pp. 68–73).

The formulation of the condition (ii) above creates some important problems. Perceptual world-lines are, according to Hintikka, the subjects of our "unedited perceptual judgements", so that they "serve the very function for which sense-data were initially introduced" (*ibid.*, p. 67).[26] However, Hintikka insists that these 'perceptual individuals' (world-lines) should be clearly separated from the physical objects existing, in the ordinary sense, in *one* world: world-lines are not "inhabitants of any world" (*ibid.*, p. 67). But how can these world-lines which are "neither here nor there" be *caused* by actual objects, as Hintikka requires? (See *ibid.*, pp. 70, 73.) If we are not willing to accept the existence of trans-world causality – i.e., causal connections stretching from one possible world to another or from one world to a set of other worlds – how can one claim that actual objects in a world w can bring about 'visual objects' which are not inhabitants of w? Does Hintikka's way of introducing a causal element into his logic of perception after all commit him – contrary to his own expressed intention – to the old sin of the traditional representative theories of perception, viz. to the assumption there is some entity *in* the actual world which functions as the object of our immediate perceptions?

I think that the answer to the last question is negative. I shall sketch the argument by illustrating the general case in Figure 1. The causal relation obtains between two events (my looking at individuals b and c; my seeing that something is the case). This relation is mediated by the

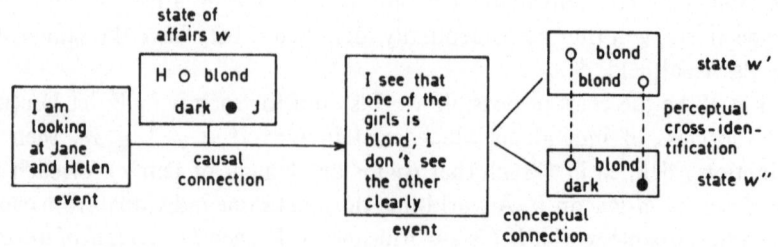

Fig. 1.

actual state of the world; this state essentially involves the individuals b and c. Only intra-world causality is presupposed here. Further, it is a conceptual feature of my visual impression, i.e., of the latter of the two events mentioned above, that its propositional content is compatible with several possible states of the world, so that the individuals b and c may play some role in each of these alternatives. One may then say, in a slightly misleading derivative sense, that the actual individuals b and c and the corresponding perceptual world-lines are 'causally related'.

We are now ready to formulate the truth-conditions for quantified intensional logic. Let $L(O_1, \ldots, O_n)$ be the language which is obtained from the ordinary predicate logic L (with individual constants $a, b, \ldots$, predicates $F, G, \ldots$, individual variables $x, y, \ldots$, propositional connectives $\sim, \vee, \wedge, \supset, \equiv$, and quantifiers (Ex), (x)) by adding a new pair of quantifiers $(\exists x)$ and $(\forall x)$ as well as the unary intensional operators $O_1, \ldots, O_n$. If formula A does not contain intensional operators, then formulae $(Ex)A$ and $(\exists x)A$ – and similarly formulae $(x)A$ and $(\forall x)A$ – have the same (i.e., the ordinary) truth-conditions. If $A(x)$ is a formula of $L(O_1, \ldots, O_n)$ where x is free, then

(6) $(Ex)O_iA(x)$ is true at w iff there is a physical world-line f on $\mathcal{O}_i(w)$ such that, for each w' in $\mathcal{O}_i(w)$, $f(w')$ satisfies $A(x)$ at w'.

(7) $(\exists x)O_iA(x)$ is true at w iff there is a perceptual world-line f on $\mathcal{O}_i(w)$ relative to w (and to the person whose propositional attitude O_i is) such that, for each $f(w')$ in $\mathcal{O}_i(w)$, $f(w')$ satisfies $A(x)$ at w'.

In these conditions, $\mathcal{O}_i(w)$ denotes the class of the O_i-alternatives to w in W. As w itself need not belong to $\mathcal{O}_i(w)$, when O_i does not satisfy the success condition, the existential quantifiers $(\exists x)$ and (Ex) in (6) and (7), respectively, need not have existential import with respect to w. However, if one considers sentences of the form $(\exists x)(A(x) \wedge O_iB(x))$ – and analogously for (Ex) – their truth in w requires the existence of a perceptual world-line f on $\mathcal{O}_i(w) \cup \{w\}$ such that $f(w)$ satisfies $A(x)$ in w and $f(w')$ satisfies $B(x)$ in w' for all $w' \in \mathcal{O}_i(w)$. In other words, the existential quantifier $(\exists x)$ (resp. (Ex)) has existential import if it binds variables which occur outside the scope of intensional operators.

The basic idea of conditions (6) and (7) is the stipulation that when we are quantifying from outside into a context within the scope of an intensional operator, the quantified variables are ranging over suitable world-lines. For example, formulae

(8) $(Ex)K_a(x = d)$
(9) $(\exists x)K_a(x = d)$

are true at world w if and only if there is some physical world-line on
$\mathscr{X}_a(w)$ (resp. perceptual world-line on $\mathscr{X}_a(w)$ relative to w and a) such
that $x = d$ in all K-alternatives to w. The natural readings of (8) and (9)
are thus

(8)* a knows who d is
(9)* a knows d,

where (9)* means that a is acquainted with d. (See Section VII, however.)
Similarly, formula $(Ex)B_a(x = d)$ says that a has an opinion as to who d is.
(See Hintikka, 1975, pp. 44, 50.)

Two general principles follow immediately from conditions (6) and (7):

(10) $(Ex)O_iA(x) \supset O_i(Ex)A(x)$
(11) $(\exists x)O_iA(x) \supset O_i(\exists x)A(x)$,

where O_i may be replaced by any of K_a, B_a, and S_a. In these cases, the
converses of (10) and (11) are not valid.[27]

VI

In this section, several formulae of $L(S_a)$ are discussed with suggestions
concerning their natural reading in ordinary discourse. It may be instructive to compare this section with Section VI in Thomason (1973).[28]
Language $L(K_a, B_a, S_a)$ will be considered in the next section.[29]

What are the simplest descriptions of perceptual situations that one can
think of? Let us consider the following formulae:

(12) $S_a(\exists x)(x = x)$
(13) $S_a(Ex)(x = x)$
(14) $(\exists x)S_a(x = x)$
(15) $(Ex)S_a(x = x)$
(16) $(\exists x)S_a(\exists y)(y = x)$
(17) $(\exists x)S_a(Ey)(y = x)$
(18) $(Ex)S_a(\exists y)(y = x)$.

Here (12) and (13) are equivalent. They are true at world w if and only if
each S_a-alternative to w is non-empty. If empty possible world are excluded

from W, as usual, then (12) and (13) are in fact valid, and they do not give any information about a's perceptual situation. In particular, (12) is a true description of a blind man's visual impressions.

(14) is true at w if and only if there is a perceptual world-line f on $\mathscr{S}_a(w)$ relative to w and a. The requirement which the individuals $f(w')$ picked up by f from the S_a-alternatives w' to w are supposed to satisfy, viz. that they make the formula $x = x$ true, is empty (trivially true). The corresponding requirement in (16) and (17), viz. that $f(w')$ exists in w', is likewise empty. Therefore, (14), (16), and (17) are all equivalent. As they require the existence of a perceptually individuated 'individual' in a's visual space, there must be something in a's visual field which appears visually discriminated from its environment. Thus, these formulae say essentially that it appears to a that there is something in front of him. The simplest reading for these formulae is therefore

$(14)^* = (16)^* = (17)^*$ a sees something.

Formulae (15) and (18) are equivalent. They are true at w if and only if there exists a physical world-line on $\mathscr{S}_a(w)$, i.e., a physically identified 'individual' existing in each S_a-alternative to w. Thus, they essentially say that it appears to a that there is some physical object (a person, perhaps) in front of him. In analogy with $(14)^*$, we have the simple reading

$(15)^* = (18)^*$ a sees some thing.

As these readings suggest, (15) implies (14):

(S15) $(Ex)S_a(x = x) \supset (\exists x)S_a(x = x)$.

If there is physical world-line on $\mathscr{S}_a(w)$, then there is at least one perceptual world-line on $\mathscr{S}_a(w)$ relative to w and a.[30]

Let us next take the formulae

(19) $(\exists x)S_a(x = b)$
(20) $(Ex)S_a(x = c)$
(21) $(\exists x)(x = b \wedge S_a(\exists y)(y = x))$.

The following readings have been proposed by Hintikka:

(19)* a sees b (*de dicto*)
(20)* a sees who c is
(21)* a sees b (*de re*)

(see Hintikka, 1969, pp. 173–174). These readings suggest that the statement 'a sees somebody' can be formalized by means of the formulae $(\exists x)(Ey)S_a(x = y)$ (*de dicto*) and $(\exists x)((Ey)(x = y) \wedge S_a(\exists z)(x = z))$ (*de re*).

The reading (20)* is natural in those cases where c corresponds to some perceptually identified individual in a's environment. For example, if c stands for 'the man sitting by the window', then (20) is true if and only if there is a physical world-line which is identical with c in all of a's S-alternatives, i.e., if a sees who the man sitting by the window is. However, (20) does not entail (8), i.e., seeing who c is does not entail knowing who c is. Moreover, (20) does not entail that c exists in the actual world.

If b stands for 'Faye Dunaway', then (19) is true if and only if Faye Dunaway is in front of a in each of a's S-alternatives. Then one may say that a sees Faye Dunaway in the weak sense that a sees something which he identifies as Faye Dunaway. Formula (19) has thus only the force of the locution 'a seems to see b'. (19) does not entail that a really sees something which exists or which is b in the actual world. Moreover, the name 'b' occurs in (19) in a referentially opaque context: if a seems to see Faye Dunaway, it does not follow that a seems to see the actress who played the role of Bonnie in *Bonnie and Clyde*.

Formula (19) is weak in the sense that it does not contain a success condition with respect to seeing. But it is quite strong in another sense: it requires success with respect to individuation. If I seem to see Faye Dunaway in the sense of (19), then in all worlds compatible with what I see Faye Dunaway is in front of me. This condition seems to put extremely severe conditions for my ability to see: as far as my visual impressions are concerned, it is quite possible that there exists some person who looks sufficiently similar to Faye Dunaway so as to deceive me. But then there should also be possible words compatible with what I see where the person whom I see is this 'duplicate' of Faye Dunaway. Here we actually see how the background knowledge and background assumption are operative (cf. the quotation from Hintikka in Section III): if we could not first exclude those anomalous possible worlds with 'duplicates' and other strange creatures, we could never claim that we seem to see some particular persons.

If a perceptual world-line relative to a is identical with a physical object b in all of a's doxastic and epistemic alternatives, then one may expect that it will be identical with b in all of a's S-alternatives, too. This amounts to the following principle

(G12) $(\forall x)(y)(B_a(x = y) \supset S_a(x = y))$.[31]

G12 justifies an inference of the form

$$(\exists x)S_a(x = b)$$
$$\underline{B_a(c = d)}$$
$$(\exists x)S_a(x = c)$$

For example, if a seems to see Faye Dunaway and if a believes (or knows) that Faye Dunaway is the actress who played the main female role in *Chinatown*, then a seems to see the actress who played the main female role in *Chinatown*.

Formula (21) is very different from (19) – a's attitude in (21) is *de re* in the sense that it is directed towards a specific physical object b. (21) is an 'object-perception statement' in the sense of Armstrong (1976). It entails that $(\exists x)(x = b)$, i.e., that b exists (in the actual world). As the world-line x in (21) is extended to the actual world, the truth of (21) further entails that the object b is the (essential part of a) cause of a's visual impression (see Section V above). Moreover, the name 'b' occurs in (21) in a referentially transparent context. In spite of the success condition with respect to existence, it is sufficient for the truth of (21) that b is "visually differentiated from its immediate environment" by a (cf. the notion of 'seeing$_n$' in Dretske, 1969, p. 20). In other words, a need not have any opinions about the identity of the object he is seeing, and it need not even appear to him that he sees b (cf. Warnock's analysis of object-perception in Warnock, 1965, pp. 50–58). A neutral reading of (21) without too strong connotations is the following:

(21)** a looks at b.

(See also the next section.)[32]

Howell (1974) has suggested that the formula

(22) $(\exists x)(x = b \wedge S_a F(x))$

is the best reconstruction of the sentence

(22)* a sees b as an F.

The reading (22)* is intuitively very satisfactory: (22) says that a is looking at b and it appears to a that b is an F. Here (22) entails (21). Moreover, a may be mistaken or not in seeing b as an F. Thus, (22) corresponds to 'object-appearance statements' in the sense of Armstrong (1976) who would read (22) in the form 'b perceptually appears F to a'.[33]

The name 'b' occurs in (22) in a referentially transparent position, while the position of predicate 'F' is referentially opaque. This gives an explanation to the puzzle of N. R. Hanson: when Kepler and Tycho Brahe are looking at sunrise, do they see the same thing in the east dawn? (Cf. Hanson, 1969, p. 5.) The answer is that they see the same physical object, the sun, but Kepler sees it as a moving body and Tycho sees it as static or fixed. As Hanson argues (*ibid.*, p. 19), this sort of seeing is 'theory-laden' or shaped by prior knowledge – if I see b as an F and if I know that all F's are G's, then (unless I see that some F's are not G's) I mediately see b as a G. In other words, principle C6 justifies the following inference:

$$(\exists x)(x = b \wedge S_a F(x))$$
$$K_a(x)(F(x) \supset G(x))$$
$$\frac{\sim S_a \sim (x)(F(x) \supset G(x))}{(\exists x)(x = b \wedge S_a G(x)).}$$

It does not follow, however, that seeing b *as* an F entails seeing *that* b is an F (cf. Achinstein's criticism of Hanson in Achinstein, 1972, p. 243), since (22) does not entail the statement $S_a F(b)$.

As special cases of (22), we obtain

(23) $(\exists x)(x = b \wedge S_a(x = c))$
(24) $(\exists x)(x = b \wedge S_a(x = b))$,

with the readings

(23)* a sees b as c
(24)* a sees b as b.

If $b \neq c$ in the actual world, then (23) describes a simple perceptual mistake – visual misidentification. Here one sees also that if we had accepted condition S9 for seeing, then formula (23) would be contradictory. Formula (24) says that a is looking at b and he correctly sees b as b, i.e.,

(24)** a correctly identifies b.

A stronger reading for (24) might be

(24)*** a recognizes b.

However, if we require – as the etymology suggests – that a can re-cognize b only if a knows b, then (24)*** is too strong (cf. Section VII).

Let us then consider formulae

(25) $S_aF(b)$
(26) $(\exists x)S_a(x = b \wedge F(x))$
(27) $(Ex)S_a(x = b \wedge F(x))$.

Here (25) says that it appears to a that b (whoever he is) is an F, i.e.,

(25)* a seems to see that b is an F (*de dicto*).

It does not follow from (25) that the individuals picked up by 'b' from a's S-alternatives are perceptually or physically cross-identified. Therefore, (25) does not entail (26) or (27), even if each of these formulae entails (25). (25) does not entail (21), either, i.e., a need not be even looking at b.

Formulae (26) and (27) express the fact that something or somebody appears both b and an F to a. This 'something' need not exist in the actual world, however. Thus, (26) and (27) do not entail

(28) $(\exists x)[(Ey)(y = x) \wedge S_a(x = b \wedge F(x))]$.

The following formula nevertheless is inconsistent:

$$(\exists x)[\sim(Ey)(y = x) \wedge S_a(x = b \wedge F(x))],$$

since it entails the contradiction $(\exists x) \sim (Ey)(y = x)$. The situation where a is only having a *hallucination* can be described by

(29) $(\exists x)S_a(x = b \wedge F(x)) \wedge \sim(\exists x)((Ey)(y = x) \wedge S_a(x = b \wedge F(x)))$,

which is equivalent to

(29)' $(\exists x)S_a(x = b \wedge F(x)) \wedge (\forall x)((Ey)(y = x) \supset \sim S_a(x = b \wedge F(x)))$.

In this case, the perceptual world-line x cannot be extended to the actual world at all, so that what appears to a as b and an F does not exist. (29) can thus be read in the following way:

(29)* a is suffering from a hallucination of seeing b who is an F.

Hallucinations should be distinguished from visual *illusions* where the percipient is looking at some object but mistakenly sees some of its properties:

(30) $(\exists x)(\sim F(x) \wedge S_a F(x))$.

Here a sees something which in fact is not an F as an F. Formula (23) also describes a visual illusion if $b \neq c$.

It is instructive to contrast with (30) two different possibilities where a sees something correctly:

(31) $F(b) \wedge S_a F(b)$
(32) $(\exists x)(F(x) \wedge S_a F(x))$.

(More complex formulae could easily be constructed.) (31) may describe a situation where a *accidentally* sees correctly that $F(b)$: a may here have the "right perceptions for wrong reasons" (cf. Hintikka, 1975, p. 70), since there is nothing in formula (31) which would guarantee that there is an appropriate causal relation between the fact that b is an F and a's perception. The situation is different in the case of (32) where a's perception is *veridical* in a strong sense.

The difference between (31) and (32) suggests that principle C4 in Section III is mistaken in that its premise allows accidentally correct seeing while it should require at least that the seeing is veridical. A more plausible principle might be the following:

(C13) $(\exists x)[x = b \wedge F(x) \wedge S_a(x = b \wedge F(x))] \wedge B_a F(b) \supset K_a F(b)$.

According to C13, if a believes that $F(b)$ and if a correctly identifies b and veridically sees that $F(b)$, then a is justified in claiming that he knows that $F(b)$.

VII

What happens when the formulae studied in Section VI are combined with epistemic operators? In answering this question, the validity of the principles adopted in Section IV will be assumed.

The interpretation of formulae of the type

(33) $K_a(\exists x)S_b(x = c)$
(34) $K_a(\exists x)(x \neq c \wedge S_b(x = c))$

is straightforward: (33) says that a knows that b seems to see c, while (34) says that a knows that b is having a visual illusion of seeing c.

We are here more interested in situations which involve a's knowledge about his own perceptions. The KS-thesis I10 entails, for example, that

(I12) $S_a F(b) \equiv B_a S_a F(b) \equiv K_a S_a F(b)$.

If it appears to a that $F(b)$, then a knows and believes that he seems to see that $F(b)$. Further consequences of I10 together with (10), (11), and K9 are:

(I13) $(\exists x)S_a A(x) \equiv K_a(\exists x)S_a A(x)$

(I14) $(Ex)S_a A(x) \equiv K_a(Ex)S_a A(x)$

(I15) $(\exists x)S_a A(x) \supset B_a(\exists x)S_a A(x)$

(I16) $(Ex)S_a A(x) \supset B_a(Ex)S_a A(x).$

For example, a sees something if and only if he knows that he sees something (cf. (16)); a seems to see b if and only if he knows that he seems to see b (cf. (19)); a sees who c is if and only if he knows that he sees who c is (cf. (20)).

On the other hand, in the following chain of formulae the implications cannot be strengthened to equivalences:

$$
\begin{aligned}
(I17) \quad (\exists x)K_a(A(x) \wedge S_a C(x)) &\equiv (\exists x)(K_a A(x) \wedge S_a C(x)) \\
&\supset K_a(\exists x)(A(x) \wedge S_a C(x)) \\
&\supset (\exists x)(A(x) \wedge S_a C(x)) \\
&\equiv (\exists x)(A(x) \wedge K_a S_a C(x)).
\end{aligned}
$$

The same result holds also for the quantifier (Ex). One may thus distinguish between

(35) $(\exists x)(K_a(x = b) \wedge S_a(\exists y)(y = x))$

(36) $K_a(\exists x)(x = b \wedge S_a(\exists y)(y = x))$

(37) $(\exists x)(x = b \wedge K_a S_a(\exists y)(y = x)).$

(37) is equivalent to (21), so that it says simply that a looks at b (cf. (21)**). It is consistent with (37) that a is only stupidly staring at b without paying much attention to what he sees – in particular, without being actively aware of the causal relation between b and his perceptions. On the other hand, (36) requires that there is an appropriate sort of a causal process in each of a's K-alternatives. (36) thus describes a situation where a is actively looking at b or observing b – for example, as in the case where 'Tom is watching a girl who is walking by'. One way of reading (36) is therefore

(36)* a is watching b.

(36), but not (37), entails that a knows that b exists. If a child looks at the sun and painfully becomes aware of its existence, we say that she at least noticed the sun.[34] One may therefore read (36), in some cases, also by

(36)** a notices b.

(35) is stronger than (36) in that it entails $(\exists x)K_a(x = b)$, i.e., a knows b (cf. (9)). It thus expresses the fact that a is watching b whom he knows. In ordinary discourse, (35) might be expressed by saying

(35)* a knows that he sees b

with a special emphasis on 'b'.

One often encounters situations where somebody says: I must have seen her sometimes, but I don't know when, since I don't know her (cf. Warnock, 1965, pp. 51–52). In view of the distinctions made above, there is nothing strange in such a situation. Indeed, as $\sim(\exists x)K_a(x = b)$ entails $(\forall x) \sim K_a(x = b)$, a cannot know that he sees b (in the sense of (35)) if he does not know b (in the sense of (9)). Therefore, the following inference is correct, as we might expect:

> a looks at b (but not anyone else)
> a does not know b
> Hence, a does not know that the person he is looking at is in fact b.

Let us next consider formulae

(38) $(\exists x)(K_a(x = b) \wedge S_a(x = b))$
(39) $K_a(\exists x)(x = b \wedge S_a(x = b))$
(40) $(\exists x)(x = b \wedge K_a S_a(x = b))$.

(40) says that a correctly identifies b (cf. (24)**). (39) entails that a knows the existence of b, and (38) entails that a knows b. One might propose here the reading

(38)* a recognizes b.

A very interesting problem appears – or rather becomes clearly visible – at this point. We have earlier proposed a principle C12 which essentially tells how background information influences perceptual cross-identification. C12 entails now that formula (38) is equivalent simply to $(\exists x)K_a(x = b)$, i.e., (9). However, we have proposed different readings for these equivalent formulae: 'a knows b' for (9), and 'a recognizes b' for (38). Something seems to be wrong here, since 'a knows b' should be equivalent to 'a is *able to recognize* b' rather than to 'a recognizes b'. In other words, if a knows b, then he recognizes b if and *only if* he looks at b.

As a first reaction to this problem, one might suggest that there is something wrong with principle C12. It seems to me, however, that the blame cannot be put on C12, but instead the trouble lies deeper in the given semantical framework. To see this, note first that by K9 formula $(\exists x)K_a(x = b)$ is equivalent to

(41) $(\exists x)(x = b \wedge K_a(x = b))$.

(41) requires that the perceptual world-line x is extended to the actual world, so that the semantics given in Section V demands that there exists a causal connection between the actual object b and the perceptual world-line relative to a – i.e., between b and a's perceptions. But such a connection cannot exist unless a in fact is looking at b. According to this argument, formula (41) implies that a in fact is looking at b. Another way of arguing for this conclusion is the following: perceptual world-lines relative to world w and person a exist only if a perceives something in w. For perceptual world-lines, *esse est percipi*. If a's perceptions are brought about by an object b in w, then there must be something in front of a in each world w' in $\mathscr{S}_a(w)$. Thus, the perceptual world-lines go through the whole $\mathscr{S}_a(w)$, even if a need not know whether they can be extended to w as well. The latter information is contained in the assumption (41) which therefore entails that a looks at b precisely in the sense of formula (21).

If these arguments are correct, then formula (9) not only says that a knows b in the potential sense of being able to recognize b, but it also says that a actually recognizes b by looking at him. While the reading (9)* for (9) is not wrong, it is definitely weaker than the more appropriate reading (38)*.

One might present an objection to this conclusion by noting that Hintikka has in fact required the use of causal methods of cross-identification only in connection with perception – but not with knowledge. Indeed, he claims that the success condition K9 which makes $\mathscr{S}_a$ reflexive "partly eliminates the special problems of cross-identifying between the actual world and its alternatives" (Hintikka, 1975, p. 72). I don't see why this should be so: if I am able to *perceptually* cross-identify, without any special problems, between the actual world w and one of its K-alternatives w' in W, then why could not I use the same method of cross-identification when w' happens to be also a S-alternative to w? After all, w' is the same possible world in both cases. If we could make sense of formulae like $(\exists x)K_aF(x)$ without introducing causal methods of cross-identification, we should not have reason to raise these special problems in connection with perceptual statements, either. Conversely, if the cross-identification between the actual world and its S-alternatives has to rely on causality, as Hintikka insists, then causality will be involved in the interpretation of formulae containing perceptual quantifiers and epistemic operators.

To be able to express in our formalism the difference between (9)* and (38)* we should stipulate that in perceptual quantification into a knowledge context the perceptual world-lines have only a *conditional* existence – that is, they 'really' exist only on the condition that the person in question is having some visual impressions. This stipulation would not change the fact that (9) and (38) are equivalent, but then we could say that, in a situation where *a* knows *b* but is not actively looking at *b*, *a* only *mediately* recognizes *b*. I shall not try to develop this suggestion further in this paper, but it is clear that many clarifying remarks concerning its details would be needed.

Let us conclude this section by considering visual illusions and hallucinations. Corresponding to the simple illusion (23), we have the formulae

(42) $(\exists x)(K_a(x = b) \wedge S_a(x = c))$
(43) $K_a(\exists x)(x = b \wedge S_a(x = c))$
(44) $(\exists x)(x = b \wedge K_a S_a(x = c))$.

Here (44) is equivalent to (23); it says that *a* sees *b* as *c*, while *a* need not be aware of the illusionary character of his seeing. On the other hand, (43) describes a situation where *a* is knowingly making a perceptual misidentification: the object *b* is the cause of *a*'s visual impression in all of *a*'s K-alternatives, so that *a* knows that he is looking at *b* and that in fact *b* is the cause of his impression of seeing *c*. Such a situation may arise, for example, when some of my friends is wearing a clever disguise and I know it.[35] However, when principle C12 is assumed, formula (42) is inconsistent.

Principle C6 allows for the possibility that even a formula of the form

(45) $(\exists x)(K_a \sim F(x) \wedge S_a F(x))$

is consistent. An example of such a 'conscious illusion' was mentioned in Section III; as other examples one can mention the case where a stick put in water looks bent and the case where a cloud is seen as a horse. If I see a cloud as a horse and if I know that all horses are animals, then I see the cloud as an animal. If I see the cloud as horse without seeing 'its' legs, then it does not follow that I see the cloud as an four-legged animal (see Section VI). Moreover, if I see the cloud as a horse and if I know that it is a cloud rather than a horse, then I don't see (believe, expect) that it has all the properties of horses. Achinstein is therefore right in criticising Hanson's claim that "seeing X as Y entails seeing that X may be expected to behave in all the ways that Y's do" (Achinstein, 1972, pp. 242–243).

The consistency of formulae (43) and (45) shows that visual illusions need not involve any false beliefs.[36] On the other hand, visual illusions may lead to false beliefs when they are created by deliberate cheating. To mention one real-life example, it happened recently in a top-class Finnish restaurant that a man, Mr. X, presented a show by pretending to be a famous artist, Mr. W, and also pretending to play harmonica, while the music actually came from a record. For some minutes, Mr. X succeeded in cheating the distinguished audience, among them Mr. V, a famous Finnish long distance runner. In this situation, it appeared to Mr. V that Mr. W is playing harmonica, and, as he did not have reason to believe otherwise, he also believed that Mr. W is playing harmonica. As a matter of fact, what he really saw was Mr. X who was not playing anything. The situation can be described by the sentence

(46) $\quad (\exists x)(x = b \wedge \sim F(x) \wedge \sim B_a(x \neq c) \wedge \sim B_a \sim F(x) \wedge S_a(x = c$
$\qquad \wedge F(x)))$.

which by C10 imp ies

(47) $\quad (\exists x)B_a(x = c \wedge F(x))$.

Let us finally note that epistemic operators can be combined with formula (29) as follows (cf. I13):

(47) $\quad K_a \sim (\exists x)((Ey)(y = x) \wedge S_a(x = b \wedge F(x))) \wedge (\exists x)S_a(x = b$
$\qquad \wedge F(x))$

(48) $\quad (\forall x)K_a \sim ((Ey)(y = x) \wedge S_a(x = b \wedge F(x))) \wedge (\exists x)S_a(x = b$
$\qquad \wedge F(x))$

(49) $\quad \sim (\exists x)((Ey)(y = x) \wedge K_a S_a(x = b \wedge F(x))) \wedge (\exists x)S_a(x = b$
$\qquad \wedge F(x))$.

Here (47) is stronger than (48) and (48) stronger than (49). Again, (49) is equivalent to (29), so that it describes a situation where a is suffering from a hallucination. In this case, a need not know whether he is hallucinating or not – "even Macbeth was not sure whether it was a real dagger or not!" (Hamlyn, 1970, pp. 168–169). On the other hand, (47) and (48) entail that a knows that the world-line x cannot be extended to the actual world, i.e., a is knowingly having the hallucination.

VIII

From a philosophical point of view, Hintikka's logic of perception has the remarkable feature that it treats propositional perception as basic and object-perception as derivative (cf. formulae (19) and (21)). According to this analysis, there is something propositional even in the simplest cases of perception of objects. Even our most primitive "unedited sense-impressions" are "already structured categorically", i.e., "organized so as to be *of* definite objects", so that their informational content has to be described by speaking of these objects (Hintikka, 1975, pp. 200–202). It is precisely this aspect of perception which Hintikka has referred to as the intentionality of perception.[37]

Is this view about perception incompatible with those analyses of perception which start from some sort of 'non-epistemic' seeing of objects and only later proceeds to the discussion of propositional seeing? How is it related to the thesis that "all seeing is seeing as"? (Cf. Vesey, 1965, p. 72.) I shall conclude this paper with a brief comment on these questions.

Dretske starts his account of perception by giving a description of man's primitive and "fundamental visual capacity" of seeing things, where this "non-epistemic seeing" is "devoid of positive belief content" in the sense that 'a sees b' does not entail 'a believes that p' for any proposition p (Dretske, 1969, p. 6). According to Dretske, a non-epistemically sees b if and only if b "looks some way to" a, so that b is "visually differentiated from its immediate environment" by a (*ibid.*, p. 20). This sense of seeing is contrasted with different forms of 'epistemic' seeing where seeing always involves "acquiring some true belief about what is seen" (*ibid.*, p. 76).

On the other hand, Hanson has argued that "seeing an object x is to see that it may behave in the ways we know x's to behave" (Hanson, 1969, p. 22). For Hanson, seeing typically involves seeing as, and seeing as is at least partly epistemic in that it in turn involves seeing that (*ibid.*, pp. 19–25). Relying on Dretske's analysis, Achinstein (1972) has argued that Hanson's attempt to reduce seeing X to seeing that *via* seeing as fails, since there exists a reasonable non-epistemic sense of seeing.

Our treatment of perception in this paper differs from Dretske's and Hanson's in some important respects. On the one hand, we have already found reason to argue against Hanson's view of seeing as-statements, especially their relation to seeing that-statements (see Sections VI and VII). It would also be a little artificial to claim that all statements concerning visual perception are at bottom seeing as-statements – unless we are willing to say that in hallucinations one sees something non-existent as an F (cf. Howell, 1972, pp. 417–418).

On the other hand, in our analysis, all seeing involves propositional seeing, but often in a quite complex way. In a scale indicating the inter-vowenness of epistemic attitudes with perception, Dretske's notions of non-epistemic and epistemic seeing seem to mark two extreme points between which most types of perceptual situations are situated. In Dretske's description, non-epistemic seeing is so purified of all epistemic associations that not only "a great variety of sentient beings", such as dogs and cats, but also such non-living things as cameras are able to 'see' things in this sense.[38] It is clear, of course, that perceptual cross-identification would not be possible at all without the existence of articulated visual fields, so that our treatment in fact presupposes something like the ability which Dretske is describing. However, his thesis that non-epistemic seeing is devoid of positive belief content is not strictly coherent with the analysis (19) and (21) of object-perception if principles C10 and I10 are assumed. If a is looking at b in the sense of (21), then I10 entails that

$$(\exists x)(x = b \wedge B_a S_a (\exists y)(y = x))$$

and C10 entails that

$$(\exists x)(x = b \wedge B_a (\exists y)(y = x)) \vee (\exists x)(x = b \wedge B_a \sim (\exists y)(y = x)).$$

Thus, a believes that he sees something, and he has an opinion on the question whether he is looking at something existing or non-existing. Even if these consequences of (21) are not of the form $B_a p$, as Dretske's definition requires, it still seems appropriate to say that (21) has some positive belief content. This belief content is considerably weaker, however, than in Dretske's epistemic seeing, since it need not include any true beliefs about what is seen.

University of Helsinki

NOTES

[1] See Hintikka (1962, 1969), especially the essay 'Semantics of Propositional Attitudes' in Hintikka (1969), pp. 87–111.

² See the essay 'On the Logic of Perception' in Hintikka (1969), pp. 151–183. For an excellent review of Hintikka's treatment of quantification and propositional attitudes, see Saarinen (forthcoming).

³ See Hintikka (1969), pp. 166–168; the essay 'Information, Causality, and the Logic of Perception' in Hintikka (1975), pp. 59–75; and the essay 'The Intentions of Intentionality' in Hintikka (1975), pp. 192–222.

⁴ As an illustrative example, one can mention the fact that in Carnap's system of inductive logic all genuine universal statements about infinite universe receive the probability zero. Hintikka has argued convincingly that this feature of Carnap's formalism is unnecessary from the technical viewpoint and undesirable from the philosophical viewpoint.

⁵ A is used here as a syntactical metavariable which ranges over the sentences of $PL(O_1, \ldots, O_n)$.

⁶ See the remark concerning 'small worlds' in Hintikka (1975), p. 195. It seems to me natural to think that these 'small worlds' are not only spatio-temporally limited, but also otherwise restricted – e.g. with respect to the *relevant* properties of the 'inhabitants' of these worlds. Note that the possible worlds w in W may, in some cases, be alternative courses of events rather than states of affairs.

⁷ This principle would follow from S9. For deontic logic, see D. Föllesdal's and R. Hilpinen's 'Introduction' in Hilpinen (1971), pp. 1–35.

⁸ Cf. Hintikka (1962). Most of these principles are relatively uncontroversial. B7 and part of B4 (implication from right to left) have been challenged by Armstrong (1973), pp. 104–107.

⁹ Ordinary language is not a reliable guide in this case, since a large part of what is today called 'scientific knowledge' is strictly speaking false. Most philosophers since Plato have nevertheless made K9 as one of the defining conditions for knowledge.

¹⁰ See, for example, the definition of 'seeing that b is P in a primary epistemic way' in Dretske (1969), pp. 78–88, and the definition of 'non-inferential knowledge' in Armstrong (1973), p. 197. Cf. also formula C13 in Section VI below.

¹¹ Hintikka has proposed one way of weakening this assumption (see Hintikka, 1975, pp. 179–191), but we shall not consider this issue here.

¹² This distinction is vaguely stated, but it is sufficient for our purposes in this paper. (For a detailed discussion of this distinction, see Armstrong, 1961, 1976. See also Dretske, 1969, p. 153.)

¹³ See the chapter 'Revolutions as Changes in World View' in Kuhn (1962), pp. 110–134, and the chapter 'Observation' in Hanson (1969), pp. 4–30.

¹⁴ The tentative adoption of C6 means that we use the S-operator even in cases which do not seem to have anything to do with visual perception. Dr. Esa Saarinen has made some good objections to this suggestion: it is unintuitive enough that Hintikka's approach to the logic of perception commits us to saying that we can 'see that p' for tautologous p; why then even multiply (by C6) such situations? For example, I know that there are more than 700 millions Chinese, and as I surely do not see otherwise I am committed, by C6, to claim that I *see* that there are more than 700 millions Chinese. I should like to make three remarks in defense of my strategy. First, one might try to argue (if one supports some form of empiricism) that the example concerning the Chinese – and all other cases dealing with factual knowledge rather than with analytical

truths – can after all be analyzed in terms of mediate visual perception: I know about the number of Chinese, since I have read (i.e., had an immediate visual perception) about it in some reliable source of information. Secondly, mediate perception has to be taken into account within a logic of perception in any case, at least if one is not a naive empiricist and if one wants to take the intentionality of perception seriously (cf. Section VIII). Thirdly, I am quite ready to accept that C6 is only a first approximation to a reasonable principle concerning the influence of knowledge to perception. Still, I think it is approximately true in the sense that a better principle cannot expressed in $PL(K_a, S_a)$ – that is, without introducing quantifiers or new perceptual operators. The situation here is somewhat analogous to the difference between principles C4 and C13; we shall also propose below another principle, C12, which concerns the effects of background assumptions within cross-identification. Within predicate logic, one can express statements of the form 'a is looking at b' (see (21) below). Thus, we could also formulate C6 in a more concrete form by the following requirement: if a is looking at b, a knows that b is an F, and a does not see that b is a non-F, then a sees that b is an F. These remarks serve to illustrate the general point that the logic of perception at the propositional level is much less interesting than at the level of quantification theory.

[15] See Howell (1972), p. 409. It should be mentioned that Howell is very well aware of the difficulties with C9, and he presents interesting comments on the possibilities of handling them (see *ibid.*, pp. 409–413).

[16] For these results, see Hintikka (1962), pp. 60–61, 103–106, 123, 158–159. Hintikka rejects the principles $B_a p \supset K_a B_a p$ and $\sim K_a p \supset K_a \sim K_a p$ (*ibid.*, pp. 51–53, 106); for the former, see however Hilpinen (1970), p. 129. The following implications are consequences of C1:

$$K_a K_b A \underset{\textstyle B_a K_b A}{\overset{\textstyle K_a B_b A}{\lessgtr}} \gtrless B_a B_b A \, .$$

Hintikka has later admitted that there are other reasonable notions of knowledge for which the KK-thesis fails (see Hintikka, 1970). I shall ignore here the possibility, noted by H.-N. Castañeda, that a does not happen to know that he himself is referred to by the name 'a'.

[17] Note that this is consistent with principle C6, since, according to I1, $\sim S_a \sim K_b A$ is equivalent to $\sim K_a \sim K_b A$ which is entailed by $K_a K_b A$.

[18] For example, Descartes says in his *Meditations*: "But it will be said that these appearances are false and that I am dreaming. Let it be so; all the same, at least, it is very certain that it seems to me that I see light, hear a noise and feel heat" (see Descartes, 1968, p. 107).

[19] A. J. Ayer has argued that I can be mistaken in judging that one line *looks* to me longer than another (Ayer, 1956, p. 69). See, however, the counter-argument referring to the indeterminateness of some sense-impressions in Armstrong (1961), pp. 38–46.

[20] See Armstrong (1961), pp. 123–124, and Dretske (1969), p. 12.

[21] Cf., however, Section VIII below.

[22] Note that S10 is equivalent to the assumption that $\mathscr{S}_a$ is transitive.

[23] An example approximating the desired situation might be the following: I am sitting in front of a window, and I see that there is a bird behind the window. Moreover, through a complex system of mirrors I am at the same time able to see myself sitting at the window, but from such an angle that it seems that the window might as well be a wall. This example is so artificial, however, that I am not willing to take it seriously.

[24] This distinction is explained and clarified at several places in Hintikka (1975). See also Saarinen (forthcoming).

[25] In Thomason's treatment, the individuals in each possible world are divided (not necessarily exclusively) into 'physical objects' and 'perceptual objects' (see Thomason, 1973, p. 271). If the latter are only physical objects which a percipient sees (cf. *ibid.*, pp. 269–270), then there is no reason to introduce this division as a semantical primitive. If 'perceptual objects' may be such entities as sense data, then there are strong philosophical reasons for not assuming them to be objects in *one* world (see below).

[26] If we say that a perceptual world-line f on V has the property P in case each $f(w)$, for w in V, has the property P, then perceptual world-lines can be said to similar to sense data in that "they do not always obey the Law of Excluded Middle" (Barnes, 1965, p. 145).

[27] A simple counter-example to the principle $S_a(\exists x)A(x) \supset (\exists x)S_aA(x)$ is given by the non-equivalent formulae (12) and (14) below. Another example is given by the case where I see two persons and I am able to see that at least one of them is a girl but I still don't see which one of them is a girl.

[28] Instead of two pairs of quantifiers, Thomason uses two sorts of variables – perceptual φ and physical x. He allows the use of φ independently of the S-operator: for example, $(\exists \varphi)\varphi = a$ says that a exists visually for the percipient (*ibid.*, pp. 269–270), even if he later suggests that this formula might describe a situation where the percipient "does not really see" a (*ibid.*, p. 274). The difference between formulae like $(\exists \varphi)\varphi = a$ and $(\exists \varphi)S\varphi = a$ remains unclear to me.

[29] Note that we might interpret the variables x, y, ... as ranging over *singular events* instead of objects (cf. Thomason, 1973, p. 183). One could then formalize statements like 'I see the man waving to his wife' and explain their difference to statements like 'I see that the man is waving to his wife' (cf. Dretske, 1969, pp. 33–34). These questions will not be discussed in this paper.

[30] Note that these world-lines need not be identical: a physical world-line on $\mathscr{S}_a(w)$ may intersect or go together with several perceptual world-lines. S15 can be proved by observing that if there are no perceptual world-lines on $\mathscr{S}_a(w)$ relative to w and a then it is surely compatible with what a sees in w that any physical individual either exists or does not exist in the world. But this shows that no physical world-line can be extended to the whole $\mathscr{S}_a(w)$.

[31] Note that C12 is equivalent, by 110, to $(\forall x)(y)(B_a(x = y) \supset B_aS_a(x = y))$.

[32] (19) and (21) both entail (16), but (19) does not entail (21), and (21) does not entail (19).

[33] Formula $(Ex)(x = b \wedge S_aF(x))$ says that there is a physical world-line which picks an F from each of a's S-alternatives and which can be extended to b in the actual world. Howell argues that this formula entails (22) (Howell, 1972, p. 407). However, this formula would be true, and (22) would be false, in the following situation: I am acquainted with two identical twins, and I see them from a distance so that I am not able to say which one of them is standing in the left. (If they were nearer, I could

distinguish them from the way they comb their hair.) Let b be the name of one of these twins, and choose '$x = b$' for '$F(x)$'. Then $(Ex)(x = b \wedge S_a(x = b))$ is true, while $(\exists x)(x = b \wedge S_a(x = b))$ is false. This example seems to disprove Howell's argument. Howell's suggestion for formalizing the sentence 'a recognizes b as an F' is

$$(Ex)(x = b \wedge S_a(x = b \wedge F(x))).$$

However, this sentence does not entail 'a sees b as an F', as he thinks (*ibid.*, p. 408). A more appropriate formalization which entails the corresponding seeing as – statement is

$$(\exists x)(x = b \wedge F(x) \wedge S_a(x = b \wedge F(x))).$$

[34] For 'noticing', see Warnock (1965), p. 54, and Dretske (1969), pp. 15, 21.

[35] A parallel situation arises for hearing when Sammy Davis Jr. is imitating Dean Martin.

[36] Cf. the statement that "sensory illusion is nothing but false belief, *or inclination to a false belief*, that we are perceiving some physical object or state of affairs", in Armstrong (1961), p. 87.

[37] The intentionality of 'raw' sensation means, among other things, that if I see the moon then the moon as a whole (not moon without back side) occurs in my perceptual alternatives.

[38] Cf. Dretske (1969), p. 4. Note that Hanson explicitly says that cameras cannot see (Hanson, 1969, p. 20).

BIBLIOGRAPHY

Achinstein, P., Review of N. R. Hanson, *Perception and Discovery*, *Synthese* **25** (1972), 241–247.

Armstrong, D. M., *Perception and the Physical World*, Routledge and Kegan Paul, London, 1961.

Armstrong, D. M., *Belief, Truth and Knowledge*, Cambridge University Press, London, 1973.

Armstrong, D. M., 'Immediate Perception', in R. S. Cohen *et al.* (eds.), *Essays in Memory of Imre Lakatos*, D. Reidel, Dordrecht, 1976, pp. 23–35.

Ayer, A. J., *The Problem of Knowledge*, Macmillan, London, 1956.

Barnes, W., 'The Myth of Sense-Data', in Swartz (1965), pp. 138–167. (Originally published in 1944–1945.)

Chisholm, R. M., '"Appear", "Take", and "Evident",' in Swartz (1965), pp. 473–485. (Originally published in 1956.)

Clark, R., 'Old Foundations for a Logic of Perception', *Synthese* **33** (1976), 75–99.

Descartes, R., *Discourse on Method and the Meditations*, Penguin Books, Harmondsworth, 1968.

Dretske, F. I., *Seeing and Knowing*, Routledge & Kegan Paul, London, 1969.

Hamlyn, D. W., *The Theory of Knowledge*, Anchor Books, Doubleday, New York, 1970.

Hanson, N. R., *Patterns of Discovery*, Cambridge University Press, London, 1969.

Hilpinen, R., 'Knowing That One Knows and the Classical Definition of Knowledge', *Synthese* **21** (1970), 109–132.

Hilpinen, R. (ed.), *Deontic Logic: Introductory and Systematic Readings*, D. Reidel, Dordrecht, 1971.

Hintikka, K. J., *Knowledge and Belief*, Cornell University Press, Ithaca, 1962.

Hintikka, K. J., *Models for Modalities*, D. Reidel, Dordrecht, 1969.

Hintikka, K. J., '"Knowing That One Knows" Reviewed', *Synthese* **21** (1970), 141–162.

Hintikka, K. J., *The Intentions of Intentionality and Other New Models for Modalities*, D. Reidel, Dordrecht, 1975.

Howell, R., 'Seeing As', *Synthese* **23** (1972), 400–422.

Hughes, G. E. and Cresswell, M. J., *An Introduction to Modal Logic*, Methuen, London, 1968.

Kuhn, T. S., *The Structure of Scientific Revolutions*, University of Chicago Press, Chicago, 1962.

Saarinen, E., 'Hintikka on Quantifying In', forthcoming.

Segerberg, K., *An Essay in Classical Modal Logic*, vols. 1–3, Filosofiska Studier, Uppsala, 1971.

Swartz, R. J. (ed.), *Perceiving, Sensing, and Knowing*, Anchor Books, Doubleday, New York, 1965.

Thomason, R. H., 'Perception and Individuation', in M. K. Munitz (ed.), *Logic and Ontology*, New York University Press, New York, 1973, pp. 261–285.

Unger, P., 'Propositional Verbs and Knowledge', *Journal of Philosophy* **59** (1972), 301–312.

Vesey, G. N. A., 'Seeing and Seeing As', in Swartz (1965), pp. 68–83. (Originally published in 1955–1956.)

Warnock, G. J., 'Seeing', in Swartz (1965), pp. 49–67. (Originally published in 1954–1955.)

Williams, B., 'Deciding to Believe', in H. Kiefer and M. Munitz (eds.), *Language, Belief and Metaphysics*, The State University of New York Press, Albany, 1970, pp. 95–111.

VI

PHILOSOPHICAL AESTHETICS

PAUL ZIFF

ANYTHING VIEWED

Look at the dried dung!
What for?
If I had said 'Look at the sunset!' would you have asked 'What for?'?
People view sunsets aesthetically. Sunsets are customary objects of aesthetic attention. So are trees rocks wildflowers clouds women leaping gazelles prancing horses: all these are sometime objects of aesthetic attention. But not everything is: not soiled linen greasy dishes bleary eyes false teeth not excrement.

Why not? It's not because they're unbeautiful or even ugly. Beautiful things are no problem for a rambling aesthetic eye but not all objects of aesthetic attention are beautiful: Grunewald's *Crucifixion* isn't neither is Picasso's *Guernica*. Breughel's rustics aren't lovely. The stark morning light in a Hopper is powerful but it is not beautiful. Not being beautiful needn't matter.

These unbeautiful objects are works of art. By chance some objects of aesthetic attention have been naturally produced. For the rest: they are products of art.

What is a work of art? Something fit to be an object of aesthetic attention. Most likely nowadays (now that didactic art is largely dead) something tailor-made for the purpose designed to be just that. If you want to attend aesthetically to something fix on a work of art as your object: that's the way it's thought to be. Is a work of art the paradigm of an object fit for aesthetic attention? What does a work of art have or lack that dung doesn't?

What is a work of art? Not everything. Leonardo's portrait of Ginevra de'Benci is. A mound of dried dung isn't. Nor is an alligator at least a living gator basking in the sun on a mud bank in a swamp isn't. A reason they are not is plain: nothing is a work of art if it is not an artefact something made by man. A gator basking a mound of dried dung are products of nature made or produced by natural forces. Not being made or produced by men they are not classed artefacts. Not being artefacts they are

285

E. Saarinen, R. Hilpinen, I. Niiniluoto, and M. Provence Hintikka (eds.), Essays in Honour of Jaakko Hintikka, 285–293. All Rights Reserved.

not classed works of art. Such is a common or the common if there is anything that is the common conception of a work of art.

Most likely there is no such thing as the common conception of a work of art: these are vague ill-defined notions. And some say that some objects that are not artefacts are nonetheless works of art. That needn't concern us: undoubtful examples of works of art are all that are wanted here and now and these are easier to come by when one considers artefacts rather than nonartefacts.

When one looks at a gator basking a mound of dried dung is one at once cognizant of the fact that not one or the other is man-made? And does such cognizance at once preclude all possibility of aesthetic attention to the gator basking the mound of dried dung? Though the gator basking is not man-made it is (to invoke the shade of Paley) remarkable in design and structure. By no stretch of the imagination can it be imagined to be less detailed rich intricate in design less complex in structure than an artefact. Given the present state of technology there's no way anyone can actually make a gator basking. But making a mound of dried dung is easy. Conjure up this image: a field in which there are two virtually identical mounds of dried dung. One was and the other was not man-made. Would that fact render the latter less accessible than the former to aesthetic attention?

Imagine this: that the Henry Moore statue at Lincoln Center was in fact not an artefact by Moore but a naturally formed that is nonman-made object found in a desert and transported to Lincoln Center. Would that matter to an appreciation of the statue? Yes enormously. Knowing that one's view of the object would be restructured: one would not in looking at the work look at it as a work. One would not look for manifestations of craftsmanship. One would not look for and see signs of the sculptor's hands: there would be none. But the object would still have shape form mass and balance. The various parts of the object would still be in the spatial relations they are in. The solidity of the volumes would remain unaltered. Nor would the expressive aspects of the object be seriously impaired if impaired at all by its lacking the status of an artefact. It would still possess those physiognomic characteristics which serve to make it an imposing impressive work. That it was not an artefact would not indicate that it was not a fit object for aesthetic attention.

That something is not an artefact does not suggest let alone establish that it is therefore unfit to be an object of aesthetic attention. And unless one has a compelling narcissistic obsession with the marks of men's endeavours one can view things in the world aesthetically without being concerned with or inhibited by their lack of status as artefacts.

If a work of art is a paradigm of an object fit for aesthetic attention it is not owing to the status of a work of art as an artefact. Not that just any artefact is classed a work of art: a garden rake a screwdriver a green paper plate are not though they are undoubtful examples of artefacts. What if the paper plate were on a pedestal displayed as a piece of sculpture? Would it then be classed a work of art? By some. Not by others. Even so: if one wanted an undoubtful example of a work of art wouldn't one prefer Leonardo's *Ginevra* to the paper plate? An undoubtful example of a work of art is a hand-made work a product of an art a craft: it is an artefact the production of which called for considerable and unmistakable craftsmanship. Look at Leonardo's *Ginevra*: that the craftsmanship displayed is remarkable is obvious. (And that is not belied by the fact that one may wonder whether the portrayed slight strabismus is rightly to be attributed to Ginevra herself.)

This exquisite portrait is incomparably more beautiful than any reproduction can suggest. The marvelous sense of atmosphere surrounding Ginevra, the harmonious unity of landscape and figure, and the incredible delicacy with which minute details are rendered can only be appreciated in the original painting.[1]

Reproductions rarely capture the quality of a work of art of an exquisite and refined craft. That a work does not lend itself to easy reproduction however may be owing either to its being remarkably ordered (so to speak) a product of great craftsmanship or to its being a clear manifestation of entropy. Leonardo's *Ginevra* would be difficult to copy and so would one of Pollock's· typically dribbled pieces: to smash an egg is easy but to replicate the appearance of the smashed egg in all perceivable details may be impossible.

A display of craftsmanship may on occasion facilitate aesthetic attention to an object. The lack of that display in no way indicates that an object is unfit for such attention. Consider a typical work by Piet Mondrian: one of black lines and white ground. Such a work displays virtually nothing of the painter's craft rightly so-called: a tolerably steady hand an ability to apply masking tape judiciously is about all the the technical skill required to produce it. Or to reproduce it: a perfect copy would be a matter of a few hours work at most.

That works of art may be artefacts that they may be skillfully hand-made objects here doesn't signify. Figuratively and on occasion literally speaking works of art are framed objects. It is that more than anything else that makes them plausible paradigms of objects fit for aesthetic attention. But both the efficacy and the necessity of a frame are something of an illusion.

Works of art are framed mounted hung illuminated displayed exhibited. The object is supplied with a milieu an environment a background. Presumably all that facilitates aesthetic attention to the works by those concerned to appreciate them. The basic idea would seem to be this: a person p performs certain relevant actions a in connection with a work of art an entity e under conditions c. The entity e is supposed to be of a kind or character to facilitate and make valuable the performance of a by p under c. If so e is then a fit object for aesthetic attention. And what if e is dried dung? Then the performance of actions a by person p under conditions c in connection with e the dried dung is supposed to be neither facilitated nor rendered valuable by the dried dung. Hence the dried dung is not supposed to be a fit object for aesthetic attention. But obviously all this depends on the person p the actions a and the conditions c.

Aesthetic value is as it were a cooperative affair. If attending aesthetically to an object is worthwhile then the object contributes its presence and possibly the conditions under which one attends to the object contribute their share while the person contributes his: what is wanted is an harmonious relation between the person and the object. It is never the case that such harmony depends solely on the contribution of the object. For despite its presence the conditions of attention may be infelicitous: who could enjoy viewing Klee's *Twittering Machine* while being tortured? (Perhaps a *roshi*.) If both object and conditions make their contribution something about the person may occasion a difficulty: a color blind person may be cut off from an appreciation of a Matisse nude and so conceivably could one psychologically disturbed about sexual matters.

To say of something that it is worth attending to aesthetically is to speak in an abstract way. For in so saying one abstracts from reference to persons actions and the conditions under which the actions are to be performed. On occasion this abstract way of speaking is somewhat fatuous. A case in point: 'Michelangelo's Sistine Chapel murals are worth viewing.' Presumably these are great works of art. Theoretically the viewing of these works is aesthetically worthwhile. In fact it is not. It would be worthwhile if the works were not where they are if the conditions of viewing were altered for example if the Chapel were turned on its side. Where they are high up and almost out of sight they are for all save presbyopes virtually inaccessible to the performance of any relevant aesthetic action. Viewing them is literally a pain in the neck. One can recline on a bench or the floor (if the guards permit and the spectators don't trample) but that position is not conducive to aesthetic attention. Here one should keep in mind the illusion of the full moon on the horizon: the apparent size of the moon is radically reduced by turning one's back to it bending down and viewing it between one's legs with one's head upside down. Evidently the positions in which one views things can serve to alter the apparent size of the things viewed. (It is said that Frank Lloyd Wright hated paintings: that would account for the sloping floors and tilted perspectives of the Guggenheim Museum which serve effectively to sabotage any delicately balanced work.)

A work of art is supposed to retain its identity from frame to frame wall to wall room to room: those who suffer from inept framers know how silly this view is. Seurat took care at times to prepare and paint his own frames. But he could do nothing about the walls floors company his works were forced to keep. Conversely is there any doubt that dried dung displayed by the lighting engineers of the New York Museum of Modern Art could prove to be a fantastically intriguing aesthetic object? With appropriately placed lights and shadows walls of the right tint in the right position of the right height carefully proportioned pedestals anything at all that could be displayed could be a fit object for aesthetic attention.

Would it be the dried dung or the dried dung under special environing conditions that would be a fit object for aesthetic attention? Certainly at least the latter is obviously true and I think also the former but let's focus on the latter for the moment for that's the way it always is anyway with any work of art. Works of art such as paintings and pieces of sculpture are best thought of as scores awaiting realization in actual performance. Viewing a yellow version of Josef Albers' *Hommage to the square* displayed in a yellow frame on a yellow stuccoed wall would be like listening to a Rossini overture performed *con sordini* with all instruments muted.

To say that an object is fit for aesthetic attention is not simply to say that there are or could be environing conditions under which the object would be worth attending to aesthetically. That seems plainly true (to me anyway) and not surprising in the light of twentieth century art and techniques of display. In saying that an object is fit for aesthetic attention one is saying much more namely that the object can be attended to and is worth attending to aesthetically in that such attention to the object is worthwhile and if it is not that it is not is attributable either to interference by the conditions or to something about persons or their actions.

When attention to an object is not aesthetically worthwhile it may be uncertain what the lack is attributable to. If aesthetic attention to a floating clump of seaweed was not worthwhile that may be owing to the fact that while contemplating the clump one was being savaged by a school of sharks. Here conditions may fairly be said to have interfered. But if on a cold dank winter's day in Venice one finds the contemplation of a Tintoretto in a dim unheated church not aesthetically worthwhile is the lack to be attributed to the conditions under which the work is viewed or to a failure of concentration on the part of the person?

As the character of the objects attended to vary the character of the actions the conditions and the requisite qualities skills and capacities of the person may also have to vary if attention to the objects is to be aesthetically worthwhile. Demands made on a person are absolutely minimal in the appreciation of the popular art of his own culture: soap operas rock and roll comic strips western flicks. No special knowledge is called for no special actions are wanted: not even the capacity for continued attention is requisite. (Which is not to deny that from an intercultural point of view these demands can be seen as prodigious: the wonderful world of *Barry McKenzie* a comic strip is not apt to be available to those who haven't lived among the kangabloodyroos.) Popular art is popular because it is so readily available to all within the culture. But traditional works of art such as Leonardo's *Ginevra de'Benci Mona Lisa* the madonna on the rocks Botticelli's Venus on the half shell are also popular and for much the same reason: from a western intracultural point of view an appreciation of these works calls for nothing special on the part of the viewer. The same is true of the appreciation of many carefully hand-crafted objects of many beautiful things in general.

When one turns to modern works demands on the person are apt to increase. Eliott Carter's *2nd Quartet* is a work of rare beauty but it is not instantly available to all. If one attempts in listening to the quartet to attend to recurring themes and variations as one would in listening to a work in standard sonata form then one is ready for Beethoven's *C minor Opus 18 No 4* but not for Carter: eighteen seconds of the opening *Allegro fantastico* should be enough to make that clear. Modern works of art often call for prolonged continuous close attention if one is to appreciate them. The same is true of a gator basking in the sun on a mud bank in a swamp. Anything viewed makes demands.

To suppose that anything that can be viewed is a fit object for aesthetic attention is not like supposing that anything one can put in one's mouth is a fit object to eat. It is more like supposing that anything that can be seen can be read. Because it can. It isn't true that one can't read just anything that one can see. Not everything has meaning but anything can be given meaning. One can read a blank piece of paper or a cloud or a sea anemone as some read palms and tea leaves and entrails. One can give meaning to stones but one can't make them edible. And one can see them as displays of solidity as expressive objects.

What's a fit subject to photograph? Anything that can be seen. Or is it not what the photographer photographs but what he makes of it? With his camera and darkroom and skills? What he does with art I can do with my (or maybe you too can with your) eyes. One can look at anything and within limits and depending on one's powers create an appropriate frame and environing conditions for what one sees.

I will describe what I call 'antiaesthetic litter clearance'. A nonaesthetic approach is a simple exercise in futility: the litter is offensive pick it up put it in trash cans sweep and tidy the area. Which owing to the unchanging propensities of the inhabitants will soon almost immediately be covered with litter again. The antiaesthetic approach is to alter one's view to see the original litter not as litter but as an object for aesthetic attention: a manifestation of a fundamental physical factor: entropy. One can look upon the disorder of litter as a form of order a beautiful randomness a precise display of imprecision. (And if you cannot look at litter in this way perhaps you can learn to do so by looking at Pollock Tobey and others.) Garbage strewn about is apt to be as delicately variegated in hue and value as the subtlest Monet. Discarded beer cans create striking cubistic patterns.

Consider a gator basking in the sun on a mud bank in a swamp. Is he a fit object for aesthetic attention? He is and that he is is readily confirmable. Go look and see if you doubt what I say. He is presently to be seen around Chokoloskee Island in the Everglades. What is in question is the American alligator (*Alligator mississipiensis*) not to be confused with a crocodile. Gators have shorter broader heads and more obtuse snouts. The fourth enlarged tooh of a gator's lower jaw fits into a pit formed for it in the upper jaw whereas a crocodile's fits into an external notch. It helps in viewing a gator to see it as a gator and not as a crocodile. But that requires knowing something about gators.

Seen from the side the gator appears to have a great healthy grin conveying a sense of well-being vitality. When Ginevra's portrait was painted by Leonardo she must have been sick for a long time. The pallor of her face conveys a 'sense of melancholy'.[1] The ossified scutes along his back forming the characteristic dermal armour constitute a powerful curving reticular pattern conveying simultaneously an impression of graceful fluidity and of remorseless solidity. Ginevra's face is "framed by cascading curls. These ringlets, infinitely varied in their shapes and movement, remind us of Leonardo's drawings of whirling eddies of water."[1] He has just come out of the water to bask in the sun. His sight is acute as is his power of hearing. But his eyes now have a lazy look being half-closed for he has upper and lower lids as well as a nictitating membrane. Ginevra too stares at us out of half-closed eyes. He is not strabismic. Her eyes are hazel. His seem green and remote despite the great grin.

Anything that can be viewed is a fit object for aesthetic attention. But not everything can be viewed just as not everything can be eaten. And in eating and in viewing the difficulties may be attributed either to the object or to the person. The former are obvious: stones can't be eaten and some gases subatomic particles and so forth can't be viewed because they can't be seen. But what cannot be eaten or cannot be viewed owing to the person is another matter. There are places where a rat foetus is considered a delicacy. The same is true of sheep's eye balls in aspic. In India warm monkey's brains are served up raw. Eskimoes are reported to munch with delight on deer droppings (perhaps only in times of stress). Many in my society could not ingest these items: they would be stricken with nausea in the attempt. And there are hideous offensive nauseating objects that one cannot bear to view. Are such objects fit for aesthetic attention?

Yes why not? That I am psychologically incapable of attending aesthetically to a certain object tells you something about me nothing about the aesthetic qualities of the object. The same could be true of a work of art. Suppose Derain had done an heroic portrait of Hitler: I could not attend aesthetically to that work. Hitler was a repulsive nauseating object. That nausea is readily evoked by any lifelike image of the person. But my nausea would not be a criticism of Derain's art. Many of us cannot bear to look at blood particularly our own: that is not to deny that blood may be of a beautiful color and form beautiful patterns as it flows. If there were something that no one was psychologically capable of viewing even though the object was available for viewing then one might wonder whether such a thing was a fit object for aesthetic attention. But as far as I know there is no such thing and even if there were there's no need in theory anyway to countenance a morbid sensitivity that makes one psychologically incapable of viewing something in the world.

If anything that can be viewed is a fit objects for aesthetic attention aren't some things more fit than others? No why think it? But granted that both a gator basking and Leonardo's *Ginevra* are fit objects for aesthetic attention isn't *Ginevra* more fit? No. In what way? It would make sense to compare the two only if there were some basis of comparison. But there isn't.

But isn't one painting better than another? In some ways and not in others. Rubens' paintings were superior to those of many of his contemporaries with respect to technique and pigmentation. Ingres' work displays finer draftsmanship than that of David. Vuillard's works have finer color than Manet's. But this isn't to say that Vuillard's works are more fit than those of Manet for aesthetic attention. If you are concerned to attend aesthetically to color then giving such attention to Vuillard's works will prove more worthwhile than giving such attention to Manet's works. But there are other things to attend to in viewing Manet's works. There are always other things to attend to.

For one can attend to anticolor: one can attend to precisely those aspects of hue value saturation of Manet's works which when standing on one's right foot adopting the stance of a judge one judges to be inferior to Vuillard's. And one can without losing one's balance adopt a different stance standing on one's left foot one judges Manet's color superior to Vuillard's. And one can stand squarely on both feet and abandon the silliness of aesthetic judgements.

In looking at *Ginevra* one can attend to the display of craftsmanship and the beauty of form and shape: in looking at the gator basking one can attend to the beautiful grinning display of life. Anything that can be viewed can fill the bill of an object fit for aesthetic attention and none does it better than any other. Granted that 2 3 5 7 11 and so forth are primes: are some more so than others? A monk asked Ummon: "What is Buddha?" Ummon answered him: "Dried dung."

The University of North Carolina
at Chapel Hill

NOTE

[1] Perry B. Cott, *Leonardo Da Vinci Ginevra de'Benci*, Washington D.C.. National Gallery of Art, March 1967.

VII

HISTORY OF PHILOSOPHY

G. H. VON WRIGHT

THE 'MASTER ARGUMENT' OF DIODORUS

1. One of Jaakko Hintikka's many brilliant contributions to the history of ideas and of logic is his study of the famous κυριεύων or 'Master Argument' of Diodorus Cronus.[1] The literature on this argument is vast and continues to grow. I shall here contribute to it with yet another paper. My interest in the argument was roused by reading Hintikka's essay many years ago. My treatment here, unlike Hintikka's, will be 'unscholarly'. The only serious reference to sources will be to the wording in which the argument has survived. I shall sometimes say of things which seem to *me* to make good sense that 'it is plausible to think' that Diodorus accepted them, and of certain other things that 'we have no reason' to doubt that Diodorus would have disagreed. If a reader finds such statements annoying because of lack of supporting evidence, I advise him to ignore them. My aim is not primarily to reconstruct a piece of past history but to say something of interest about things which continue to puzzle *us*.

2. *The sources.* We know the argument from Epictetus, who lived and wrote, it should be remembered, about four hundred years after Diodorus. According to Epictetus[2], Diodorus held the following three propositions or theses to be mutually incompatible:

(a)   Everything that is past and true is necessary.
(b)   From the possible the impossible does not follow.
(c)   Something is possible which neither is nor will be true.

Diodorus, we are told, accepted (a) and (b). Therefore he thought that (c) had to be rejected. Or to put it slightly differently, Diodorus thought that from the conjunction of (a) and (b) the negation of (c) follows logically. The negation of (c) is the thesis

(d)   If something is possible then it either is or will be true.

One could paraphrase this thesis as follows: Every possibility will sooner or later become real, actualize. The purpose of the Master Argument

297

E. Saarinen, R. Hilpinen, I. Niiniluoto, and M. Provence Hintikka (eds.), *Essays in Honour of Jaakko Hintikka*, 297–307. *All Rights Reserved.*
Copyright © 1979 by D. Reidel Publishing Company, Dordrecht, Holland.

was evidently to 'establish' or 'prove' (d). Ancient authors also refer to it, or some equivalent formulation, as Diodorus's 'definition' of possibility.[3] Hintikka has argued that it also corresponds with Aristotle's view of possibility. (See below p. 307)

3. *Preliminaries on propositions, time, and truth.* When discussing these matters it is important to keep clear a distinction between what I have elsewhere called *generic* and *individual* propositions. (I am not sure that the terms are altogether felicitous.) A generic proposition is not, by itself, true or false. It 'gets' a truth-value, 'becomes' or 'turns out' true or false when it is individualized or instantiated in space and time. Since different instantiations may yield different truth-values for the same generic proposition, a generic proposition is sometimes said to have a 'variable' truth-value. For present purposes only instantiation in time matters.

The schematic letter '*p*' will here be used to represent a sentence which expresses a generic proposition. The proposition could be, for example, that it is raining in Athens. This proposition is not, by itself, true or false – like the individual proposition that it is raining in Athens at time *t*, for example on 5th January 1499.

To say that it is true at *t* that *p* is equivalent with saying that it is true that *p* at *t* and this again with saying that *p* at *t*. There is an important sense in which the phrase 'it is true that' or 'it is the case that' is otiose. To say of something that it is true (false) at a certain time is tantamount to saying that a generic proposition (the 'something') instantiated at that time turns into a true (false) individual proposition. An individual proposition is true or false timelessly. ('Truth is eternal.')

To say of something that it was or will be true is to maintain that there is an instant *t* in time, past or future in relation to some designated present (a 'now'), such that a certain generic proposition (the 'something') is true when instantiated at *t*. One can also make combined uses of past and future tense when speaking about truth. Thus, for example, one may say that it will be true that it was true that *p*. This means that there is an instant *t* in time, future in relation to a chosen *now*, and another instant *t'*, anterior to *t*, such that (it is true that) *p* at *t'*. Locutions like this are important when discussing Diodorus's argument.

4. *Preliminaries about modality.* Another thing which is important to note in a discussion of Diodorus's argument is that the modal status of possibility, necessity, and impossibility can belong both to generic and to

individual propositions. 'It is possible that p' expresses a different thought from 'it is possible that p at t'.

A proposition to the effect that something is possible, necessary, or impossible can be either generic or individual. In the former case it has no truth-value by itself; it may turn out to be now true, now false when individualized relative to different stations in time. It may, for example, be true at t that it is possible that p, but false at t' that it is possible that p. In such a case one can speak of a *change or shift in modal status* of the *generic* proposition that p, consequent upon a shift in truth-value of the generic proposition that it is possible that p.

One should note that the locution that it is true at t that it is possible that p is equivalent with saying that it is possible at t that p. ('It is true that' is otiose.)

The attributions of possibility so far mentioned have been of the *generic* proposition that p. Attributions of possibility, however, can also be of the *individual* proposition that p at t. Such attributions can in their turn yield either generic or individual propositions. This means the following:

There is a tenseless sense of the phrase 'it is possible (necessary, impossible) that'. When it is conjoined with an individual proposition we get another individual proposition, true or false. Then the words 'at t'' in the sentence 'it is possible at t' that p at t' are otiose and can be omitted. The sentence means the same as 'it is possible that p at t'.

There is, however, also a tensed sense of the phrase 'it is possible (necessary, impossible) that'. When it is conjoined to an individual proposition, we get a generic proposition which is not yet true or false. In order for it to acquire a truth-value we must qualify the attribution of possibility (necessity, impossibility) temporally *e.g.* by appending the words 'at t'' to 'possible' ('necessary', 'impossible'). Then it may happen that it is true that it is possible at t' that p at t but false that it is possible at t'' that p at t. In such cases we can speak of a change in modal status of the *individual* proposition that p at t – due to a change in truth-value of the (generic) proposition that it is possible that p at t.

Saying that it was (will be) possible that p at t is saying that there is a moment in time, t', which is past (future) in relation to a designated present and which is such that it is possible at t' that p at t.

These distinctions, or some equivalent way of making them, are not useless pedantry. They are necessary for an understanding of the use of modal words in relation to temporality and truth. They are also needed for an understanding of Diodorus's argument. And this is one reason why

I find this argument challenging and interesting.

5. *Prospective possibility*. In the Diodorean view, it is possible that *p* if, and only if, it either is or will be the case that *p*. Assume that it was the case that *p* at some time in the past, but that it neither now is nor will in future ever be the case that *p*. Then, according to Diodorus, it is not possible *now* that *p*, although there was a time when it *was* possible that *p*. The modal status of the (generic) proposition that *p* has changed.

Diodorus's notion of possibility could be called a *prospective* possibility. Possibility is defined in terms of what is or will be the case; the modal status which a (generic) proposition now has may be different from the status it might have had in the past.

I think that often when we say that something is possible we mean that this thing either is there already or will *perhaps* materialize later. This too is a prospective idea of possibility. I shall call it *F*-possibility. The word 'perhaps' can, of course, be exchanged for 'possibly'.

That it will perhaps (possibly) be true that *p* means that it is possible now that *p* will, at some future time, be true. It does not mean that it will, at some future time, be possible that *p*.

Diodorus's notion of prospective possibility, however, is stronger. In his view, something is now possible if, and only if, it either is the case already or will be the case later. I shall call this *D*-possibility. Whether something is *D*-possible, or not, can of course, be a matter of conjecture. One would then entertain the proposition that *perhaps* this thing is or will be the case. But if something *is* *D*-possible, then it is *certain* that this thing is or will be the case. For the words 'it is certain that' one could also put 'necessarily'.

6. *Retrospective necessity (impossibility)*. Thesis (a) mentions the idea of necessity. Implicitly, it also involves the idea of impossibility, if we assume that to hold that everything past and true is necessary is equivalent to holding that everything past and false is impossible. This is simply the 'orthodox' relation between necessity and impossibility, and there is no reason to think that Diodorus had not accepted it.

The notions of necessity and impossibility which are involved in thesis (a) are what I propose to call *retrospective*. Unlike the Diodorean notion of prospective possibility which is an attribute of generic propositions, they are attributes of individual propositions. Assume that it is true (false) that *p* at *t*. Then, according to thesis (a), it is necessary (impossible) at any time after t that p at t. We can also express this by saying that if, in relation to an arbitrarily designated present, it *was* true (false) that p at t, then it is necessary (impossible) *now* that p at t. This is how I understand the meaning of the thesis that everything that is past and true is necessary – and I cannot think of any other plausible way of understanding it.

Prospective possibility, too, can be an attribute of an individual proposition. Assume that it is true that p at t, then, according to Diodorus, it is at least possible at any time not later than t that p. This is *D*-possibility of the generic proposition that p. But we could also say that it is possible at any time not later than t that p at t. This is not Diodorean possibility in the sense of thesis (c) but it is a perfectly respectable attribution of prospective possibility to the individual proposition that p at t.

The notions of possibility and impossibility which are mentioned in thesis (b) are less clear than those at stake in (c) and (a). It is by no means certain that they have to be understood in either a prospective or a retrospective sense. Nor is it clear whether the two modal attributes, as contemplated in (b), pertain to generic or to individual propositions or perhaps to both. But it cannot be regarded as excluded, I should say, that Diodorus in maintaining (b) was thinking of possibility in the prospective sense of thesis (c) and of impossibility in the retrospective sense implicit in thesis (a).

7. *The necessity of the past.* I have already said how I think we ought to understand the meaning of thesis (*a*). The question then arises whether we should think the thesis acceptable or not.

The idea that what is past and true is also necessary was, it seems, commonly entertained by ancient and medieval logicians. However, I am not myself aware of any recent study of its significance.

Thesis (a) does not mean that everything which actually happened, turned out to be true, did so 'of necessity'. That it is necessary (now) that it was true that p at t is logically fully compatible with holding that it *might not* have been true that p at t. This last is then understood to entail that, at some time prior to t there were two possibilities or alternative developments ahead of us, *viz.* that it was going to be the case that p at t and that it was not going to be the case that p at t. Factual developments, let us assume, 'annihilated' the second possibility. That is: at t it was (became) true that p. And, when looked at in retrospect, what was, was – helplessly and without possibility of change. *This* is what is meant by saying that it is now necessary that it was true that p at t.

Thesis (a) is not a principle of what would nowadays be called modal logic. One would rather call it a 'metaphysical' idea relating to time, causation, and necessity. To say that what is true and past is also necessary is another way of affirming what is sometimes called the *closedness* of the past. This again is related to the idea that causal relations are *directed* in time so that there cannot exist *retrospective causation*. Nothing which happens later can make a difference to what happened before; but whether what happens now is this or is that can make a great difference to what happens later.

Understood as a metaphysical idea in the above sense, thesis (a) may be debatable. But one can hardly deny that it has great intrinsic plausibility.

In our interpretation, thesis (a) is equivalent to a thesis that retrospective possibility as an attribute of an individual proposition is tantamount to retrospective necessity. One could also express this by saying that facts of history are never retrospectively contingent – except, of course, in the epistemic sense that we may not know whether it was the case that p at t and therefore wonder whether it was perhaps the case that not-p then.

8. *The problematic thesis (b).* If we interpret thesis (a) as above we run, I think, a minimal risk of not doing justice to the thoughts of Diodorus. It is much more difficult to conjecture what he might have meant by saying that from the possible the impossible does not follow. The interpretation of thesis (b) is the crux of every attempt to reconstruct and evaluate the Master Argument.

There is a straightforward interpretation of (b) as a thesis of normal modal logic. Then it says that a proposition which follows logically from a possible proposition is itself a possible proposition. Hence, if a proposition which follows from another one is impossible, the proposition from which it follows is impossible, too. This corresponds to the formula of modal logic $Mp \ \& \ N(p \rightarrow q) \rightarrow Mp$ and its equivalent variations.

There is no reason to think that Diodorus had *not* subscribed to this modal principle. He may even have (tacitly) relied on it in his proof. But it does not seem to me certain that his thesis (b) was meant to be an expression of it. Perhaps (b), like (a), should not be understood to express a law of pure modal logic at all, but rather as a 'metaphysical' idea concerning possibility and the other modalities. If that is so, then it is natural to understand (b) as a principle concerned with what I called above *change in modal status*. It does not seem to me unreasonable to think that thesis (b) should be understood to say the following: If, at any time, the proposition that p at t is prospectively possible, then this same proposition cannot, at some other (later) time be retrospectively impossible.

An interpretation of (b) *roughly* to this effect was suggested by E. Zeller in a paper from the year 1882.[4] Zeller's interpretation has been contested on philological grounds. The verb ἀκολουθεῖν, scholars maintain, is never used in the Stoic tradition to denote *temporal* succession. I have no competence to contradict them and side with Zeller, himself a formidable scholar in these matters. But I should like to add that interpreting (b) as suggested above does not necessarily mean giving to the verb ἀκολουθεῖν a temporal connotation. On the suggested interpretation, thesis (b) only says that one cannot from assumptions which involve that something is (prospectively) possible establish *by logical argument* that this same thing is (retrospectively) impossible.

9. *Reconstruction of the argument.* The aim of the argument is clear. It is to establish that prospective possibility as an attribute of a generic proposition must amount to D-possibility. An attempted reconstruction of the proof should, I think, proceed in two steps. The first step is relatively uncontroversial, the second is less so.

Assume thesis (c). It says that, for some 'p', it is now possible that p but that it neither is nor will be the case that p. Then it follows by virtue of thesis (a) that at any time t after the present time it will be impossible that p (was true) at a time not earlier than the present time but earlier than t. (If we assume that time is a discrete ordering of moments, we could simplify the formulation 'a time not earlier than the present time but earlier than t' to '$t - 1$'. But I prefer not to make this simplifying assumption here.)

In order to complete the proof we shall have to invoke thesis (b). How this is to be done is not clear, since the interpretation of (b) is controversial. This much is certain, however: Diodorus must have thought that the assumption that it is now possible that p contradicts the conclusion that at any future time t it will be impossible that p (was true) at a time not earlier than the present but earlier than t. Because had he not thought this, he could not have thought that his Master Argument was valid either.

If thesis (b) was meant to exclude a change in modal status from prospective possibility to retrospective impossibility, Diodorus could have established the contradiction as follows: If it is now possible that p then, for some t not earlier than the present moment, it is (now) possible that p at t. But we have shown that, for any moment after such a moment

t, it is impossible that p (was true) at t. Hence there will exist a moment t in time such that it is possible before this moment t in time that p at t but impossible after it that p at t. This, by virtue of (b), is a contradiction, something which cannot be. Consequently, we must reject thesis (c), *i.e.* we must accept that either it is not possible now that p or it is not the case that it neither is nor will be the case that p. This means: if it is possible now that p, then it either is or will be the case that p.

If again we interpret thesis (b) as the principle of modal logic which says that a possible proposition cannot entail an impossible one, then we could attribute to Diodorus the following piece of reasoning: It is impossible that it is now possible that p but at any later time t impossible that p at a time between now and t. But this impossibility followed, by sound logical argument, from assuming (a) and (c). Hence, by thesis (b), the conjunction of (a) and (c) must be an impossibility, too. If the conjunction is impossible, it is *a fortiori* false. This means: if it is possible now that p then it either is or will be the case that p.

To this piece of reasoning Diodorus would in all probability have subscribed. He would perhaps have thought it trivial and not worth spelling out in detail. The non-trivial part of the reasoning is the starting point, *viz.* that the conclusion drawn 'by sound logical argument' from (a) and (c) is an impossibility. Either Diodorus took this simply for granted when conducting his argument, or he did not do this. If the former, the role of (b) was merely that of the law of modal logic in question, *i.e.* it was a rather trivial role. If the latter, then the role of (b) was not just that of a principle of pure logic but was that of some more substantial 'metaphysical' principle concerning change in modal status. The second alternative seems to me intrinsically much more likely than the first.

10. *Examination of the argument.* To assume (c) is to assume two things, *viz.* that it is possible that p *and* that it neither is nor will be the case that p. I shall refer to these two assumptions as the two 'halves' of the thesis (c).

In the proof we derive from the second half of (c) with the aid of (a) the conclusion that, for any t not earlier than the present moment, the proposition that p at t will be retrospectively impossible after t. This conclusion is thought to contradict the first half of (c).

Let us for a moment forget about the Master Argument. Assume that, as a matter of fact, it *is false* that p at t, but that it *might have been true* that p at t. There is nothing *logically* wrong with this assumption. It can also be expressed by saying that it is *contingently false* that p at t. Speaking in our terminology of 'prospective' and 'retrospective' modal status, our assumption entails that, at some time anterior to t, it was prospectively possible that p at t. Accepting the thesis that what is past and true is necessary, our assumption further entails that, at any time after t, it will be retrospectively impossible that p at t.

A similar argument can be used for showing that the assumption that it is *contingently true* that p at t entails a change from prospective possibility to retrospective impossibility in the modal status of the proposition that $\sim p$ at t. This is a characteristic of contingent propositions. If such a change is excluded, the proposition that p at t cannot be contingently true or contingently false. It can only be either necessarily true or necessarily false (impossible).

We may now return to Diodorus. To think that the conclusion drawn from the second half of (c) with the aid of (a) contradicts the first half of (c) is equivalent to thinking that any proposition of the form 'p at t' is (prospectively) either necessary or impossible. To think this is tantamount to subscribing to a version of *Universal Determinism*.

Shall we say then that the Master Argument, as it stands, is inconclusive and that it depends for its conclusiveness on a concealed or suppressed assumption (premiss) about determinism? One must be cautious in saying this. For what is the argument 'as it stands'? As long as the meaning and role of thesis (b) in the argument remains obscure we cannot tell this. If (b) is simply the law of modal logic saying that a proposition which follows from a possible proposition is itself possible, then the argument is indeed inconclusive and dependent upon a concealed deterministic premiss. But I do not think it implausible to understand (b) in such a way that in fact it supplies the needed premiss. *On this interpretation thesis (b) is an oblique way of stating a thesis of determinism.* And, for all we know, Diodorus *was* a convinced determinist.

11. *Determinism, Contingency, and Diodorean Possibility.* There are many versions of the idea of determinism. But it is feasible to think that on any such idea the proposition that p at t is at any time either prospectively necessary or prospectively impossible. If this is so and if it neither is nor will be the case that p, then the proposition that p at t is prospectively impossible at any time before t. If follows that, if the generic proposition that p is, at some time, prospectively possible, then it either is at that time or will be at some later time the case that p. If determinism reigns, prospective possibility is Diodorean, D-possibility.

If it is *D*-possible that *p* then *perhaps* it is already the case that *p*, or perhaps not, or perhaps it will tomorrow be the case that *p*, or perhaps not, ... *certain* (necessary) is only that at some time, either now or in the future, is it the case that *p*. Here 'perhaps' is an *epistemic* notion of possibility (contingency) which we employ when we want to indicate that we do not *know* whether something is or is not the case. In a deterministic universe contingency can only be epistemic.

Determinism commits to *D*-possibility. The converse, however, does not hold. One can adhere to the view that prospective possibility is (must be) *D*-possibility without adhering to determinism. If it is *D*-possible that *p*, then it is now necessary that, for some not-past *t*, *p* at *t*. Let *t'* be such a not-past *t*. Then it is true that *p* at *t'*. But it does not follow that it is now prospectively necessary that *p* at *t'*. It can be prospectively possible both that *p* at *t'* and that ~*p* at *t'*. It is necessary only that for *some* not-past *t*, it will be true that *p* at *t*.

12. *Comparison with Hintikka's treatment.* I shall end with a comparison between Hintikka's reconstruction of the Master Argument and my own. In order to make the comparison perspicuous, I shall first paraphrase my reconstruction from Sections *9* and *10* above as follows:

(1) Assume that it is possible that *p*. (2) From this it follows that, for some future time *t*, it is possible that *p* at *t*. (3) But from Diodorus's third thesis it follows that it is false that *p* at *t*. (4) From his first thesis again it follows that, for any time after *t*, it will be impossible that *p* at *t*. (5) By virtue of his second thesis, (4) contradicts (2) and *a fortiori* (1) too. Hence the set of the three theses or propositions (a)–(c) is inconsistent, and (a) and (b) conjunctively imply (D).

Hintikka's reconstruction goes as follows: (1) Assume that it is possible that *p*.[5] (2) Assume further that, for some future time *t*, (it is *true* that) *p* at *t*. (3) From Diodorus's third thesis it follows that it is false that *p* at *t*. (4) From his first thesis again it follows that, for any time after *t*, it will be impossible that *p* at *t*. (5) Since (2), or the assumption that the possibility mentioned in (1) is realized, thus leads to an impossibility, (4) contradicts (1) by virtue of Diodorus's second thesis.

I hope that my paraphrase does justice to Hintikka's chain of reasoning. As seen, there is one substantial difference between the two reconstructions. It lies in the passage from (1) to (2). I pass from the assumption that it is possible that *p* to the consequence that, for some future *t*, it is *possible* that *p* at *t*. Hintikka proceeds from the assumption that it is possible that *p* to the further assumption that this possibility is realized at some future *t*. As a consequence, the second thesis of Diodorus is assigned a somewhat different role in our respective reconstructions.

Hintikka's move from (1) to (2) and the role it plays in his argumentation is motivated by his reading of the passage 32a18–20 in the *Prior Analytics*. If Hintikka's reading is correct, then the passage in question amounts, practically speaking, to a proof that Aristotle's notion of possibility was Diodorean. This is a view for which Hintikka has argued forcefully in several publications. I shall not here discuss whether he is right. Let it only be said that my own reconstruction of the argument is neutral with regard to disputed questions of interpreting Aristotle.

Academy of Finland

NOTES

[1] Jaakko Hintikka, 'Aristotle and the "Master Argument" of Diodorus', *American Philosophical Quarterly* 1 (1964). As the title indicates, the paper is also concerned with interpreting Aristotle. The present essay is not concerned with this.
[2] Epictetus, *Dissertationes* II, 19: Κοινῆς γὰρ οὔσης μάχης τοῖς τρισὶ τύτοις πρὸς ἄλληλα,
(a) τῶ Πᾶν παρεληλυθὸς ἀληθὲς αναγκαῖον εἶναι,
(b) καὶ τῷ Δυνατῷ ἀδύνατον μὴ ἀκολουθεῖν,
(c) καὶ τῷ Δύνατον εἶναι ὃ οὔτ ἔστιν ἀληθὲς οὔτ ἔσται.
The letters *a*, *b*, *c* are my insertion. Translations vary slightly; the correct rendering of ἀκολουθεῖν in (b) has been a matter of scholarly dispute.
[3] Alexander, *Commentarium in Aristotelis Analyticorum Priorum Liber I*, ed. Wallies, p. 183–184. Boethius, *Commentarii in Librum Aristotelis Περὶ ἑρμηνείας*, ed. secunda, Meiser: "Diodorus possibile esse determinat, quod aut est aut erit."
[4] E. Zeller, 'Über den κυριεύων des Megarikers Diodorus', *Sitzungsberichte der Königlichen Preussischen Akademie der Wissenschaften*, 1882.
[5] The steps (1)–(5) in my paraphrase of Hintikka's reconstruction answer, with minor changes, to the steps on p. 6 of his paper from (10) through (10)[+], (11)[+], and (13)[+] to the conclusion that the set of three theses (a)–(c) is inconsistent.

SIMO KNUUTTILA AND ANJA INKERI LEHTINEN

PLATO IN INFINITUM REMISSE
INCIPIT ESSE ALBUS

*New Texts on the Late Medieval Discussion on the Concept of
Infinity in Sophismata Literature*

I

In the history of western philosophy there has never been as long a period
of intensive study of logic and conceptual analysis as there was in the
Middle Ages. In recent years the end results of this tradition, as they are
to be seen in fourteenth-century philosophical logic and theory of science,
have become an object of lively interest. The first results of this new interest
have shown a perhaps surprisingly high standard of analysis of problems
which have much in common with some of the basic problems of modern
research on logic and theory of science. It suffices to mention as examples
the problems, much discussed in the fourteenth century, of modal logic,
epistemic logic, and deontic logic, questions of intensional identity and
methods of individuation, the explanatory nature of scientific sentences
and so on. Until recently many of these topics have been studied only
occasionally, if at all.[1]

One traditional working method of medieval philosophy was to discuss
sophisms, *i.e.*, sentences the analysis of which caused exceptionally great
problems. It was customary to separate true and false readings of sophisms
and especially in the fourteenth century the different readings were
accepted or refuted on basis of elaborate syntactical and semantical rules.
Thus analytical languages were created for different topics.[2]

One collection of sophisms typical for the first part of the fourteenth-
century English philosophy was written by Richard Kilvington, a member
of the group now known as the Oxford Calculators.[3] The first part of
Kilvington's collection was concentrated on problems caused by concepts
of beginning and ceasing (*incipit, desinit*). Moreover, the problems Kilvington discussed were chosen in such a way that in their treatment an
analysis of problems involving comparative terms, infinitesimals, limits,
and the continuum was needed. These logico-mathematical concepts
were much discussed in that time by other even better known authors, too. A
good introduction to this exciting area is offered by Curtis Wilson in his book
William Heytesbury: Medieval Logic and the Rise of Mathematical Physics.[4]

*E. Saarinen, R. Hilpinen, I. Niiniluoto, and M. Provence Hintikka (eds.), Essays in
Honour of Jaakko Hintikka, 309–329. All Rights Reserved.*

In the second part of this paper we present a new series of medieval texts connected with the problematics just described. To this series, preserved in the hitherto unanalysed MS *Uppsala*, Universitetsbiblioteket C 640, ff. 87 ra – 102 vb[5], belongs also a commentary on the Sophisms of Richard Kilvington with the heading: *Declaratio sophismatum climitonis*. An edition of this text is published for the first time in the third part of this paper. Before that we shall discuss the first sophism of Kilvington and some later commentaries on it, or more exactly, a corollary to it presented by Kilvington in his second sophism. Three different treatises in the *Uppsala* MS take up the same problematics. In addition, the discussion concerns also the comments given by Gregory of Rimini in his Commentary on the Sentences.

Even if we cannot give a definite identification for the anonymously preserved *Declaratio*, there are some reasons to believe that this text has either been influenced by the now lost *Sophismata abbreviata Kylmynton*, composed by the well-known Merton School logician Richard Billingham, or it might even be a fragment or an epitome of Billingham's lost work itself.[6] In fact, there are no *a priori* reasons for denying the latter alternative. Anyway, the author of the *Declaratio* devotes to the first sophism of Kilvington a detailed discussion and shows how the original question can be answered in terms of the medieval theory of supposition. This was one of the most frequently used tools of logical analysis in the fourteenth-century philosophical logic. When a distinction is made between a merely confused supposition and a determinate supposition of a term in sophisms, in the former case the sentence is analysed into a form in which the problematic term is substituted in the sentence by a term-disjunction. In the latter case, where we deal with a determinate supposition, the sentence is analysed into a disjunction of singular sentences. The different readings explicated in this way often resemble those which in modern logic are obtained by changing the order of operators.[7]

The first sophism of Kilvington is rather uncomplicated. It has been discussed by Francesco Bottin (with an edition of the Latin text on basis of 12 MSS) as well as by Norman Kretzmann (with an English translation of the text).[8] The sophism runs as follows: 'Socrates is whiter than Plato begins to be white'. It is here assumed that Socrates is white in the highest degree, Plato now first begins to be white, and there is some time before Plato will be as white as Socrates now is. It is also supposed that the increase of whiteness as well as time are continuous. The task then is to compare Socrates' whiteness to the incipient whiteness of Plato. Because of the continuous latitude of whiteness any determinate proportion between Socrates' whiteness and Plato's incipient whiteness can be greater and therefore Kilvington says that the proportion cannot be finite. One could perhaps say that it is no determinate finite proportion and that it increases towards infinity when we go towards the incipient whiteness of Plato. Then Kilvington would have had the Aristotelian concept of potential infinity in his mind when he says that Socrates is infinitely whiter than Plato begins to be white. Because in this analysis there is no determinate infinitely small degree of whiteness, Plato's incipient whiteness cannot be identified with any certain degree of whiteness. Thus the verb 'begin' in this case must be analysed so that Plato now is white and after the present instant Plato will be white and there will be no instant such that between that instant and the present instant Plato will not be white. If the limit of Plato's whiteness were intrinsic then he would now have the first determinate degree of whiteness, which would be in contradiction with Kilvington's claim that there is not first, least degree of whiteness.

In his second sophism 'Socrates is infinitely whiter than Plato begins to be white' Kilvington discusses the problem in what sense the solution of the first sophism implies the reverse sentence, which gives the incipient whiteness of Plato in the sense of infinitely smaller whiteness than that acquired by Socrates. He accepts the sentence in the form: 'Infinitely Plato begins to be less white than Socrates now is white' but denies it in the form: 'Plato begins to be infinitely less white than Socrates now is white'. Kilvington says that the latter reading is not acceptable because there is no first degree of whiteness by which Plato will be white. The former reading is correct because more than two times less Plato begins to be white than Socrates now is white, more than three times less, and so on *in infinitum*.

This solution of Kilvington's seems to follow the usual distinction between the syncategorematic sense and the categorematic sense of the term 'infinite'. If the term is in the beginning of the sentence, it has a syncategorematic sense, *i.e.*, it is a term which has no signification taken by itself, but which has a consignification together with the subject and the predicate. In the above sentence it has the effect that the whiteness which Socrates begins to be has a merely confused supposition in the sense that it can be made smaller *in infinitum*. The term 'infinite' belongs to the terms which have a force of confounding (*vis confundendi*) a term which follows it and thus gives to it a merely confused supposition, in this case to the whiteness which Plato now has not and immediately after now will have. When the term 'infinite' is preceded by the term it modifies, it is used in a categorematic sense and then it gives a determinate supposition to the term it modifies, in this case to the effect that there is a certain infinitely small degree of whiteness, which is the first whiteness Socrates will have.[9]

Kilvington is not so explicit in his discussion of the sophism, but we can find this explanation in the distinction 17 of the first book of Gregory of Rimini's Commentary on the Sentences, where he discusses several of Kilvington's sophisms. Gregory says that from the sentence '*a* is infinitely whiter than *b* begins to be white' it does not follow '*b* begins to be infinitely less white than *a* is white' but it does follow 'infinitely less white *b* begins to be than *a* is white'. When the term '*in infinitum*' in the first sentence is placed after the expression '*b* begins to be white' it, according to Gregory of Rimini, refers to a certain degree of whiteness labelling it as the incipient whiteness infinitely less than *a* is white. If the term '*in infinitum*' is in the beginning of the sentence, it means that *b* begins to be more than two times less white, more than three times less white, and so on *in infinitum*. In this case the whiteness implied in the explication of *b*'s beginning to be white is preceded by the term '*in infinitum*' which has a *vim confundendi* the incipient whiteness which thus has a merely confused supposition. In Gregory of Rimini's next example this is stated as follows:

Et ad probationem, posito casu, concedenda est hec: in infinitum minor pars albedinis quam est hec tota albedo completa acquisita in *c*, fuit acquisita post *b*. Ista tamen non est concedenda: aliqua pars albedinis in infinitum minor est quam hec tota albedo complete acquisita in *c*, fuit acquisita post *b*. Prima enim est vera, quia plus quam in duplo minor pars quam est hec tota albedo fuit, et cetera, et plus quam in triplo minor, et sic in infinitum. Unde ista est universalis, et iste terminus minor pars supponit confuse tantum, sed secunda est particularis et ly "minor pars" supponit determinate, eo quod non sequitur terminum confundentem, sicut in prima.[10]

In the above interpretation we have supposed that Kilvington as well as Gregory of Rimini make use of the concept 'infinite' in the potential sense. It should, however, be noticed that Kilvington was taken by his contemporaries as a representative of the doctrine of actual infinity and that Gregory of Rimini is known as the most radical infinitist of the fourteenth century writers.[11] Although Gregory in this context does not explain how his solution is compatible with his doctrine of infinitely small entities, the point can be seen in the following way. If Gregory thought that the explanation of the first reading mentioned above can be thought as an actually infinite disjunction of degrees of whiteness of which *b* begins to be less white, then it seems that the incipient whiteness is less than all finite degrees of whiteness, although it cannot be zero. Now we know that Gregory of Rimini really had such a theory of infinitely small parts of a *continuum* and that he denied them to be any determinate *minima*.[12] These transfinite parts can be of different size depending on the proportion of the division which is continued *in infinitum*. Thus the concept of an infinitely small whiteness must in a sentence have only a merely confused supposition.

Some closely parallel ideas are developed in the question *Utrum universalis affirmativa debet exponi per eius indefinitam affirmativam et eius universalem negativam* (MS *Uppsala* UB C 640, ff. 87 ra – 88 vb). The writer says that if you are speaking about an actually infinite multitude, you must add the term *finitum* to the context in order to show that it exceeds all finite multitudes:

Pro quo nota, quod si tenetur, quod aliqua multitudo sit actu infinita, tunc oportet addi illam particulam finita data, verbi gratia, infinita multitudo est partium in continuo, id est, qualibet multitudine data finita, adhuc maior est in continuo multitudo ipsarum partium. . . .

Correspondingly he explains the problem of the incipient (or desinient) whiteness of Socrates:

Eodemmodo de ista: Sortes in infinitum remissam albedinem habebit, id est, infra omnem gradum albedinis adhuc remissiorem habebit, vel quantolibet gradu albedinis dato, et cetera.

It might be convenient to mention here that the way in which the infinitists of the fourteenth century are speaking about transfinite entities has some salient similarities with the intuitive ideas involved in the modern non-standard analysis developed by Abraham Robinson. Correspondingly the way in which fourteenth-century supporters of the potential infinite are operating with the order of operators while discussing the problems connected with infinity seems to be parallel to the usual ε, δ-approach.[13]

If we compare these treatments to that presented by the author of the *Declaratio* in the *Uppsala* MS, it seems that what Richard Kilvington and Gregory of Rimini took as a correct reading is here refuted and what they took as a wrong reading is here accepted as a correct one. The author first discusses the sentence: *Plato in infinitum remisse incipit esse albus*, which is resolved as follows: *Quantalibet albedine data remissionem Plato incipit habere.* This seems to be very similar with what Gregory of Rimini offers as an explanation of what he takes to be an acceptable reading: more than two times less white *b* begins to be white than *a* is white, more than four times less white, and so on *in infinitum*. In the *Declaratio* the above reading is however refuted as well as the following: *In infinitum remissam albedinem Plato incipit habere*. On the other side, the reading accepted in the *Declaratio*, namely: *Plato incipit habere in infinitum remissam albedinem* is very close to such a sentence which is refuted by Kilvington and Gregory of Rimini.

As a reason for his analysis different from that of Richard Kilvington and Gregory of Rimini the author of the *Declaratio* refers to the order of the terms 'whiteness' and 'begin'. He thus wants to put forward a rule to the effect that similarly as the term 'infinite' has a *vim confundendi* the term which follows it, so also 'begin' has a *vim confundendi* the term it precedes which then has a merely confused supposition. If the term in question precedes the verb 'begin', it has a determinate supposition. This rule seems to be in conflict with that concerning the term 'infinite' in sentences, in which the expression *in infinitum remissa albedo* is placed before the verb 'begin', because in that case a certain degree of whiteness would be, after all, designated as the incipient whiteness. This is why the author refutes the sentences *In infinitum remissam albedinem Sortes incipit habere* and *Plato in infinitum remisse incipit esse albus*, which he analyses into the form *Quantalibet albedine data remissionem Plato incipit habere*.

The author thinks that this difficulty is avoided if the sentence is written as follows: *Sortes incipit habere in infinitum remissam albedinem*. He says that the whiteness now has a merely confused supposition, because it comes after the verb 'begin'. And apparently the term 'infinite' is thought to be used in the syncategorematic sense, so that it is in the beginning of one of the sentences in which the original sentence is exposed: Socrates is not white and immediately after now he will be white and infinitely there will be a lesser degree of whiteness than any achieved degree of his whiteness.

One could ask, of course, whether this solution is correct, because it *prima facie* seems that the term 'infinite' now has a categorematic sense at least on the surface level and thus the problem is only pushed elsewhere. One could also ask, whether the analysis of the first sentence *Plato in infinitum remisse incipit esse albus* is correct, because the whiteness does not occur in the sentence before the verb 'begin'. Literally taken the analysis offered in the *Declaratio* would then suit only to the sentence *In infinitum remissam albedinem Plato incipit habere*. If the rules concerning the confounding force of the terms 'infinite' and 'begin' are taken in their face value, then one should, perhaps, guarantee the syncategorematic sense of the term 'infinite' by placing it in the beginning of the sentence and in order to avoid a determinate supposition of 'whiteness' let the verb 'begin' precede it. Then the correct form would be *Infinitam remissam Sortes incipit habere albedinem*. Now these doubts and this solution are presented by the anonymous author of the treatise *De immediate*, occurring on ff. 93 rb – 94 vb in the *Uppsala* MS.

Infinitam remissam albedinem Sortes incipit habere. Falsum est istud sophisma, quamvis multi concedunt ipsum, et non apparet prima vice ex eo, quia ly 'albedinem' stat confuse tantum in prima, sed tamen si exponitur primo et secundo modo resolvitur vel inducitur, tunc liquet quod est falsum. Sortes incipit habere infinitam remissam albedinem. Falsum est ut primum. Infinitam remissam Sortes incipit habere albedinem. Istud est verum et non differt a primo, nisi quod ly 'albedinem' in tertia postponitur ly 'incipit' et in prima preponitur illo termino 'incipit' et in prima oportet dare certum suppositum post expositionem et in tertia non oportet, igitur.

The difficulties which late medieval logicians encountered when analysing the sentence '*Plato in infinitum remisse incipit esse albus*' may shed light to the problem why the promising tradition of medieval philosophical logic was broken so drastically at the end of the medieval period. In its attempt to formulate the semantics of natural language medieval logic provided interesting analyses of problems which only recently have started to be studied by extending the methods of modern logic into the treatment of natural language.[14] Instead of applying abstract models to natural language late medieval logic tried to extend the domain of formal logic directly inside of natural language. Thus the study of the semantical presuppositions of natural language did not have any barrier against the tendency of the rules to become more and more complicated.

II

The MS *Uppsala*, Universitetsbiblioteket C 640 originates in late fourteenth-century Prague and contains in its 110 ff. several series of texts on medieval logic and natural philosophy.[15] It belongs to the material brought to Sweden by some northern students at the University of Prague in the turn of the fourteenth and the fifteenth century. Three dates occur in the first part of this MS in the end of the works of Marsilius of Inghen: fol. 16 r 'et sic est finis. anno domini 1388° in proximo sabbato ante festum sancti iohannis baptiste', fol. 19 r 'et sic est finis ampliationum Marsilii anno domini 1388° in vigilia sancti iohannis baptiste' and on fol. 23 r giving also the place 'et sic est finis liber(!) appellationum. finitus prage anno domini M°C°C°C°88° post festum sancti iohannis baptiste'. Because texts written by the same hands are found also in the later parts of this MS, it seems safe to date the complete MS in 1388/90, with the possible exception of some of the notes and divisions added on vacant ff. This places for instance the works of Thomas Maulefelt clearly on the side of the fourteenth century[16] and gives some grounds for dating the works of such unresearched logicians as Thomas de Clivis and Hugo (Capellanus?)[17].

The *Declaratio sophismatum Climitonis* belongs to a series of texts which show a strong influence of the English logic in Prague. The following is an analysis of the contents of this series, occurring on ff. 87 ra – 102 vb of this MS:

Ff. 87 ra – 88 vb Questio
Inc. Utrum universalis affirmativa debet exponi per eius indefinitam affirmativam et eius universalem negativam. Quod sic dicit magister byligam. Quod non, probatur, quia sequitur, quod quelibet universalis affirmativa implicaret contradictionem. . .
cit. 'byligam' (= Richard Billingham)

Ff. 88 vb – 89 vb De gradibus comparationis
Inc.: SEQUITUR DE GRADIBUS comparationis. Notandum est primo, quod quandocumque positivus ponitur primo loco, non reddit propositionem exponibilem, sed resolubilem. . .
cit. 'mawluelt' (= Thomas Maulefelt)

Ff. 89 vb – 90 va De comparativo
Inc.: SEQUITUR DE COMPARATIVO. Notandum est primo, quod comparativus habet tres exponentes. Prima ponit primum comparatorum in positivo, secunda ponit secundum comparatorum in positivo, tertia abnegat equalitatem. . .
cit. 'biligam' (= Richard Billingham)

Ff. 90 va – 91 vb De superlativo
Inc.: ⟨U⟩ TRUM SUPERLATIVUS DISTRIBUITUR. Quod sic patet ⟨secundum⟩ auctorem confusionum. Et probatur ratione, quia ad propositionem. . .
cit. 'auctor confusionum', 'anglici'

Ff. 91 vb – 92 va De maximo et minimo
Inc.: SEQUITUR DE MAXIMO et minimo. Nota primo, quod duplex

est potentia, scilicet activa et passiva. Activa ⟨est⟩ potentia transmutandi principium alterum ut alterum, sed passiva est transmutandi alterum ab altero...

Ff. 92 vb – 93 rb De incipit et desinit

Inc.: ⟨C⟩IRCA ILLA VERBA Incipit et desinit primum notabile est illud, quod primum instans esse et ultimum instans non esse...

Ff. 93 rb – 94 vb De immediate

Inc.: ⟨S⟩EQUITUR DE ILLO TERMINO immediate. Et ponitur linea per quam designatur tempus et per litteras instantia et est presens instans .a.b.c.d.e. Nota primo, quod immediate...

In the end of this text there is a reference to the sophisms of Richard Kilvington: 'Istis habitis magister declarabit aliqua sophismata climitonis que patent in textu suo'.

Ff. 94 vb – 95 rb De aliud

Inc.: ⟨S⟩EQUITUR DE ALIUD. Sequitur de istis dictionibus 'aliud', 'non idem', 'differt'. Sortes differt a Platone exponitur sic: Sortes est et Plato est et Sortes non est Plato, igitur...

Ff. 95 rb – 96 rb De exclusivis

Inc.: ⟨D⟩E EXCLUSIVIS DICENDUM EST. Nota, quod propositiones de signis exclusivis sunt in quadruplici differentia. Primo, quod negationem habent ante signum exclusivum, ut non tantum homo currit...

cit.: 'maweluelt' (= Thomas Maulefelt)

Ff. 96 rb – 96 vb De exceptivis

Inc.:⟨S⟩EQUITUR DE EXCEPTIVIS. Omnis homo preter Sortes currit. Exponitur communiter sic: Omnis homo alius a Sorte currit et Sortes non currit, igitur...

Ff. 96 vb – 97 vb De reduplicativis

Inc.: ⟨D⟩E REDUPLICATIVIS. Nota primo, quod propositiones reduplicative sunt in quadruplici differentia, quia alique habent negationem ante signum, alique habent negationem post signum, alique utroque modo...

Ff. 97 vb – 98 vb De termino officiali

Inc.: ⟨D⟩E TERMINO OFFICIALI. ET capiam textum byligam pro me. Contra primum terminorum modum officialium sequitur, quod illi termini videre, stare essent termini officiales...

cit. 'byligam' (= Richard Billingham), 'burlaus' (= Walter Burley)

Ff. 99 ra – 101 ra ⟨De obligationibus⟩
Inc.: Circa obligatoria primo queritur, quid sit subiectum in obligatoriis. Non ly 'obligatio' cum obligatio est actus in disputationem obligisticam ortus tamquam in finem, igitur non est subiectum in illa arte. . .
The text ends with the name 'clapstert', probably referring to the name of its author.
cit. 'marsilius' (= Marsilius of Inghen), 'hollandrinus' (= Johannes Hollandrinus), 'anglici'

Ff. 101 ra – 102 vb Declaratio sophismatum Climitonis
Inc.: SEQUITUR DECLARATIO sophismatum climitonis. Sortes est albior quam Plato incipit esse albus stante casu sophismatis quod Sortes sit summe albus et quod Plato incipiat dealbari. . .

Fol. 102 vb Two short notes
Inc.: A parte disiunctivi ad totum valet consequentia. Et a tota disiunctiva ad partem non valet consequentia. . .
Inc.: Item. De numero exclusivarum quedam est simpliciter affirmativa. . .

III

The *Declaratio sophismatum Climitonis* deals with the following sophisms of Richard Kilvington: 1–14, 17, and 19–23. These numbers correspond with those given by Curtis Wilson in his list of Kilvington's sophisms.[18] The transcription is based on the MS *Uppsala* UB C 640. No other MSS of this work have so far been discovered.

SEQUITUR DECLARATIO SOPHISMATUM CLIMITONIS.

Sortes est albior quam Plato incipit esse albus stante casu sophismatis quod (*fol. 101 rb*) Sortes sit summe albus[1] et quod Plato incipiat dealbari per remotionem de presenti.

Tunc conceditur sophisma et exponitur sic: Sortes est albus et Plato incipit esse albus et Plato non incipit esse ita albus sicut Sortes est albus. Et sic exponitur ratione primi termini mediati, scilicet ratione ly '*albior*' et prima et secunda ex casu patent, et si negatur tertia, arguitur: da oppositum: Plato incipit esse ita albus sicut Sortes est albus, igitur Plato immediate post hoc habebit summam albedinem, igitur.

Contra istam expositionem arguitur sic: Sequitur, quod per comparativum posset comparari futurum tempus ad presens et per consequens non presupponit suum positivum in utroque extremorum, quod est contra prius dicta, quia ibi solum ponitur Sortem esse album et non Platonem.

Respondetur, quod comparativus in utroque extremorum presupponit suum positivum, prout copula propositionis requirit sibi unum de presenti et aliud de futuro.

Sed contra. Si Sortes est albior, et cetera, vel hoc est finite vel infinite. Si finite, sequitur, quod Plato inciperet esse albus in aliquo certo gradu, quia si Sortes est, exempli gratia, albus ut quattuor, tunc sic Plato inciperet esse albus sub quadruplo, quod est falsum, et igitur dicitur quod infinite Sortes est albior, et sic Sortes in infinitum est albior quam Plato incipit esse albus. Conceditur secundum casum predictum[2] et resolvitur sic: Quantalibet albedine data ultra istam Sortes est albior quam Plato incipit esse albus. Et exponitur sic: Ultra aliquam proportionem Sortes est albior quam Plato incipit esse albus et nulla proportio albedinis est, quin ultra istam Sortes est albior quam Plato incipit esse albus.

Contra. Si in infinitum Sortes est albior, et cetera, igitur ibi non est comparatio certi gradus albedinis ad certum gradum. Tunc arguitur sic: Vel hoc est quod ista proportio est infinita ex eo, quod Sortes infinite vel quod Plato in infinitum remisse incipit esse albus.

Respondetur, quod ex parte secundi, scilicet remissionis. Et exponitur: Quantolibet gradu et remissius, et cetera.

Plato in infinitum remisse incipit esse albus.

Sophisma est falsum et resolvitur: Quantalibet albedine data remissionem Plato incipit habere. Et exponitur sic: Aliquanta albedine data remissionem Plato incipit habere et non aliquanta albedo est, quin adhuc, et cetera. Et tunc primo inducitur sic: Ista albedine data remissionem incipit habere et tunc oportet dari certa albedo, quia ibi demonstratur aliquid per ly '*istud*' et per consequens Plato habet vel habebit aliquando in[3] infinitum[4] remissam albedinem. Et si[5] dicitur quod numquam habebit, tunc numquam incipit habere.

In infinitum[6] remissam albedinem Sortes incipit habere.

Sic resolvitur ut prius et exponitur ut prius et iterum est falsum, quia oporteret, quod daretur albedo infinite remissa per deductionem prius factam.

Sortes incipit habere in infinitum remissam albedinem.

Conceditur et resolvitur et exponitur ratione ly '*infinitum*' et semper ly '*albedinem*' stat confuse tantum.

Sed contra. Sicut in propositionibus precedentibus confusio istius quod est '*albedinem*' est evacuata per deductionem, quod tandem deveniebatur (*fol. 101 va*) ad aliquod determinatum, sic etiam hoc potest evacuari.

Respondetur, quod posset evacuari, sed tamen nullum sequitur inconveniens quod in primis fuit inconveniens, quia in istis ly '*albedinem*' sequitur ly '*incipit*' et sic semper stat confusio, et si evacuatur, difficultas illius quod est incipit. Tunc tamen ista albedo est danda per remotionem de presenti et positionem de futuro, et sic numquam potest tam diu descendere quod veniatur ad singularem in quo ly '*albedinem*' stat determinate.

Sortes in infinitum est albior quam Plato incipit esse albus, igitur Sortes in infinitum est albus.

Consequentia tenet, quia omnis comparativus presupponit positivum. Negatur consequentia, quia solum est summe albus quasi gradus summus. Forte est ut octo.

Ad probationem dicitur, quod quando comparativus ponitur inter duo comparata sic quod ipse sit primus terminus probabilis, tunc presupponit suum positivum. Sed quando ponitur in ultimo loco, tunc non comparat extrema adinvicem, quia non sequitur: Sortes solum in tribus pedibus est longior Platone, igitur Sortes solum trium pedum est longus, quia stat antecedens esse verum et consequens falsum.

Sortes incipit esse albior quam Plato incipit esse albus, posito, quod Sortes et Plato incipient esse eque albi et semper equaliter augmentantur in albedine usque ad desinitionem[7] eorum.

Tunc semper sophisma est verum, quia eius exponentes sunt vere, igitur. Antecedens probatur, quia sic exponitur: Sortes nunc non est albior quam Plato incipit esse albus et Sortes immediate post hoc erit albior quam Plato incipit esse albus, igitur. Prima exponens est vera, quia tunc nullam albedinem habet. Secunda etiam est vera, quia si non, da oppositum: Sortes non immediate post hoc erit albior quam Plato incipit esse albus, igitur non ante quodlibet instans post hoc Sortes erit albior quam Plato incipit esse albus, igitur non ante illud instans futurum Sortes erit albior, et cetera, et non ante illud, et cetera, et tandem demonstrando instans in quo Sortes erit summe albus. Tunc est manifeste falsum, quia ita albus Plato non incipit esse[8]. Et consequentia tenet, quia ly '*erit*' stat confuse distributive.

Sortes incipit esse albior quam ipsemet[9] incipit esse albus.

Conceditur in eodem casu et probatur sic: Iam non est albior quam ipsemet incipit esse albus et immediate post hoc erit albior quam, et cetera, quia si non, da oppositum: Immediate post hoc Sortes non erit albior quam ipsemet, et cetera, igitur ante quodlibet futurum instans non erit albior quam ipsemet, et cetera, igitur ante illud instans demonstrando instans sue desinitionis non erit albior quam, et cetera. Consequens falsum est, quia ante illud instans Sortes erit multum albior quam ipsemet incipit esse albus, et ante illud, et cetera. Et sic de aliis, quia si non, da aliud instans futurum ante quod non erit albior. Sit igitur illud *d* quod non est primum instans non[10] esse *a*. Tunc arguitur sic: Vel *d* est in *a* vel post *a* vel ante *a*. Non ante *a* nec in *a*, quia tunc non fuit albus, sit igitur quod post *a*. Tunc sic: Vel hoc est divisibiliter vel indivisibiliter. Non indivisibiliter, ut prius probatum est, igitur divisibiliter. Igitur tunc *d* non facit instans in quo Sortes primo fuit albus, et cetera.

Sortes (*fol. 101 vb*) erit ita albus sicut ipsemet erit albus, posito, quod albedo Sortis continue intendatur usque ad eius desinitionem ita quod primum instans non esse Sortis sit primum instans non esse dealbationis.

Tunc sophisma est falsum et resolvitur sic: Sortes tunc erit ita albus sicut ipsemet, et cetera. Et tunc fore est aliquando fore. Tunc exponitur sic: Sortes tunc erit albus et ipsemet erit albus et ipsemet non erit albior quam ipsemet erit albus,[11] modo ista tertia exponens erit falsa, quia illud non est ultimum instans sui esse, quia quocumque tempore dato verum est dicere quod[12] Sortes est albior, quia ly '*erit*' in tertio loco stat confuse distributive et igitur nullum certum tempus est dandum, et in primo stat determinate, quia refertur per ly '*tunc*' et non est dandum certum tempus. Et ex ista provenit, quod sophisma est falsum.

Nota, quod omnis positivus tali modo exponitur: Sortes erit ita albus sicut ipsemet erit albus. Sortes erit albus et ipsemet erit albus et ipsemet non erit albior quam ipsemet erit albus.

Sortes erit ita albus sicut Plato erit albus, ponendo casum, quod eque similiter albedines eorum intendantur. Tunc negatur ut prius.

Contra. Ipsi sunt eque albi et erunt eque albi, igitur Sortes erit albus sicut Plato.

Respondetur negando consequentiam, quia antecedens est verum et ly '*erit*' refertur ad idem tempus, sed sic non fit in consequente, quia in consequente primum 'erit' stat determinate, sed secundum confuse distributive. Sed hec bene est vera: Sortes tunc erit ita albus sicut Plato tunc erit albus, quia ibi[13] ambo comparata respiciunt idem tempus, sed sic non fit in sophismate, quia primum refertur ad primum comparatum et secundum ad secundum, ut patet, et cetera.

Nota, quod refert dicere ita albus et precise ita albus, quia quando Sortes est albus ut sex et Plato ut quattuor, tunc ista concedatur: Sortes est ita albus sicut Plato, sed non precise, ut patet.

Sortes erit albior quam Plato erit albus in aliquo istorum, posito, quod Sortes et Plato simul dealbantur, intantum quod vivunt et simul desinunt esse et dealbari.

Tunc sophisma est falsum, quia nec in g nec immediate post g Sortes erit albior quam Plato erit albus. Et resolvitur sic: Sortes tunc erit albior quam, et cetera. Et tunc fore est aliquando fore, et cetera. Et exponitur sic: Sortes tunc erit albus et Plato erit[14] albus in aliquo istorum et Plato non erit ita albus in aliquo istorum sicut Sortes tunc erit albus, igitur. Et si negatur minor debet probari sicut prius. Vel melius exponitur sic: Sortes tunc erit albus et Plato in quolibet istorum erit albus et Plato in nullo istorum erit ita albus sicut Sortes tunc erit albus, igitur.

Contra. Sortes erit eque albus Platoni, igitur erit ita albus sicut Plato erit albus. Negatur consequentia, quia antecedens est verum et consequens falsum. Et si dicitur, quod Sortes erit eque albus Platoni, quando est hoc, vel sicut Plato est vel sicut erat.

Respondetur, quod nullum istorum, sed sicut tunc erit.

Sortes erit (*fol. 102 ra*) precise ita albus sicut Plato erit albus in aliquo istorum.

Varietur iam casus modicum, scilicet[15], in g instanti Plato desinit esse et Sortes survivit. Tunc sophisma est falsum, ut patet per resolventes et exponentes, quia primo sic resolvitur: Sortes tunc erit precise ita, et cetera. Et exponitur sic: Sortes tunc erit albus et Plato erit albus in aliquo istorum et Plato non erit albior nec minus albus quam Sortes erit tunc albus in aliquo istorum, modo tertia exponens est falsa.

Contra sophisma arguitur: Sortes erit magis albus quam Plato erit albus in aliquo istorum et Sortes erit minus albus quam Plato erit albus in aliquo istorum, quia in inceptione erit minus albus et post desinitionem erit magis albus, igitur aliquando erit similis Platoni, quod est contra prius dicta.

Respondetur, quod consequentia non valet, quod igitur erit equalis. Sed bene sequitur quando secundum comparatum esset terminus singularis, sed quando est terminus communis distributivus,[16] ut est ¦ly *'erit'* secundum tunc non valet. Ad declarandum illud primo modo pono casum, quod non sint nisi tres homines, scilicet *a*, *b*, *c*, unus bipedalis, scilicet *a*, et *b* tripedalis et *c* quadrupedalis, et sit unum lignum unius pedis, scilicet *d*, et crescat hoc usque ad sex pedes. Tunc hec est vera: *d* est minus quam aliquis homo et erit maius quam aliquis homo, et tamen non sequitur, quod in medio tempore erit precise ita magnum sicut aliquis homo, quia tunc pro medio tempore esset ita magnum sicut *a* et sicut *b* et sicut *c*, igitur esset simul et semel bipedalis et tripedalis, quod implicat contradictionem. Et si arguitur: tamen minus non potest fieri maius, nisi fiat tripedale, dicitur, hoc est verum, nisi singularia comparantur adinvicem pro eodem tempore vel unum certum datum ad unum certum datum.

Sortes erit ita albus sicut[17] Plato desinit esse albus.
Stat casus precedentis sophismatis. Tunc est falsum, quia vel hoc esset in ultimo instanti vel post vel ante. Non post,[18] quia tunc non erit in ultimo, quod[19] non est dandum, nec ante ultimum, quia tunc non desinit esse albus. Et resolvitur et exponitur sicut patet in predictis.

Sortes erit in duplo albior quam Plato erit albus in *b* instanti.
Ponatur, quod in *a* instanti incipiunt alterari usque in *b* instans, sed in *b* instanti Sortes desinit esse et alterari, sed Plato solum desinit alterari, et quod per totum tempus Sortes in duplo plus alteratur quam Plato, ita quod quando Plato habet unum gradum et tunc Sortes habet duos.[20] [Quia][21] tunc patet, quod sophisma sit falsum, quia nec in *b* nec post nec ante Sortes erit in duplo albior, et cetera.

Plura similia sophismata formantur, et cetera.
Aliquid egit ⟨*b*⟩ gradum albedinis, posito, quod in *b* gradu albedinis desinit motus et sit *b* gradus et igitur ita, quod quando verum sit (*fol. 102 rb*) dicere iam est et prius numquam fuit. Tunc sophisma est falsum.
Contra. *B* gradus albedinis est et prius numquam fuit, igitur aliquid egit eum. Negatur consequentia. Detur oppositum. Est productus, igitur est aptus. Negatur assumptum.
Pro quo nota, si ly *'egit'* tantum valet sicut complevit et determinate complevit, tunc negatur sophisma. Si autem valet tantum sicut fuit in agendo, tunc conceditur sophisma.

Contra. Utroque modo falsum est, igitur antecedens pro secunda parte probatur, quia *a* gradus albedinis consistit in indivisibili, igitur non potuit esse in agendo nec complete agetur. Negatur consequentia, quia fuit in agendo et successive producebatur. Cetera quere in sophismate.

Sortes pertransivit *a* spatium.
Ponatur idem casus in isto sophismate. Sophisma conceditur, inquantum pertransit tantum valet sicut fuit in pertranseundo. Sed negatur si valet tantum, id est, complete pertransivit.
Sed contra. Idem est dicere ⟨quod⟩ pertransivit et per omnes partes transivit. Antecedens probatur, quia ista propositio[22] per hoc importat scilicet totalitatem in comparatione vocabuli.
Nota, quod ad placitum est ut[23] pertransivit ita resolvitur, id est, per omnes partes pertransivit vel transivit, id est, fuit in pertranseundo, et simili modo de futuro simul, scilicet transire.
Sortes est in pertranseundo *a* spatium, supposito, quod *a* sit spatium hinc et Rome. Tunc dubitanda est propositio, quia est una de presenti cuius veritas dependet a futuro, quia si compleverit transitum, tunc fuit in pertranseundo. Si autem non pertransivit complete, tunc non erit pars eius.
Nota, quod Sortes simul et semel incipit ambulare et ambulavisse et non citius unum quam reliquum, quia iam non est verum dicere ipsum ambulavisse et immediate post hoc erit verum, et cetera.

Sortes in *d* instanti incipiet pertransire *a* spatium et in eodem instanti incipiet pertransivisse[24] *a* spatium et non citius nec tardius incipiet pertransire quam pertransivisse[25]. Conceditur totum. Similiter de movere et movisse. Similiter de dealbari et album esse.

A et *b* erunt vera, supposito, quod *a* et *b* erunt duo contradictoria contingentia, sicut 'Sortes currit, Sortes non currit' et successive verificentur per unam diem et numquam simul.
Tunc conceditur sophisma.
Contra. *A* et *b* erunt vera, igitur aliquando conceditur pro[26] eodem tempore quod ista sunt vera communiter, quia[27] in singularibus de subiecto obliquo quelibet de futuro vera habebit aliquam de presenti veram.
Respondetur, quod sic determinando antecedens, quia vel ly '*ista*' capitur divisive, tunc conceditur, vel collective, tunc negatur.
Contra.[28] Divisive ista erunt vera, igitur in aliquo instanti erunt vera.
Respondetur negando consequentiam, quia ly '*erunt*' stat confuse tantum, igitur non potest dari aliquod certum suppositum.

Sortes desinit (*fol. 102 va*) moveri post *b*, posito, quod *b* sit instans presens et[29] Sortes incipiat moveri in *b*.

Tunc probatur sophisma. Sortes potest desinere moveri post *b*. Ponatur igitur in esse, igitur. Antecedens probatur. Sortes immediate post *b* potest desinere moveri, igitur Sortes potest desinere moveri immediate post *b*. Consequentia est bona et antecedens est verum, igitur et consequens. Negatur sophisma. Ad probationem: Negatur consequentia, quia in prima ly '*potest*' stat confuse tantum et in alia determinate. Et si dicitur: ponatur prima in esse, negatur hoc, quia confusio cadit super ly '*potest*'.

Sortes potest ita cito desinere moveri sicut movebitur.

Sophisma est falsum.

Contra. Immediate post hoc potest desinere moveri et immediate post hoc movebitur, igitur Sortes potest ita cito desinere.

Respondetur negando consequentiam, quia ly '*potest*' sic in antecedente stat confuse tantum et in consequente determinate.

Sortes erit ita cito corruptus sicut ipsemet erit generatus, posito, quod Sortes sit unum contingens futurum quod cras poterit[30] generari.

Tunc sophisma est falsum.

Contra. In infinitum cito Sortes potest generari et in infinitum cito Sortes potest corrumpi, igitur Sortes est ita cito corruptus, sicut, et cetera. Antecedens probatur. Ante quodlibet instans futurum Sortes potest corrumpi et etiam ante quodlibet instans futurum Sortes potest generari, igitur ante quodlibet tempus Sortes potest generari et ante idem corrumpi.

Respondetur concedendo antecedens et negando consequentiam et ulterius concedendo totum argumentum, et etiam conceditur, quod quam cito Sortes potest generari, citius potest corrumpi, immo in decuplo citius, et non debet poni in esse, igitur.

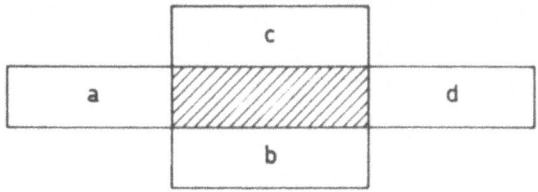

Fig. 1.

Immediate post hoc aliqua pars ipsius *b* erit sub *a* et[31] tamen nulla[32] pars ipsius *b*[33] immediate post hoc erit sub *a*, ut patet in figura[34].

A[35] incipit intendere albedinem in aliqua parte ipsius *b* et tamen quelibet pars proportionalis ipsius *b* post hoc remittetur.

Conceditur sophisma, quia aliqua parte stat confuse tantum in principio et non est dandum aliquod certum, sed in ultimo stat confuse distributive, et est dandum, et cetera.

Sophisma. *A* incipit albefacere[36] aliquam partem ipsius *b* et tamen nulla pars in *b* erit albior quam (*fol. 102 vb*) nunc est alba.
Conceditur sophisma.

(*Fol. 102 va*) *A* generabit albedinem usque ad *c* punctum et nulla albedo erit immediate *c* puncto.
Sophisma conceditur tenendo ly '*usque*' exclusive et non inclusive.
Et sic est finis. Deo gratias.

University of Helsinki
and Academy of Finland

NOTES

1 albus] *add. interlin.*   2 predictum] *add. marg.*   3 in] *add. interlin.*   4 infinitum] *corr. ex* infinite   5 si] *add. interlin.*   6 infinitum] *corr. ex* infinitam   7 desinitionem] *corr. ex* descisitionem   8 albus] *del.*   9 ipsemet] *corr. ex* platomet   10 non] *add. interlin.*   11 consequentia] *del.*   12 quod] *corr. ex* quia   13 in] *del.*   14 erit] *rep.*   15 in aliquod] *del.*   16 tunc] *del.*   17 sicud plato et sicud] *del.*   18 post] *corr. ex* potest   19 quod] *corr. ex* quia   20 tunc in duplo albior] *del.*   21 Quia] *add. marg.*   22 propositio] *rep.*   23 ut] *corr. ex* utrum   24 pertransivisse] *corr. ex* pertransisse   25 pertransivisse] *corr. ex* pertransisse   26 pro] *corr. ex* per   27 quia] *add. marg.*   28 ista] *del.*   29 capiat] *del.*   30 voluit] *del.*   31 et] *add. interlin.*   32 nulla] *add. interlin.*   33 et] *del.*   34 *figura occurrit post verba* 'Deo gratias'.   35 *A* incipit intendere ... Conceditur sophisma] *add. post textum manu eiusdem scriptoris.*   36 albefacere] *corr. ex* albescere.

We wish to thank Professor Norman Kretzmann and Mrs Barbara Kretzmann for their valuable suggestions and corrections to the first draft of this edition.

NOTES AND REFERENCES

[1] For an introduction to modern research on medieval logic see Paul Vincent Spade, 'Recent Research on Medieval Logic', *Synthese* **40** (1979).

[2] Cf. Curtis Wilson, *William of Heytesbury: Medieval Logic and the Rise of Mathematical Physics* (The University of Wisconsin Publications in Medieval Science, vol. 3), The University of Wisconsin Press, Madison, 1956, and John E. Murdoch, 'From Social into Intellectual Factors: An Aspect of the Unitary Character of Late Medieval Learning' in J. Murdoch and E. Sylla (eds.), *The Cultural Context of Medieval Learning*, D. Reidel Publishing Company, Dordrecht and Boston, 1975, pp. 271–339.

[3] A. B. Emden, *A Biographical Register of the University of Oxford to A.D. 1500* Oxford 1958, vol. II, coll. 1050a–1051a. On previous research on Kilvington, see Francesco Bottin, 'Un testo fondamentale nell' ambito della 'Nuova Fisica' di Oxford: I Sophismata di Richard Kilmington' *Miscellanea Mediaevalia* **9**, De Gruyter, Berlin – New York, 1974, pp. 201–205, and Norman Kretzmann, 'Socrates is Whiter than Plato begins to be White', *Noûs* **11** (1977), 3–15.

[4] See note 2 above.

[5] An analysis of the contents of this MS is presented in the second part of this paper.

[6] In his *Index Britanniae Scriptorum* (ed. from the 1552 edition by R. Lane Poole with the help of Mary Bateson, Oxford 1902, p. 346) John Bale lists among the works of Richard Billingham an *Abbreviata G. Kylmynton* with the *incipit* 'Sortes est albior quam Plato'. The work is also mentioned in Bale's *Scriptorum illustrium Maioris Britanniae quam nunc Angliam et Scotiam vocant Catalogus*, Basle 1559, p. 460, and in Pits' *Relationum Historicarum de Rebus Anglicis* I, Paris 1619, p. 489. The *incipit* given in these catalogues corresponds with the beginning of the first sophism of Richard Kilvington as well as with the beginning of the *Declaratio* in the *Uppsala* MS. In his recent article 'Richard Billingham's Works on Logic', *Vivarium* **XIV** (1976), pp. 121–138, L. M. De Rijk does not present any MSS of this work of Billingham. Judging from the content of the other texts in the *Uppsala* MS, it seems that the authors of these texts were well acquainted at least with Billingham's *Speculum puerorum*. In the end of the treatise *De immediate* (fol. 94 vb) there is a remark probably referring to the *Declaratio*: "Istis habitis magister declarabit aliqua sophismata climitonis que patent in textu suo". It remains so far uncertain, whether this unknown master presented his own comments on Kilvington's sophisms, or gave the text directly on the basis of Billingham's now lost abbreviation.

[7] See *Guillelmi de Ockham Summa Logicae*, ed. by P. Boehner, G. Gál, St. Brown (Editiones Instituti Franciscani Universitatis S. Bonaventurae), St. Bonaventure, N.Y., 1974, *Pars I, cap.* 70–74.

[8] Francesco Bottin, 'Analisi linguistica e fisica Aristotelica nei "Sophismata" de Richard Kilmyngton' in *Filosofia e Politica e altri saggi*, a cura di Carlo Giacon (Università di Padova, Pubblicazioni dell' Istituto di Storia della Filosofia e del Centro per Ricerche di Filosofia Medioevale, N.S. 14), Antenore, Padova 1973, pp. 125–145, Norman Kretzmann, see note 3 above.

[9] For the different senses of the term 'infinite' see, e.g., Curtis Wilson, *op. cit.*, p. 16.

[10] Gregorii Ariminensis, O.E.S.A., *Super Primum et Secundum Sententiarum*, Reprint

of the 1522 Edition (Franciscan Institute Publications, Text Series 7), St. Bonaventure 1955, *Liber* I, *dist.* 17, *q.* 2, *a.* 2.

¹¹ Anneliese Maier, *Die Vorläufer Galileis im 14. Jahrhundert. Studien zur Naturphilosophie der Spätscholastik* (Storia e Letteratura 22), Roma 1949, pp. 172–176, 214.

¹² *Ibid.*, pp. 212–215.

¹³ Abraham Robinson, *Non-standard Analysis* (Studies in the Logic and the Foundations of Mathematics), North-Holland Publishing Company, Amsterdam, 1974, pp. 279–282.

¹⁴ Cf. Ernest A. Moody, 'The Medieval Contribution to Logic' in his *Studies in Medieval Philosophy, Science, and Logic. Collected Papers 1933–1969*, University of California Press, Berkeley, Los Angeles, London, 1975, pp. 371–392.

¹⁵ Of the texts contained in this MS only *De consequentiis* of Marsilius of Inghen has so far been identified, see M. Andersson-Schmitt, *Manuscripta Mediaevalia Upsaliensia. Übersicht über die C-Sammlung der Universitätsbibliothek Uppsala* (Acta Bibliothecae R. Universitatis Upsaliensis, vol. XVI), Uppsala 1970, no. 708, p. 46. The other texts have been characterized as "Verschiedene philosophische Stücke und Notizen", *ibid.*, no. 1722, p. 95. Because this MS contains hitherto unresearched texts and early information of the reception of the English logic in Prague, it seems convenient to give a short analysis of its contents here: Marsilius of Inghen, *De consequentiis* (*pars prima*), ff. 1 r – 12 v, Marsilius of Inghen, *De suppositionibus* (incomplete), ff. 13 r – 16 r, Marsilius of Inghen, *De ampliationibus*, ff. 16 r – 19 r, Marsilius of Inghen, *De appellationibus*, ff. 19 v – 23 r, Marsilius of Inghen, *De consequentiis* (*pars secunda*), ff. 23 r – 31 v, Petrus de Alliaco (?), *Tractatus de arte obligandi*, ff. 32 r – 45 v, *inc.*: Tractaturus de obligationibus dei gratia tractatum in eisdem in sex capitula distinguebam. Primo diversorum terminorum huic arti pertinentium descriptiones ponam ..., *Notabilia circa materiam appellationum*, f. 45 v, *Expositio super tres libros De anima*, ff. 46 r – 57 r, *inc.*: ⟨B⟩onorum honorabilium. Liber de anima cuius subiectum est ly "*anima*" tamquam signum ⟨in⟩ principali sui divisione dividitur in tres libros partiales. In primo ostendit de ipsa secundum opinionem antiquorum. Et habet duos tractatus ... *De conversionibus*, ff. 57 v – 61 v, *inc.*: Est notandum quod quilibet qui vult debite convertere propositiones primo debet videre quid sit subiectum et predicatum propositionis ... (*cit.*: Marsilius, Buridanus), *De suppositionibus*, ff. 62 ra – 74 rb, *inc.*: Circa librum Suppositionum queritur primo utrum notitia libri suppositionum est scientia. Quod sic patet. Contra. Omnis scientia ... (*cit.*: f. 62 va Marsilius, Thomas de Clivis, Maulefelt, f. 63 ra Maulefelt, Marsilius, f. 64 va Marsilius, Buridanus, f. 65 rb Marsilius, Buridanus, f. 67 va "parisienses", "anglici", f. 71 vb Albertus, Marsilius, f. 72 vb Wiclef, f. 73 ra Marsilius, Hugo ⟨Capellanus?⟩, f. 73 va Marsilius, Buridanus), *De confusionibus*, ff. 74 rb – 76 rb, *inc.*: ⟨Q⟩ueritur circa librum ⟨Confusionum⟩. Queritur quid sit subiectum in libro confusionum. Dicitur quod terminus confundens. Contra. Terminus qui confunditur est hic subiectum, igitur non terminus confundens ..., *De consequentiis*, ff. 76 rb – 80 ra, *inc.*: Circa librum consequentiarum queritur quid sit subiectum in libro consequentiarum. Arguitur quod ly "*consequentia*". Contra. Si iste terminus "*consequentia*" est subiectum ... (*cit.*: f. 77 rb Buridanus, Marsilius et alii, f. 79 rb Hollandrinus), *De exclusivis*, ff. 80 ra – 80 va, *inc.*: Queritur de exclusivis utrum solum sunt due dictiones exclusive. Quod sic patet. Contra. Aliqua est dictio exclusiva que non est aliqua istarum duarum..., *De exceptivis* (incomplete), ff. 80 va – 82 va, *inc.*: Utrum

omnis exceptio. Utrum omnis exceptio propria debet fieri a toto, et cetera. Contra. Ab universali ad suam particularem est bona consequentia. Sed quandocumque in universali est propria exceptio, tunc etiam in particulari..., ff. 82 vb – 83 rb *vacant, Nota de descensu propositionum,* ff. 83 va – 85 ra, *inc.*: Nota. Ad videndum primo de descensu propositionum predictarum... (*cit.*: f. 84 ra Marsilius, f. 84 rb Billingham), *Nota de restrictionibus,* ff. 85 ra – rb, *inc.*: Item. Nota circa restrictiones quod subiectum est ly 'restrictio'..., *Divisiones de restrictionibus,* f. 85 v, *inc.*: Restrictio capitur dupliciter. Uno modo prout est terminus prime impositionis..., *Cantus (cum notis musicis),* f. 86 r, *inc.*: Rumpetur exercitium vocentur ad convivium convive et sorores..., *Nota de appellationibus,* f. 86 r, *inc.*: Item. Nota de appellationibus. Et nota quod hec est appellatio simplex..., *Regule et divisiones,* f. 86 v. The *incipits* of the following series of texts are given on pp. 316–318. On ff. 103 r – 107 r there is a series of logical exercises, *inc.*: Domina mea sancta maria. Supposito, quod aliquis dicat aliquod dictum ut "homo est asinus" vel "deus est diabolus" esse falsum..., ff. 103 r–103 v, *inc.*: Omnis propositio est falsa, igitur et tua conclusio. Tenet consequentia et antecedens probatur..., ff. 103 v – 104 r, *inc.*: ⟨O⟩mnis propositio si est impossibilis est falsa. Consequentia tua est propositio..., ff. 104 r – 104 v, *inc.*: ⟨C⟩onclusio tua est impossibilis, igitur est falsa. Consequentia est bona et antecedens probatur..., f. 104 v, *inc.*: ⟨S⟩ortes incipiendo scire tuam conclusionem non incipit scire conclusionem veram, igitur ipsa est falsa..., ff. 104 v – 105 v, *inc.*: ⟨N⟩ulla propositio est vera, igitur tua conclusio est falsa. Consequentia tenet dum conclusio tua sit quedam propositio..., ff. 105 v – 106 r, *inc.*: ⟨S⟩olum hominem esse hominem est verum, igitur tua conclusio non est vera..., ff. 106 r – 106 v, *inc.*: ⟨O⟩mnis propositio non est vera, ergo nulla propositio est vera. Consequentia probatur, quia negatio postposita..., f. 106 v, *inc.*: ⟨N⟩on quedam propositio est vera, ergo nulla propositio est vera..., ff. 106 v – 107 r, *inc.*: ⟨Q⟩uelibet conclusio necessaria est vera. Conclusio tua non est conclusio necessaria, igitur non est vera..., f. 107 r, *inc.*: ⟨A⟩liqua propositio preter hanc "deus est" est falsa. Conclusio tua est aliqua propositio preter hanc.... The MS ends with a German chant. The MS has belonged to the library of Vadstena, cp. the note inside the cover "Liber monasterii Vadsteniensis intitulatus anno domini MCD90 die Blasii".

[16] The earliest date connected with Thomas Maulefelt has so far been the year 1413 in the MS *Prague* UB 2065, see Neal Ward Gilbert, 'Ockham, Wyclif, and the "Via Moderna"' (*Miscellanea Mediaevalia* 9). De Gruyter, Berlin – New York, 1974, pp. 113– 114, n. 75. His name is mentioned several times in the treatise *De suppositionibus* on ff. 62 ra – 74 vb in the *Uppsala* MS. This text seems to be a reportation of a commentary on Maulefelt's respective work. In addition, his name occurs many times in the texts of the series on ff. 87 ra – 102 vb.

[17] A reference to the Suppositions of Thomas de Clivis is found on f. 62 va of this MS: ".... quod probatur, quia Marcilius, Thomas de Clivis in suppositionibus suis ponunt alias diffinitiones quam magister facit hic...". The name "Hugo" occurs also in connection with Marsilius of Inghen.

[18] Wilson, *op. cit.*, pp. 163–168.

ROBERT HOWELL

A PROBLEM FOR KANT

I. INTRODUCTION

Kant's basic account of knowledge in the *Critique of Pure Reason* is well known. Objects, which exist in themselves in independence of their relations to our human sensibility, affect that sensibility in a quasi-causal manner. As a result of this affection, we acquire intuitions which represent these objects to us as single, individual entities. By means – and only by means – of our mental awareness of these intuitions, in conjunction with rule-governed synthetic activities that we carry out on the manifolds of the intuitions, do we come to know the objects in question. We know the objects simply in the forms that the intuitions represent them to us as having. But, Kant argues, these forms are necessarily spatiotemporal in character, and space and time are simply entities in the mind that structure objects as those objects are represented to us as being by our intuitions. Hence, according to Kant, objects, as they exist in themselves, are non-spatiotemporal. And when we come to know objects via our mental contemplation of our intuitions, we know these objects only as they appear to us (or are represented to us as being) by our intuitions and not as they are in themselves.

Crucial to this Kantian picture of knowledge is the idea that the objects which we know are entities that *both* exist in themselves *and also* appear to us via our intuitions. This idea Kant himself clearly and frequently emphasizes. Thus Kant writes, for example, of "the distinction, which our Critique has shown to be necessary, between things as objects of experience and *those same things* as things in themselves" (Bxxvii, my italics); he says that "the object is to be taken *in a twofold sense*, namely as appearance and as thing in itself" (Bxxvii); he notes that "the things which we intuit are not *in themselves* what we intuit them as being" (A42/B59, my italics); and he speaks of distinguishing "the mode in which we intuit [objects] from the nature that belongs to them in themselves" (B306).[1] Moreover, it is not just that Kant clearly and frequently emphasizes this idea. He

331

E. Saarinen, R. Hilpinen, I. Niiniluoto, and M. Provence Hintikka (eds.), *Essays in Honour of Jaakko Hintikka*, 331–349. *All Rights Reserved.*
Copyright © 1979 by D. Reidel Publishing Company, Dordrecht, Holland.

also assigns to it an important role within his overall treatment of knowledge. For instance, Kant appeals to such an idea, in part, to distinguish his form of idealism from Berkeley's.[2] And, as I think that one can show, Kant uses this idea in a certain way in his Transcendental Deduction of the Categories.

Many Kant scholars have in fact recognized the significance to Kant of the idea that we are now considering.[3] But in their commendable efforts to present Kant's theory of knowledge accurately, these scholars do not sufficiently stress the severe philosophical problems that attend this view. In the present paper I examine one central such problem. This problem is that Kant really cannot, given his conclusion that things as they exist in themselves are nonspatiotemporal, consistently accept the view in question. Kant really cannot hold with consistency, that is, that the objects of our knowledge are things as they appear to us in spatio-temporal form, things which also have an existence in themselves, in a nonspatiotemporal form, outside our knowledge.

In Section II below, I provide the background needed to consider this problem. Then in Section III I present the problem itself. In the concluding Section IV, I observe that this problem is really not resolved by shifting over to Kant's additional account of the objects of our knowledge in terms not of things as they appear but of appearances. And I note for future investigation the bearing of the problem on two further Kantian topics: first, the question of how Kant can be sure that there is a one-to-one correlation of the objects of knowledge with the members of some set of things existing in themselves; and, second, the Transcendental Deduction of the Categories. The upshot of my Section III and Section IV discussion is that the problem for Kant which we here consider is both centrally important to his theoretical philosophy and apparently irresolvable (in a sense that I explain later) within the framework of that philosophy.

My recognition of this problem results from my application to Kant's writings of the possible-worlds techniques that Hintikka has developed in various papers on perception and the propositional attitudes.[4] Some of this application I present below. My questions about Kant do not, of course, depend for their existence – or even for their proper statement – on the success of Hintikka's work. But it is a tribute to that work that its use leads one to uncover such questions about the thought of a deep, subtle, and great philosopher like Kant.

II. KANTIAN OBJECTS OF KNOWLEDGE AS THINGS APPEARING
TO US VIA OUR INTUITIONS

For the sake of concreteness, let us focus our discussion henceforth on the specific case of my knowledge of the round cup before me. Then, according to Kant, this knowledge amounts to – or at least prominently involves – my mental contemplation of the cup as it is displayed to me by a sensibly given intuition. Yet, as we have just seen, Kant holds that the object which is thus displayed to me in the form of this cup has also an existence in itself, as a nonspatiotemporal thing, outside my knowledge of it. Our main task, in the present section of this paper, is to understand the sense in which the single object that appears to me as the cup does also have such an existence in itself. That is, we must understand the sense in which, for Kant, an identity obtains between an object that, in the form of the spatiotemporal cup, I know, and an object that, as it exists in itself in a nonspatiotemporal form, I do not know.

In carrying out this task, we should concentrate on the fact that, for Kant, the intuition via which I know the cup is an intuition that represents to me the object, existing in itself, that we have just mentioned. The intuition in question is in fact *of* this object existing in itself, and the intuition represents this object to me *as being* the round cup before me. But then that round cup simply *is* the object which the intuition is *of*, in the form that that object takes in its representation by the intuition. And so the sense in which Kant holds there to be an identity between (i) an object that, in the form of the spatiotemporal cup, I know, and (ii) an object that, as it exists in itself in a nonspatiotemporal form, I do not know, is evidently just the sense in which there is an identity between (iii) the round cup which the intuition represents the object that it is *of* to me *as being* and (iv) the object that, as it exists in itself in a nonspatiotemporal form, this intuition is in fact *of*. Hence we can understand the identity that Kant asserts between (i) and (ii) simply by examining the identity that holds between (iii) and (iv).

Now the identity that holds between (iii) and (iv) is obviously the sort of identity that is presupposed between an *F*-thing and an object *b* in our use of sentences of the form

(1)   Intuition *i* represents object *b* as being an *F*

where sentences of this form are interpreted as claiming

(1a)   *Of* the object that in itself is in fact *b*, intuition *i* represents
          that that object is an *F*

So we can understand the above (iii)–(iv) identity if we can understand
this sort of *F*-thing–*b* identity.

However, in speaking – as in (1a) – of an object that, in itself, is in fact
such-and-such a thing, we seem clearly to be speaking of an object that
has what Kant would call (transcendently) actual existence, namely
the sort of actual existence that things have insofar as they exist in independence of any relations to our human sensibility.[5] In consequence,
any sentence of form (1a) can be understood as claiming

(1b)   *Of* the object that has (transcendently) actual existence and
          in that (transcendently) actual existence is *b*, intuition *i*
          represents that that object is an *F*

But then any sentence that claims what (1b) claims seems clearly equivalent
to an existential claim of the form[6]

(2)   ($\exists x$) (*x* has actual existence & *x* = *b* & intuition *i* represents
          that *Fx*)

where by '*x* has actual existence' I mean in (2) (and hereafter) '*x* has
(transcendently) actual existence.' And so the sort of *F*-thing–*b* identity
that is presupposed in our use of sentences of form (1), where those
sentences are interpreted as claiming something of form (1a), must be the
sort of strict identity that is evidently presupposed in (2). That is, it
must be the sort of strict identity that obtains between the value of the
variable '*x*' in the third conjunct of (2) and the value of that variable in the
first two conjuncts of (2).

Given this result, however, we can isolate immediately the exact sense
in which the above (iii)–(iv) identity – and hence the above (i)–(ii) identity –
holds. It is simply that sense in which a strict identity holds between the
value of '*x*' in the third major conjunct of (3) below and the value of '*x*'
in the first two major conjuncts of (3):

(3)   ($\exists x$) (*x* has actual existence & *x* is nonspatiotemporal &
          intuition *i* represents that (*x* = the cup & *x* is round))

where for convenience I use '*i*' in (3) (and hereafter) as a constant designat-
ing the particular intuition that does display to me the round cup before
me that I know. This strict identity between the values of these various
occurrences of '*x*' in (3) is thus precisely that identity which Kant takes
to hold between an object that, in the form of this spatiotemporal cup, I

know, and an object that, as it exists in itself in a nonspatiotemporal form, I do not know.

Although we have thus succeeded in specifying the sense in which this last identity holds for Kant, we cannot yet claim really to understand this identity, for we have not yet provided well-defined truth conditions for (3). But to this end we can argue, as I now do, that the expression 'intuition *i* represents that,' as that expression occurs in (3) (or, for that matter, in any claim of form (2)), functions as an intensional operator. Given this argument, we can supply an intensional semantics that states truth conditions for claims like (3) that contain this operator. Using our grasp of these truth conditions, we can then understand the strict identity that is presupposed between the values of '*x*' in (3). And so we can understand the above identity that Kant affirms. (The fact that an intensional semantics proves appropriate to (3) indeed shows – what I do not argue further in this paper – that such a semantics is of great use in understanding the intensionality-involving claims that permeate Kant's philosophy.)

To demonstrate that 'intuition *i* represents that' functions in (3) as an intensional operator, I observe as a preliminary that when we and Kant talk of intuition *i*'s representing some object existing in itself as being a round cup – that is, when we talk of the truth of (3) – we clearly have in mind an activity of representation that occurs in the realm of the sort of (transcendently) actual existence that we have introduced in claims (2) and (3) above.[7] We clearly have in mind such an activity of representation – and so an activity of representation that itself has a (transcendently) actual rather than a merely phenomenal existence – for the following reason. The reason is that Kant's view certainly is not that an intuition like *i* merely *appears to be* representing an object existing in itself as being a round cup. Nor, certainly, does Kant suppose (or write as though) it is merely that some other representation *represents i to me* as thus representing this object existing in itself. (In fact, Kant would fall into an unacceptable regress were he to pursue any such position consistently.) Rather, Kant's view obviously is that *i actually does* represent an object existing in itself as being a round cup.

But were *i*'s representation of such an object as being a round cup to have a merely phenomenal existence – as does, of course, this round cup itself – then it *would* be the case that *i* merely appears to be representing an object existing in itself as being a round cup. Or it *would* be the case that some other representation merely represents *i* to me as thus representing this object. Since, however, Kant does not describe in any such terms the representation that is effected by i, we must conclude that this representation does not have a merely phenomenal existence. Instead, we must attend to the fact that i actually does represent an object as being a cup. And we must take this fact to mean that i's activity of representation occurs within the realm of (transcendently) actual existence and so itself possesses such an existence.

If, however, i's activity of representation has (transcendently) actual existence, then the truth value of (3) is to be evaluated at the world or realm of such existence; and this truth value (evaluated at that world) is in fact true. But then from this result we can now argue directly for the conclusion that 'intuition i represents that' functions in (3) as an intensional operator. To do so, we need merely recall that (i) no things, as they exist in themselves, are, according to Kant, spatiotemporal; (ii) round and square things are spatiotemporal; and (iii) we are counting things, as they exist in themselves, as being those things that have (transcendently) actual existence.

Given points (i) to (iii), it follows immediately that, in the world of (transcendently) actual existence, the set of round things is identical to the set of square things; for both these sets are, in that world, identical to the null set. Hence the predicates 'x is square' and 'x is round' are, at the world of (transcendently) actual existence, coextensive. But (we can quite properly imagine) intuition i does not, in its (transcendently) actual representation of the (transcendently) actual thing that it is *of* as being round, also represent that thing as being square. And so, despite the coextensiveness of 'x is square' and 'x is round,' we cannot substitute the former of these predicates for the latter in (3) and preserve the truth, at the world of (transcendently) actual existence, of (3). The expression 'intuition i represents that' therefore does function as an intensional operator in (3) (and, by similar reasoning, in all claims of form (2)).

Because this expression functions in this way in (3), we must, as I have intimated earlier, supply an intensional semantics for claims like (3) if we are really to understand the identity that is presupposed in (3). For this purpose, an intensional semantics in Hintikka's style proves simple and perspicuous.[8] As all those acquainted with such a semantics know, we are required by it, in stating the truth conditions for claims like (3), to appeal not only to the world of (transcendent) actuality (that is, to the world of transcendently actual existence). We are required to appeal, also, to the various worlds that are compatible with what intuition i represents.

Since this type of appeal should be familiar to most readers, I note here only that its effect is to make (3) hold true at the world W of (transcendent) actuality just in case there is a single object o which satisfies two conditions: First, this object o occurs in (and so has transcendently actual existence in) W and, as it thus occurs in W, is a nonspatiotemporal thing. And, second, this selfsame object o also occurs in each of the worlds W_1, W_2, W_3 (and so on) that are compatible with what i represents and, as it thus occurs in each of those worlds, is a round cup. The situation that obtains when these two conditions are satisfied is shown in Figure 1, where the dotted line represents the fact of there being a single object o that occurs, as just described, in W and in W_1, W_2, and W_3.

In the last paragraph, we have of course been stating well-defined truth conditions for (3). Our possession of these truth conditions enables us,

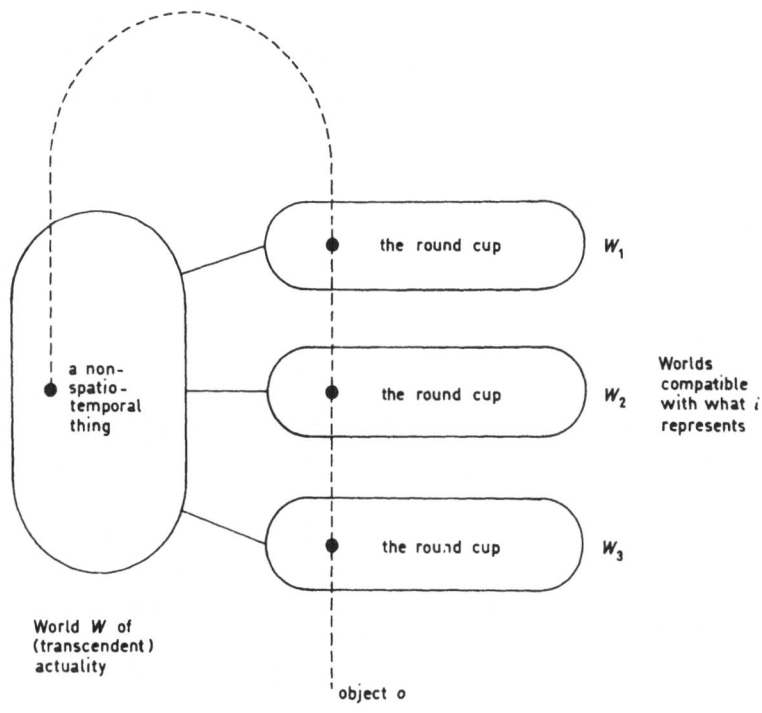

Fig. 1.

finally, to understand the exact force of the strict identity that is presupposed in (3) between the values of the various occurrences of the
variable 'x.' And so we now can finally understand the identity that Kant
takes to hold between the object that, in the form of the spatiotemporal
cup, I know, and the object that, as it exists in itself in a nonspatiotemporal
form, I do not know. This identity (and the identity between the values
of 'x' in (3)) clearly is *not* an identity that, contrary to all logic, somehow
obtains in a single world (say the world of transcendent actuality) between
the spatiotemporal cup that I know and a nonspatiotemporal entity that
I do not know, both of these objects being considered to exist in that
single world. Rather, as our truth conditions for (3) bring out, the identity
in question is simply the strict identity that holds between the object o
that occurs in the world W of (transcendent) actuality as a nonspatiotemporal thing and that selfsame object o that occurs in the worlds W_1,
W_2, W_3 (and so on) as a round cup.

Or, to express this result in Kant's own terminology, we can observe
that the world W of (transcendent) actuality is just the world of things
as they exist in themselves. And the worlds W_1, W_2, W_3 (and so on) – the
worlds compatible with what intuition i represents – are, taken together,
equivalent to the Kantian phenomenal world or "field of appearances."[9]
Consequently the exact force of the identity that we are discussing is this.
The identity at issue is simply the strict identity that holds between the
object o that exists, as a nonspatiotemporal thing, in the world of things
as they exist in themselves, and that selfsame object o that exists, as a
round cup, in the phenomenal world or field of appearances. The identity
at issue is *not* an identity that – contrary both to Kant and to logic –
somehow obtains in a single world (say the world of things as they exist
in themselves) between a nonspatiotemporal entity and a spatiotemporal
cup, both this entity and that cup being considered to exist in this single
world.

The results that we have arrived at in the preceding paragraph can be
seen to agree extremely well with the sorts of things that, in Section I, we
saw Kant claiming about objects as objects of experience and those
selfsame objects as things having existence in themselves. But we have
reached our above results simply by translating into Kant's preferred
terminology our own intensional-semantics account of the identity that
Kant's Section I claims imply. And so the agreement of our above results
with Kant's Section I claims provides a strong confirmation for our own
account of this identity.

III. THE PROBLEM

The identity that we have just been explicating is one of which we have spoken repeatedly above. We have so spoken simply because, as we have observed, this identity is clearly affirmed by Kant. But, in fact, within the general context of his overall account of knowledge, Kant cannot consistently maintain such an identity in conjunction with another central point in that account, namely his claim that things as those things exist in themselves are nonspatiotemporal. As we see below, this result, which I now demonstrate, constitutes a most severe problem for Kant.

Given our Section II account of the identity between the round cup that I know and an object that, as it exists in itself, I do not know, it is a straightforward matter to show that Kant cannot in consistency affirm both this identity and the nonspatiotemporality of things as they exist in themselves. To establish this result, let us observe three points about my knowledge of a phenomenal object like the round cup.

First, let us observe that, for Kant, such knowledge takes a propositional (or judgmental), that-clause form (compare A50–52/B74–76, A67–70/B92–95, B141–42, and similar texts).[10] Let us in fact suppose – as we may without any loss of generality – that my knowledge of the cup takes the specific form

(4) H knows that the cup is round

where 'H' designates myself. And let us remark that Kant clearly holds that the knowledge that I have of phenomenal objects does encompass such a fact about shape as is expressed in (4). (Kant explicitly asserts in B69–70 note, for example, that "the predicates of space ... are rightly ascribed to the objects of the senses." In conjunction with his discussions of geometric knowledge, such an assertion shows clearly that Kant accepts (4) as a satisfactory example of my knowledge of a phenomenal object.)

Second, let us observe that one evident and familiar principle about knowledge is that all claims of the form

(5) H knows that $p \supset p$

hold true. We should, indeed, observe that this principle is so deeply embedded in our conception of knowledge that we would, I think, simply and rightly refuse to count as knowledge any cognitive states for which this principle fails. Moreover, we should observe that Kant himself seems obviously to accept what is for him the equivalent of this principle. At

A58/B83, for instance, Kant avers that "truth consists in the agreement of knowledge with its object ... knowledge is false if it does not agree with the object to which it is related." And elsewhere Kant makes many similar, (5)-accepting claims.[11] Hence Kant as well as we must take it that my knowledge of the cup, as expressed in (4), is subject to the principle that is embodied in (5).

Third, let us observe that my knowledge, as it is expressed in (4), is in fact a state of mind that has (transcendent) actuality. And let us remark that this observation needs explicit justification, for – as Kant scholars are well aware – Kant is hardly forthcoming about the exact ontological status of such knowledge as (4) expresses. Yet let us remark, also, that such justification can be readily produced. In particular, we can note that both ordinary reasoning and various of Kant's own texts are readily seen to require that (4) express a knowledge-state that is (transcendently) actual – that is, a knowledge-state that has existence in itself, not mere existence in the phenomenal world.[12]

Ordinary reasoning in fact requires us to accept this conclusion on essentially the grounds already given, in connection with the question of the ontological status of intuition i's representation, in Section II. Kant's view, after all, certainly is not that I merely appear to know that the cup before me is round. Nor, clearly, does Kant suppose (or write as though) it is merely that some representation represents me as having this knowledge. Instead, Kant obviously supposes (and writes as though) I *actually do* know that the cup is round. But then, for the sorts of reasons already developed in Section II, the only coherent way of understanding these Kantian suppositions about my knowledge is to regard Kant as accepting the conclusion that this knowledge is something that has (transcendently) actual existence. We must, in fact, regard Kant as accepting the conclusion that my knowledge is a state of me, the (transcendently) actual although unknowable thing to which, as their subject, my mental states belong.[13] We must not regard Kant as taking my knowledge to be a mere set of representations that is contained within that temporally ordered string of inner-sense sensations and representations that constitutes my phenomenal or empirical self.

I lack space here to show in detail how Kant's own texts also require the above conclusion. But we can note the following three points. (a) Kant takes knowledge to require synthetic activities of the understanding, and he does not attribute such activities to the phenomenal or empirical self, which is essentially a passive, associatively organized string of sensations and representations of the sort just mentioned. Rather, Kant seems clearly to regard these synthetic activities as belonging to the above-mentioned (transcendently) actual although unknowable thing that is the subject of my mental states. And so Kant seems clearly committed to regarding such synthetic activities as themselves having (transcendently) actual existence.[14]

(b) Texts like B68, B158, and B277 make it clear, I think, that, for Kant, the entity who in fact knows, when I am rightly said to have knowledge, is that entity that is the I of which I am made aware via my representation of apperception 'I think' (or 'I am'). These texts also make it clear, I take it, that this I of apperception has, in its very state or process of knowing, a (transcendently) actual existence, not a mere phenomenal or empirical existence. (Such texts, in conjunction with those texts cited in point (c), surely also indicate that this I is the just-mentioned transcendently actual – although unknowable – subject of my mental states.)

(c) In the second edition of the first *Critique*, Kant attempts in several places to answer an important criticism made to the first edition by the writer Pistorius. For reasons that we can ignore here, Pistorius argues that Kant really cannot sustain his whole crucial notion that our knowledge concerns the appearances, to an other-than-appearance self, of things existing in themselves. Instead, Pistorius objects, on Kant's theory even the self reduces to an appearance, and so "there will be nothing but illusion, for nothing remains to which anything can appear."[15]

As Kant scholars know, Kant's various answers to this objection raise extremely subtle and far-reaching questions about his view of the self, the range of application of the categories (including especially the category of existence), and the extent of our knowledge. But, however these questions are to be treated in detail, it seems clear to me that, at a minimum, Kant in his response to Pistorius affirms that a (transcendently) actual existence does attach, in the self's state of knowing, to the self who knows. It seems clear that Kant in fact attributes such an existence to the self who knows even while Kant also holds, as his overall theory requires him to, that this self, in our awareness of it via the representation 'I think' (or 'I am'), is given to us only in an indeterminate manner and cannot in any way be known by us. (Here see B157–58, B158 note, B277, B422–23 note, and B427–30.[16])

The consequence of the above points (a) to (c) is that a (transcendent) actuality attaches both to the synthetic activities of the understanding that are required in knowledge and – in its very state or process of knowing – to the self who knows. But seems evident that if something that is required in our knowledge has (transcendent) actuality – and if (as we argued in Section II) the representation that is effected by our intuitions in our knowledge has itself such actuality – then our state of knowledge must have (transcendent) actuality. It also seems evident that if the self who knows has (transcendent) actuality in its very state or process of knowing, then such actuality must attach also to that state or process of knowledge. And so I conclude that Kant's texts do require us to take the knowledge that is expressed by (4) to be (transcendently) actual. That is, Kant's texts do require us to take (4) to express a state of knowledge that has existence in itself, in the world W of (transcendent) actuality. Since both Kant's texts and – as we saw above – ordinary reasoning thus require this conclusion, we have consequently justified the third point that we have observed above about my knowledge of a phenomenal object like the round cup.

Notice now, however, where our observation of these three points has got us. By the first point that we have observed above, this knowledge takes a propositional form of a sort specified in (4); and by the second point that we have observed, this propositional form of knowledge is subject to the principle that is embodied in (5). Hence the conjunction of our first two points implies that any situation or world at which (4) holds true must also be a situation or world at which

(6)   The cup is round

holds true. Yet, by the third point that we have observed, (4) expresses a state of knowledge that exists in the world W of (transcendent) actuality. And so claim (4) holds true at this world W. But then if (4) holds true at W, the conjunction of our first two points requires that claim (6) also hold true at W.

The consequences for Kant of the truth of (6) at W are, however, horrendous. In order for (6) (regarded as a consequence of (4) via our first two points) to hold true at W, it must be the case that (i) the singular term 'the cup' in (6) designates, at W, an object that can be truly said, at W, to be round. And it must be the case that (ii) this object is in fact the round cup that I know via intuition i.

But then from points (i) and (ii) trouble follows immediately. The round cup in question of course exists in the worlds W_1, W_2, W_3 (and so on) that are compatible with what i represents to me. And so points (i) and (ii) imply that 'the cup' in (6) designates, at W, the cup that exists in W_1,

W_2, W_3 (and so on). Yet according to our Section II account of Kant's view of the objects of our knowledge, as Figure 1 has graphically illustrated, this cup, as it exists in those worlds, is one and the same thing as a certain object o, as that object o exists in these worlds. And that object o, as it so exists, is *strictly identical to* an object (namely, o itself) that has (transcendently) actual existence in W. Hence, by the transitivity of identity and the intersubstitutivity of identicals, 'the cup' in (6) must designate, *at* W, the object o that has (transcendently) actual existence *in* W.[17] But then by point (i) this object can be truly said, at W, to be round (and to be a cup). And so Kant is committed, by the truth of (6) at W, to there existing, in this world W of (transcendent) actuality, an object which is round (and a cup). But this result is of course flatly inconsistent with Kant's central position that objects existing in themselves are nonspatiotemporal, for this round cup, existing as it does in W, is a thing that has existence in itself.

We thus see that, given the three points that we have observed above, Kant is required, in logic, to accept the truth of (6) at W. But the truth of (6) at W in turn implies a result that is flatly inconsistent with the above central position of Kant's, when that truth is taken in conjunction with our Section II account of the identity that Kant affirms between an object that, in the form of a round cup, I know, and an object that, as it exists in itself, I do not know. Yet we have seen cogent reasons for accepting both our three above points and our Section II account of that identity. Hence, arguing by Modus Ponens, we can detach our reference to these three points and to our Section II account. We can speak merely of the commitment by Kant's view of knowledge to the truth of claims like (6) at W, and we can note that this commitment exists alongside Kant's affirmation of the sort of identity in question. Then, regarding this commitment as itself subsumed within the overall Kantian view of knowledge, we can conclude bluntly that, considered in the general context of that view, Kant's affirmation implies a result flatly inconsistent with his central position that objects existing in themselves are nonspatiotemporal. And from this conclusion it follows immediately that, within the general context of his overall account of knowledge, Kant cannot consistently maintain the identity at issue while also adhering to that central position.

Thus we arrive at the main result that we have set out to establish above. Because Kant's treatment of things existing in themselves as nonspatiotemporal is so crucial to the overall conclusions of the first *Critique*, this result does constitute a most severe problem for Kant, given the importance

ance of the identity that we have been discussing to his view of knowledge. Various attempts may be made to solve this problem by rendering that identity consistent with the nonspatiotemporality of things existing in themselves. But no such attempt that I can construct has seemed to me at all successful.[18]

IV. APPEARANCES DO NOT SOLVE THE PROBLEM

We should note one last point. Apparently without realizing the fact, Kant really accepts two different views of the objects of our knowledge. First, he accepts the view (focused on exclusively in this paper) that these objects are things that have an existence in themselves, as those selfsame things appear to us in spatiotemporal forms via our intuitions. Second, he accepts the view that these objects are, instead, appearances, mental entities (in fact, synthesized intuitions) that occur before our consciousness.[19] Because this second view allows the object known, the appearance, to be distinct from the object existing in itself that has produced this appearance in us, one might suppose that, by adhering consistently to this second view, Kant could abandon the identity that we have been discussing and so could escape (if not literally solve) our above problem.

However, proceeding in this way, Kant would only in a very literal sense escape that problem. To take the object of my knowledge (as in (4)) to be an appearance commits him, after all, to the spatiotemporality of appearances. Yet it seems clear that Kant must count these appearances or synthesized intuitions as (transcendently) actual. (Evidently he has essentially the same sort of reason for so proceeding as the reason that, we saw above, he has for counting as transcendently actual the representation that is effected by – and the knowledge that we gain through – our intuitions.[20]) Hence Kant's idea that the objects of knowledge are appearances leads to an inconsistency similar to the inconsistency that we have just seen to arise for his former view of those objects.

Moreover, to escape this new inconsistency, it seems that, within his overall framework, Kant can only revert to that former view. Kant cannot, after all, abandon his firm position that (transcendently) actual objects are nonspatiotemporal; and yet he must somehow make room for the fact that the objects of our knowledge are spatiotemporal. So he clearly must give up the idea that these objects are appearances. Instead, Kant must say, the objects of our knowledge are spatiotemporal simply in this sense: (transcendently) actual things, which in themselves are non-spatiotemporal, *appear* to us via our intuitions to be spatiotemporal.[21] And it is these objects, as they thus appear to us, that are the objects of our knowledge.[22] But of course this last view is just Kant's former view. And of course that former view commits Kant to the strict identity that we have been discussing, and hence to the problem that we have uncovered in Section III.

We therefore see that the present attempt to escape the problem of Section III by consistently adhering to the view that the objects of our knowledge are appearances does not, in the end, succeed. In a straightforward sense that I have just illustrated, our Section III problem seems insuperable within the framework of Kant's thought. No successful solutions to the problem present themselves, and attempts to escape this problem by appeal to the notion of an appearance raise similar problems whose resolution leads directly back to this problem.

A study of the familiar question of how Kant can be certain that there is a one-to-one correlation of the objects of our knowledge with objects existing in themselves would only serve to reinforce our above problem for Kant.[23] And a careful examination of the Transcendental Deduction of the Categories would show that this problem significantly affects the Deduction's treatment of a category as a concept of an object in general[24] – and so significantly affects the argument of the Deduction as a whole.

Not only may we be sure, therefore, that the problem that we have been here discussing is apparently insuperable (in the sense that I have just explained) within the overall framework of Kant's philosophy. But, also, we may expect that further investigations will show in detail that this problem receives support from other important questions about Kant's framework. And we may expect that these investigations will demonstrate that this problem has significant consequences for the most central part of that framework. In the anticipation of completing such investigations elsewhere, I close the present discussion here.

*State University of New York
at Albany*

NOTES

The research for this paper was supported, in part, by the National Endowment for the
Humanities, the SUNY Research Foundation, and the American Council of Learned
Societies. I thank these organizations for their help. An early version of the paper was
read and usefully discussed at the University of Utah and in a Stanford University Kant
seminar. Material from the present version will appear, in a somewhat different form,
in a book on Kant's Transcendental Deduction of the Categories that is to be published
in the Synthese Historical Library.
[1] I follow the standard A-, B- references to pages of, respectively, the first and second
editions of the first *Critique*; I use 'Ak.' to refer to the Academy edition of Kant (1902
to date).
[2] See B69 and B70; Kant to Beck, 3 December 1792 (Ak. vol. 11, 395). And compare
Prolegomena, Section 13, Note III; Section 32; and Section 49.
[3] Recent English-language examples are Paton (1936), vol. 1, 61–63; Schrader (1967);
Dryer (1966), 513–514; and Barker (1969). See also Prichard (1909), 73–75; Weldon
(1958), 195; and many others.
[4] See especially Hintikka (1969) and (1975); also Hintikka (1974) and Howell (1973).
The present paper modifies some specific points in this last paper, but the overall
framework remains the same.
[5] Talk of (transcendently) actual existence may seem to infringe Kant's restriction to
phenomena of the knowledge-providing applications of the categories, including the
category of existence and the Postulates' related notion of actuality [*Wirklichkeit*].
But (a) Kant's theory requires us to assign some type of actuality to objects as they exist
in themselves (here compare the claims in his reply to Pistorius, mentioned below);
(b) on what is, I believe, its most plausible interpretation, Kant's account of the categories
allows us to use the categories (and so, in particular, the category of existence or
actuality) to *think* objects existing in themselves, although only in an indeterminate,
general manner (here see A251–253, B306–307, B309); and (c) Kant himself frequently
uses the notions of actuality and existence to apply to other than phenomenal objects.
(Here see, among other passages, Bxx, "the thing in itself as . . . real *per se* [*für sich
wirklich*]"; A37/B53, "time is . . . the real [*wirkliche*] form of inner intuition"; B157,
"I am conscious of myself . . . only that I am"; B158, "I exist as an intelligence . . .";
also B422–423 note, B429, A504 = B532, and the various texts like A42/B59 and
A492/B520 where Kant talks of things as they "are" or "exist" in themselves.)
[6] Compare the discussion of the pictorial analogues of (1a) in Howell (1974). In the
present paper, I regard single quotes enclosing symbols as quasi-quotes. But I proceed
as though the symbols within these single quotes occurred in the object-language rather
than in the meta-language. I take '=' to mean genuine, strict identity, rather than
(as in Howell, 1974) simple coincidence at or in a world. And I employ a
(transcendently)-actual-existence predicate – as in (2) – in a way that is done neither in
Howell (1973) nor in Howell (1974). (I introduce such a predicate for a technical
reason, not because I now attribute to Kant the view that existence is a real predicate!)
Other quirks of usage explain themselves.
[7] There are of course serious problems for the view that i's representation has
(transcendently) actual existence. (Here compare Strawson, 1966, 39 and 249.) But,

despite these problems, the view nevertheless is one to which compelling reasoning shows Kant committed. (Note that I talk only for convenience of i's "activity" of representation; and I abstract from the very important role that, for Kant, our thought plays in referring intuitions to objects.)

[8] But any adequate intensional semantics could of course be employed.

[9] A40/B57. Of course to arrive at the phenomenal world, the worlds compatible with what is represented by intuitions besides i must also be considered.

[10] Actually, as I argue in Howell (1973), Kant seems to run together that-clause and direct-object knowledge (at least of a perceptual sort). But this point is unimportant here, since on any adequate interpretation of his theory Kant certainly accepts (4) as expressing knowledge that I may have. And example (4) generalizes easily to all objects of my knowledge.

[11] B142, last sentences; A150/B189–190; A821 = B849; *Logik*, Introduction, Section VII (Ak. vol. 9, 51–53); and compare Reflexionen 2144, 2155, 2161, and 2177 in Ak. vol. 16.

[12] This point does of course create serious intellectual problems for Kant. (Compare note 7.) But the grounds below show that the point is nevertheless one that Kant must accept.

[13] That I, the (transcendently) actual thing, am the (logical) subject (although the unknowable subject) of my mental states is borne out by, for example, A346/B404, A349, A358, B407, B411–412 note, B419, B422, B429, and A492/B520. See also the references for point (c).

[14] Note the implications of texts like A51/B75, A76/B102 ff., A106–107, A108, A111, A117 note, A121–122, B129–130, B131–135, B139–142, B153–156, B157–159, and B407.

[15] As translated in Kemp Smith (1962), 323, from a quotation that Kemp Smith cites in Erdmann's *Kriticismus*, 107. See 305, 307–308, 323–331, and 467 of Kemp Smith (not all of whose points, however, I accept).

[16] Kant's position seems clearly to be that via the 'I think' or 'I am' I am aware *that* I have (transcendently) actual existence. But via this representation I have (and can have) no awareness or knowledge of *how* I am as I thus exist.

[17] By themselves, the three points that we have observed above show Kant committed to the truth of (6) at W. But this truth would cause him no difficulties as long as he could maintain that the object of knowledge – the designation at W of 'the cup' in (6) – can be truly described *at* W as being round (and a cup) even though this object does not exist *in* W. (Kant – or a Kantian – might try to maintain such a position on the grounds that the object of knowledge truly is, where it does exist – that is, in W_1, W_2, W_3, and so on – a round cup and so can be truly described, even where it does not exist, as being a round cup. For example, one might suppose that, for Kant, the true object of my knowledge, when I know the cup, is not the object o of Figure 1 but is, rather, an object o^* that coincides with o in W_1, W_2, W_3, and so on, yet diverges from o at W by not existing in W. One might urge that, despite its not existing *in* W, this object o^* can still be truly described *at* W as being round.) Kant is, however, prevented from maintaining any such position about the object of knowledge by our Section II account of the identity that he affirms between that object and an object existing in itself (and so in W). Hence it is precisely that identity, in conjunction with the three points that we have noted above, that plunges him into the problem that we are laying out here. Hence also

if (as I suggest in Section IV) Kant were to seek to escape this problem by abandoning that identity and taking the object of knowledge to be an appearance, then Kant would manage (at least momentarily) to avoid this exact problem. (But, as we see in Section IV, he would then be forced into a similar problem whose resolution within his framework leads directly back to just our present problem.)

[18] One might, for example, suggest that not (4) but rather, say, 'H knows that some (transcendently) actual thing appears to be a cup and to be round' properly expresses my knowledge of the cup. Such a suggestion escapes the above problem. But it is not a suggestion that Kant can accept, for, as we have seen, Kant himself would certainly take (4) as it stands adequately to render my knowledge. (Moreover, my ordinary knowledge of the cup is, in any case, clearly not a knowledge that includes within its content a claim to the effect that some transcendently actual thing is appearing to me.) Another attempt to solve the above problem would be to hypostatize the object o, as o appears to me via i, by treating as a single entity the various manifestations or "world slices" of object o within worlds W_1, W_2, W_3 (and so on). But then when (6) holds true at W about this hypostatized o-as-o-appears-via-i, either this hypostatized object exists in W or (as in the case of the object o^* discussed in note 17) it does not so exist. If it does exist in W, the above problem arises again. If it does not, the identity that Kant affirms between the cup that I know and an object existing in itself is abandoned. But since to solve the above problem *is* to render this identity consistent with the non-spatiotemporality of objects existing in themselves, that problem is now ignored rather than solved.

[19] See the lucid and valuable discussion of these two views in Barker (1969). That appearances (in contrast to things, as they appear) are, for Kant, simply synthesized intuitions is a most natural interpretation for me to offer here. But certain complications exist in this interpretation (due in part to Kant's frequent use of the *term* "*Erscheinung*" to mean, indifferently, a mental entity before our consciousness – that is, an appearance proper – and an object, as that object appears via an intuition). I ignore these complications in this paper, however, as they do not significantly affect the argument below.

[20] Understood as a mental entity – in fact, a synthesized intuition – occurring before our consciousness, an appearance certainly has (transcendent) actuality as much as does an intuition's activity of representation and the knowledge that we get from that activity.

[21] Compare Barker (1969), 289, last paragraph.

[22] One might suggest that Kant could at this point take the object of my knowledge to be a thing, as that thing appears to me via my intuition, without Kant's thereby also having to take this thing to possess any sort of existence in itself (that is, any transcendently actual existence). (Compare the idea that when I see "in" a picture some object that I know to be purely fictional, I can still properly describe that object as appearing to me via the picture, even though I know that this object has no actual-world existence.) Kant does not, however, explicitly consider any such suggestion. But the tenor of his thought, as texts like Bxxvi–xxvii, B306–307, and A251–252 indicate, is that if we adopt the view that the objects that we know are things appearing to us via our intuitions, then we *must* think of these objects as also having an existence in themselves. (Furthermore, Kant's frequent emphasis on the role, in my knowledge, of the affection of my sensibility by objects existing in themselves argues against his being able to proceed as in the above suggestion.)

[23] The possibility that one thing existing in itself could appear to me as being two distinct phenomenal objects reinforces also a subtle distinction between a thing, as it exists in itself (which, in terms of Figure 1, is the entire o-object, as that object occurs in W) and what we can call a thing-in-itself (which is the manifestation or 'world slice' only of the o-object in W). For simplicity, I have ignored this distinction – and its implications for my talk of "things existing in themselves" – in this paper. (Here compare Prichard, 75, note 1, and 78, note 1; and remark the parallel with the related distinction between the thing, as it appears, and the hypostatized thing-as-it-appears of note 18 above.)

[24] Note, for example, the description at A93/B126 of a concept of an object in general as a "concept of an object as being thereby given, that is to say, as *appearing*" (italicization mine).

BIBLIOGRAPHY

Barker, Stephen F., 'Appearing and Appearances in Kant', in Beck, Lewis White, ed., *Kant Studies Today*, Open Court, La Salle (Illinois), 1969, pp. 274–289.

Dryer, D. P., *Kant's Solution for Verification in Metaphysics*, Allen & Unwin, London 1966.

Hintikka, Jaakko, *Models for Modalities*, Reidel, Dordrecht, 1969.

Hintikka, Jaakko, '"Dinge an sich" Revisited', in Hintikka, Jaakko, *Knowledge and the Known*, Reidel, Dordrecht, 1974, pp. 197–211.

Hintikka, Jaakko, 'Information, Causality, and the Logic of Perception', in Hintikka, Jaakko, *The Intentions of Intentionality and Other New Models for Modalities*, Reidel, Dordrecht, 1975, pp. 59–75.

Howell, Robert, 'Intuition, Synthesis, and Individuation in the *Critique of Pure Reason*', *Nous* 7 (1973), 207–232.

Howell, Robert, 'The Logical Structure of Pictorial Representation', *Theoria* 40 (1974), 76–109.

Kant, Immanuel, *Kants Gesammelte Schriften*, Prussian (later, German) Academy of Sciences, ed., de Gruyter, Berlin, vols. 1–28, 1902 to date.

Kant, Immanuel, *Critique of Pure Reason*, Kemp Smith, Norman, trans., Macmillan, London, 1929.

Kemp Smith, Norman, *A Contemporary to Kant's 'Critique of Pure Reason'*, Humanities, New York, 2nd edition, 1962.

Paton, H. J., *Kant's Metaphysic of Experience*, Allen & Unwin, London, vols. 1–2, 1936.

Prichard, H. A., *Kant's Theory of Knowledge*, Oxford University Press, Oxford, 1909.

Schrader, George, 'The Thing in Itself in Kantian Philosophy', in Wolff, Robert Paul, ed., *Kant*, Doubleday, Garden City, 1967, pp. 172–188.

Strawson, P. F., *The Bounds of Sense*, Methuen, London, 1966.

Weldon, T. D., *Kant's Critique of Pure Reason*, Oxford University Press, Oxford, 2nd edition, 1950.

HIDÉ ISHIGURO

SUBJECTS, PREDICATES, ISOMORPHIC REPRESENTATION, AND LANGUAGE GAMES

This paper attempts to investigate whether we could in any meaningful way say of predicate expressions (whether monadic or relational) that they refer and, if they refer, what their reference could be. In semantics based on model theory (e.g., of Carnap's state-descriptions or Hintikka's model-sets), there are entities to which predicate expressions correspond (Carnap's attributes, Hintikka's predicates). What kind of entities are they?

I will carry out this investigation through an examination of a view which Wittgenstein stated in the *Tractatus* – a view which is commonly referred to as the 'Picture Theory' of propositions – and in the light of some comments which Professor Hintikka has made about the picture theory[1] in its relation to the idea of a language game and to model theory. Unexpected though this may be, I have found that some of the things that Professor Hintikka has recently said about language games in later Wittgenstein are more relevant to the Picture Theory than the model theoretical analyses of the Picture Theory which Professors Stegmüller and Hintikka have offered. The difficulty of applying a model theoretic reconstruction of the picture theory is not entirely due to the 'absolutism' of the *Tractatus* (i.e., the view that there is ultimately only one language, and we cannot step out of our language, to give its semantics) as Hintikka has suggested. It comes more directly from the *Tractatus*' view about the nature of what it is that is represented by predicate expressions; a view which I hope to show, is harder to disagree with or to ignore.

(a) THE DIFFERENT ROLES GIVEN TO SUBJECT EXPRESSIONS AND PREDICATE EXPRESSIONS IN THE TRACTATUS; PREDICATE EXPRESSIONS DO NOT STAND FOR ANYTHING

When I assert that Nero burned Rome or that the air is chilly, is there any need to make a logical or ontological distinction between the kind of entities designated by the names 'Nero', 'Rome' and the noun phrase

351

E. Saarinen, R. Hilpinen I. Niiniluoto and M. Provence Hintikka (eds.), *Essays in Honour of Jaakko Hintikka*, 351–364. *All Rights Reserved.*
Copyright © 1979 by D. Reidel Publishing Company, Dordrecht, Holland.

'the air' on the one hand, and what is ascribed to them by the expressions
'- - - burned - - -' or '- - - is chilly'? And if one insists on making a
distinction, what does the distinction amount to?

The problem which is familiar from the time of Plato and Aristotle
has in different ways been raised again in more recent years by Frege,
Wittgenstein, Ramsey, Geach and Strawson amongst others. The obscurity
of Wittgenstein's view in the *Tractatus* suggests the difficulty of the problem
which faces us. I agree with one part of Professor Hintikka's defence of the
Tractatus picture theory. It seems to me, however, that in order to see
what is interesting about the picture theory and to see how it resembles
or does not resemble semantics based on model theory it is necessary to
understand (and make much clearer than Professor Hintikka has done)
the essential difference the *Tractatus* emphasizes between two different
roles which are played by expressions. One is the role of standing for
(*vertreten*), and the other is that of representing (*darstellen*). Unfortunately
Hintikka denominates both roles by the word 'represent'. A name typically
stands for objects, whereas a predicate or a proposition typically *represents*.
According to Wittgenstein, a function [expression] or predicate represents
a proper concept (4.126) ('concept' is obviously used in the Fregean
manner and roughly corresponds to what we mean by 'property') and
propositions (which are functions of names) represent 'material' properties
(2.0231) ('material' contrasts here with 'formal'). Every proposition or
picture represents a possible state of affairs (2.202, 4.021, 2.031) or its
sense (2.221).[2]

Standing for is a relation between object and object: one deputising
for the other (2.131, 3.22, 3.221). If a name is in use, it is already linked by
convention to a bearer. Representing is said by Wittgenstein to be a
relation between one fact and another. *How* things are (e.g., how words are
arranged in a sentence, or how people are situated in a tableau-vivant)
represents how other things are. The difference between the role which the
expressions play in 'standing for' and 'representing' is made clear also by
what is said *not* to stand for something or *not* to represent another thing.
For example logical connectives do not stand for anything, and should not
be called logical constants (4.0312). A contradiction or tautology cannot
represent any possible situation (4.462), and no proposition can represent
its logical form, and [expression for propositional] functions cannot
represent formal concepts [i.e., properties such as that of being an object,
or of being a function].

It may be thought that the distinction between standing for and representing is an esoteric detail which one can ignore. Hintikka in his attempt to bring out the resemblance between model theory and the *Tractatus* view takes very little notice of the difference and writes that the atomic sentence '$R(a, b)$' is true if and only if "the relation represented by 'R' holds between the individuals represented by 'a' and 'b' (in this order) . . . that is . . . when the sentence '$R(a, b)$' is a true picture of the entities represented by 'R', 'a' and 'b' in Wittgenstein's sense".[3] Now Hintikka himself has written that strictly speaking what represents the relation is not the symbol 'R' itself but the relations which hold between the *names* 'a' and 'b' in the sentence.[4] But according to Wittgenstein it is not only *what* represents that is different – the relationship which links what represents to the represented is quite different from the name-object relation. This relation corresponds to what has been called 'isomorphic representation' by Stenius, and following him by Hintikka. They have assumed however that this relation obtains only between sentences and facts, whereas in the *Tractatus* this also holds between predicate and properties, and *hence* between propositions and fact. But more important for our purpose is that as a consequence of this the nature of the kind of thing which is represented (i.e., a possible situation or how things could be) is different from the nature of any object stood for. What is represented is a possibility. It need not obtain in the world. Representing is not a simple word-world relation. (There is no assignment from the represented to what represents.) I believe that this difference is important for the picture theory, and to blur it and consider both relations under the general cover of 'representation' seems to me to obscure certain crucial differences between subject and predicate.

To understand the point of Wittgenstein's doctrines here, it is necessary to see what Wittgenstein was doing as directed at certain difficulties in Frege's distinction between object and concept. In his first work after the *Tractatus*, Wittgenstein wrote "concept and object, but that is subject and predicate".[5] The distinction was one he inherited from Frege's treatment of propositions as made up of function and argument. Wittgenstein accepted this (as Professor G. E. M. Anscombe has rightly suggested), even if he rejected Frege's view that a proposition was a name of a truth-value. What then was a concept (*Begriff*) for Frege? It was something introduced as the reference (*Bedeutung*) of a concept-word. It is one special kind of function (a function from objects to truth values). Just as Hintikka uses 'represent', Frege used the same word '*bedeuten*' to express both the relationship which holds between name and object, and the relationship between predicate-word and concept, even the relationship between sentence and truth-value which Frege claimed to be one kind of name and one kind of object. Whether '*bedeuten*' is really the same kind of relation in each case or not has been discussed by Dummett, Tugendhat and others. I will not go into detail here, but I think that using the same word covers up at least two problems: an obvious problem and also a deeper problem, to which I will come back later.

Frege made a very important distinction between proper names which occur as arguments of propositional functions and concept words. Unlike a proper name a concept word is strictly speaking not a word at all, but a feature common to a set of sentences. Geach and Dummett have stressed this point. A first level concept word is obtained by removing one or several singular terms (Frege's proper names) from a sentence. For Frege a concept word always has a built-in number of gaps in it which are to be filled in by expressions to make a sentence. Indeed all function names are said to have gaps. The concept word is therefore a form rather than a mere word, and is said to be 'in need of completion' (*ergänzungsbedürftig*) or 'unsaturated' (*ungesättigt*). Even if we were to believe that this is not a special feature of concept-words, it is clear enough what he means by this.

What is much more difficult to grasp is Frege's assumption that this unsaturatedness of concept-words and other function expressions entails a corresponding unsaturatedness in concepts and in functions in general. Frege is often praised for clarifying the notion of function and for analysing propositions as function and argument. What he says about function giving the law of correlation between two values is clear, though this was said before by Cauchy. This clarity also extends to the claim that we grasp the law of correlation by seeing the unsaturatedness of the function *name* as in '(—)'.[2] But unless we understand what it is for such an expression to refer and what it is for a function to be unsaturated, the clarity does not extend to Frege's breezy claim that 'naturally' the unsaturatedness of function signs has something corresponding to it in the functions themselves – including concepts. And why cannot the thing which the *predicate* of one proposition refers to be designated by a singular term which is the *subject* of another proposition? It is true that in Frege's system we cannot even sensibly raise this question. But is this as it should be?

I believe that Wittgenstein's picture theory results from thinking through this and several other difficulties which result from the doctrine of Frege. First Wittgenstein drops talk of saturated and unsaturatedness.

For, so far as expressions are concerned, both subject words and predicates require other words to make up a sentence. Thus relative to a sentence both kinds of expressions are in need of completion. Needing completion does not then distinguish concept words playing the role of function-expressions from names playing the role of occupying the argument place. Names are no more self-subsistent (*selbständig*) than predicate expressions are. As the *Tractatus* says, *all* expressions can be expressed as propositional functions – and not only concept words as Frege thought. We can treat any expression as a constant conjoined with variables such that the result of replacing the variables with arguments is a proposition (3.312–3.35). There is still however an essential difference in this theory between subject and predicate. For although both are incomplete relative to a sentence, *how* it is incomplete determines the identity of a predicate, whereas *how* it is incomplete does not determine the identity of a name. Thus in the *Tractatus* what distinguishes a predicate expression from a name is not that it is incomplete, but that it is a fact. In other words the difference between function and argument can be made clearer by treating the expression for a function as consisting not in the function sign itself, but in some fact about the names in the argument place (either a spatial fact if it is an inscription or a temporal fact if we have uttered words). We usually use an additional sign which contributes towards differentiating the fact about the arguments – but this is not necessary. $fxy = x^y$ and $fxy = x + y$ equally express functions with two variables although the first uses no additional sign.[6] When we add an extra expression, as we do in expressing propositional functions (e.g., when we ascribe relations or properties to things in ordinary language), we can identify the predicate expression as a fact about the added word *in relation to the arguments*. (This is most clearly expressed in the *Notebooks*.) I shall call this a predicate fact. Thus the occurrence of '- - - grows', and '- - - grows - - -' in a sentence correspond to different facts and hence to different predicate expressions. (The predicative fact '- - - grows' can occur in a more concealed form as in 'All who sleep grow', or 'Alfred neither sleeps nor eats'. We should therefore say that the predicative fact '- - - grows' is identified by how these words occur in elementary propositions.) A name on the other hand is identified as a word and not as a fact. Thus it is the same name 'Caesar' which occurs in 'Brutus killed Caesar' and 'Caesar was a Roman'. Once we establish the connection between the name and the object which it stands for it is the *same* name whether it occurs as subject or direct object of a sentence. Thus there is a basic asymmetry between predicates and names in subject position.[7]

If we think not of the expressions but the properties or relations which are meant by these expressions, talk of gaps and saturation are not merely inadequate. It simply does not help. First, although properties or relations do not exist on their own and exist only as properties or relations *of something*, neither do objects exist without properties nor exist without standing in some relation to other objects. So far as self-subsistence is concerned, none of them is self-subsistent. The important distinction concerning the mutual dependence or independence of entities lies rather between two kinds of property. One is what we may call sortal property or a "property whose absence would reduce the existence of the object itself to nothing" and the other is a property which is ascribed to objects by descriptions of them. Here the property may or may not hold of the object without detriment to the identity of the object itself.[8] (Wittgenstein's main criticism of Frege's theory of concept and object in *Philosophical Remarks* is that it is too general and fails for example to distinguish these different kinds of properties which make propositions of different logical forms. A sortal property does not add to how a thing could be.) But neither of these properties could be grasped as something unsaturated – something with gaps.

Second it is misleading to think of relations or properties as any kind of entities, saturated or unsaturated, that are constituents of facts. I agree with Wittgenstein when he wrote in 1931, "To say that a red circle is composed of redness and circularity, or is a complex with these component parts, is a misuse of these words and is misleading. (Frege was aware of this and told me.) It is just as misleading to say the fact that this circle is red (that I am tired) is a complex whose component parts are a circle and redness (myself and tiredness) . . . A chain too is composed of its links not of these and their spatial relations. The fact that these links are so connected isn't '*composed*' of anything at all."[9] Thus, according to him, monadic or relational predicates do not refer to any component of a fact – if by component we think of an entity which makes up the fact.

It is important to notice that the denial that in the state-of-affairs described by '*fa*' or '*gbc*', there are any entities which are designated by '*f*‑‑‑' and '*g*‑‑‑' and which are constituents of the state-of-affairs, is not to deny that there may be something in reality which is described by these predicative facts, and hence correspond to it. *How* things are is an objective feature of reality. The world is not a set of things, but the totality of facts – of how things are. The denial does however entail that the truth of propositions of the form '*fa*' or '*gbc*' does not automatically allow us to quantify over properties or relations represented by the predicative facts '*f* - - -' or '*g* - - -'. There may be other arguments to justify second-level quantification, but these must be further arguments. Whereas quantification over objects, or (since Wittgenstein firmly believes that first level propositional functions have restricted domains) quantification over certain kinds of objects, immediately follows from the truth of elementary propositions, and vice versa. This direct connection between first-level quantification and true elementary propositions is also a part, as we know, of Hintikka's game theoretic treatment of quantifiers.

(b) FREGE AND WITTGENSTEIN'S CONCEPTION OF 'OBJECT' AND THAT OF STENIUS AND HINTIKKA

Contrary to what Professor Stenius (and Hintikka?) have claimed, Wittgenstein follows Frege quite clearly in thinking that in '*aRb*' or 'The case is brown', '- - - *R* - - -' and '- - - is brown' do not refer to objects. Wittgenstein goes further and shows that since they do not refer to any component of these facts, as Frege himself is said to have acknowledged, one should not think of them as any kind of entity, and hence one should not think of the relation of words to them as that of standing for the other one.

Wittgenstein did not deny that one can talk *about* properties or relations. Properties or relations can be made to correspond to the argument rather than the function. Indeed Frege had already acknowledged this when he made the often-quoted controversial admission, 'The concept horse is not a concept'.[10] Frege was saying that when one talks *about* the concept horse, then the concept horse is given by the argument of the propositional function, and is therefore *in Frege's technical sense an object*, even if an abstract one. Similarly it seemed for Wittgenstein, and he is unequivocal about this in his early years. When he wrote in the *Notebooks*[11] that "Relations and properties are objects too" he precedes this claim by the comment "the concept 'this' is identical with the concept of object". In other words, whatever it is that one talks about, or thinks about as an argument of a propositional function is an object. An object is that of which we make descriptions – of which we make judgments. In the *Tractatus* Wittgenstein gives an example of a proposition *about* a property, i.e., when a property becomes an 'object': "This shade of blue and that

one stand *eo ipso* in the internal relation of lighter to darker. It is un-
thinkable that *these* two objects should not stand in this relation."[12]
(Here however, Wittgenstein qualifies the remark by commenting that the
words 'object' and 'relation' have a loose use.) As it was introduced by
Frege, the concept of object is relative to propositions. It is that which
is designated by a singular term which occupies the argument place of a
first level propositional function, that about which we can make identity
claims. The singular term itself can be a highly complex one made from
other concept words and class forming operators or description operators.

In the *Tractatus* not all apparent occurrences of singular terms in the
argument place of a propositional function designate objects. If what
appears to occupy the argument place is a definite description (whether it
purports to refer to an individual or a class) it is always to be analysed
in the Russellian manner – so that they turn out *not to occupy the argument
place* (3.261, 4.0031). Wittgenstein also held the view (which seems
obviously wrong) that 'a complex' can be given only by its description and
not named (3.24). It followed in the *Tractatus* that only names in a completely analysed proposition designate objects. They are given by logically
simple names that occupy the argument place of an elementary
propositional function.

This is why the expression 'this shade of blue' in the above proposition,
which seems to be in the argument place (and would thus refer to an
object for Frege), should after logical analysis be taken as a definite
description of the form 'the shade of blue of this flower (or some object)',
and the whole proposition will then be analysed out in Russellian fashion.
Thus it does not strictly speaking designate an object.

But by saying that relations and properties can be objects, Wittgenstein
is not saying that in 'Nero burned Rome' or 'the air is chilly', '- - - burned
- - -' or '- - - is chilly' refer to objects. Wittgenstein quite clearly follows
Frege. Objects or constituents of a proposition correspond only to what is
given by the argument and not by the function. The *Tractatus* thus says that
'*fa*' shows that the object *a* occurs in its sense (4.1211), and that one expresses
the proposition 'There are two objects which . . .' by '$(\exists x, y) \ldots$' (4.1272).

(c) SO-CALLED 'REFERENCE' OF PREDICATE EXPRESSIONS
AND THEIR EXTENSION

Wittgenstein goes further than Frege, or perhaps we should say that he makes much more explicit what was implicit in Frege himself. The pattern which the names make in relation to the function sign or predicate expression (which, as we have said can in principle be expressed without the use of a function sign) does not stand for anything as names do. But if, as I said, this predicative fact represents features of other possible facts, then does it not stand in an isomorphic representational relationship to a feature of reality? That is to say, even if it did not *stand for* properties or relations as Stenius and Hintikka have suggested, doesn't it represent, or correspond to, properties or relations which belong to the individuals? Isn't a relation or a property just a feature of reality which cannot exist on its own – independently of the individual? And isn't this what Frege meant by the unsaturated entity which the function expression names?

No. If we take such a view of isomorphic representation we immediately run into two kinds of difficulty. First, as Hintikka says, such correspondence to features of reality occur at most within *true* propositions. Think of the proposition 'Nero burned London', which happens to be false. The name 'Nero' stands for or names the historical Roman emperor and the name 'London' stands for or names the city. What, according to the above interpretation of the picture theory, could the fact that the word 'burned' is written to the right of the name 'Nero' and to the left of the name 'London' be correlated to? It is not a feature of historical reality that Nero stood to London in the relation of one burning the other. What kind of thing is the relation which, in this case, does not belong to Nero and London? For Frege it was quite clear that in this proposition the relation word has a reference. This is because for Frege the reference of a concept word is not in any simple way a constituent of reality. A concept-word having a reference is defined in a technical way: '- - - burns - - -' is a name of a first-level function of two arguments, and so long as the sentence which results from putting proper names in the argument places has a truth-value (i.e., has a reference in Frege's jargon), the relation word has a reference. That is all there is to reference in this case. Thus a concept word or a relation word having reference in a sentence for Frege does not entail that the word corresponds to a feature or an element one would be able to find in reality. Frege does make a sense-reference distinction for concept words too, but this distinction does *not* correspond to the intension-extension distinction in the way in which many have assumed.[13] As Dummett has pointed out, even concept words expressing contradictory properties which can have no extension must have reference in Frege's theory.[14] And the only concept words which are said by Frege to have no reference are those which are not defined for every object – i.e., those whose domain is vague.

I wonder whether the predicate which predicate expressions represent in Professor Hintikka's model theory will evade these problems. If 'Nero burned London' is false, the relation which corresponds to '- - - burned - - -' is said not to belong to Nero and London in that order. Thus, there has to be for Hintikka something which corresponds to the relational expression whose existence is invoked even to explain false elementary propositions. In model theory this is normally identified with a set of ordered pairs of objects, the first of which burns the second. It is an abstract object and hardly a constituent of reality as individuals are.

The picture theory of the *Tractatus* avoids this difficulty. There is something in reality in virtue of which an elementary proposition is true – when it is true. And there is something common to how things are when '*A* is blue' and '*B* is blue' are both true, or when 'Nero burned Rome' and 'Cromwell burned the Abbey' are both true. There may even be criteria for determining for any *a* that *a* is blue, or for any two objects that the former burned the latter, but we do not invoke any extra entity to do this. We do not need this and it would not help if we had it.[15]

This brings me to the second difficulty. If we make a model theoretical definition, in the manner of Tarski, of what logicians call an interpreted language L (and all our languages are interpreted language) then a formula of the form $P_n(Vi_1, \ldots, Vi_n)$ of L is satisfied in a model M if the sequence of individuals in the domain A of M specified by a valuation of the variables, belong to some set (or ordered n-tuples) which is a relation on A. This has the consequence that the truth-conditions of a sentence of the form fa and ga are identical if f- and g- are co-extensive. fab and gab are identical if $f(- - -)$ and $g(- - -)$ are co-extensive. And as many have pointed out, Frege can hardly avoid this difficulty either, since although his concept is an unsaturated entity which is *not* the extension of the concept-word and not even the course-of-values which is a set of ordered pairs, he claims that a relation *analogous to identity* holds between two concepts just in case they are co-extensive. We can see that such a view fails to reflect the contention of the *Tractatus* that distinct thoughts are expressions of distinct truth-conditions. And such a view would certainly not provide a theory of meaning. Thus as Professor Stegmüller suggested, one has to extend the models into possible worlds and take the extension of the predicate in all possible worlds in the manner of Carnap. Such a highly theoretical construct is hardly an element of reality of the kind hinted at by Hintikka. Also, unless we split every predicate expression and relation word into indefinitely many time-indexed predicate expressions and relation words we cannot begin to assign a set of individuals or a set of n-tuples of individuals to them, since the extension of predicate expressions changes in time. Thus making predicate expressions 'represent' some entity leads to confusion rather than clarity.

(d) METHOD OF PROJECTION AND LANGUAGE-GAMES

How then do we understand the truth-condition of elementary propositions, and what is represented by the predicates in them? The agreement people have in distinguishing types of states of affairs (in forming propositions with common predicate expressions with common syntax) is the only basis of the *correspondence* between what one might call the predicative fact (i.e., how names occur in a proposition and how they are related to the predicate word) and the property or relation which this fact represents. We will see that in his later works, Wittgenstein extended to objects what is implied in his talk about external properties and relations in the *Tractatus*.

In *Zettel* Section 291 Wittgenstein comments on the fact that the phrase 'link (*Verbindung*) between the picture and what it depicts' might mean the lines of projection (lines correlating objects to correlated objects), or might mean the technique of projection. Wittgenstein is saying that he is there concerned with the latter and not the former. Similarly in the *Tractatus*, so long as one has a method of projecting from a fact about a predicate expression and names, one can be said to depict a property or a relation of the objects and so one has a link between the predicative fact and the property or relation.

But this does not mean that the semantics of predicate expressions is based on a vertical link between them and other entities (i.e., on what Wittgenstein calls lines of projection or the existence of a mapping function which maps the predicate expressions to an entity called predicate, attribute, relation or whatever). The *Tractatus*' purpose in identifying that which represents relations or properties as a fact about signs rather than as a verbal object, is to register that facts about names can represent possibilities, i.e., possible facts about the entities which the names stand for without invoking the existence of any possible entities. In the *Tractatus* a name in the argument place of a proposition stands for a bearer which

exists whether the proposition is true or false. A name is correlated to a
bearer. In contrast to this a predicative fact represents *possible* relations
in which the objects *may* stand, or properties which they *may* have, quite
independently of whether the proposition is true or false.

We can of course construct an abstract entity $\phi(P)$ which we may call
property or whatever, such that when '*Pab*' is true we say that $\phi(P)$ belongs
to the individuals *a* and *b* in that order. (This would be like the *Principia*
view about classes, i.e., the no-class view where classes are contextually
defined as a *façon de parler*, so that to say that an individual *a* belongs to
the class $\hat{x}.fx$ means nothing more than *fa*.) But there is no entity $\phi(P)$
such that we can *independently* identify it and correlate it to an expression
P. Thus introducing such an abstract entity does not *help* us in understanding the truth-condition of propositions. To use Hintikka's own idiom,
there are no vertical links between language and reality for predicate
expressions. The meaning of predicate expressions is not based on such
links. This does not mean that the predicative fact does not represent
properties or relations which may hold of the objects denoted in reality, or
that there may be no objective criterion for deciding that this is so. I
learn such criteria by learning through action and by use of language
related to our activity when some particular kind of propositions which
contain this predicative fact can be asserted – is agreed to be true; i.e., by
catching onto whatever they represent. The *Tractatus* thus says "To
understand a proposition means to know what is the case if it is true"
(4.024), and this is to acknowledge the situation it represents (4.021).

The 'Picture Theory' of the *Tractatus* therefore naturally develops into
Wittgenstein's views given by his metaphor of 'language-games' in the
Philosophical Investigations. Hintikka has rightly emphasised the import-
ant fact that the language games invoked by Wittgenstein, as well as by
Hintikka himself, are not a games we play in language, but ones which
gives us rules of activities we carry out with language in the world. As
Hintikka says, "one does not learn meanings by being told what they are.
One learns them by learning to play the associated games . . . one learns
them in living the associated forms of life."[16] But it seems to me wrong for
Hintikka to suggest that to thus explain meaning by invoking language
games is to extol the "crucial role of vertical relationships between
language and reality". On the contrary Wittgenstein is showing how we
can talk about and describe reality without invoking such meaning-
giving vertical links between word and the world. In the *Tractatus* it was
shown that to understand what is represented by a predicative fact (so

that we can see whether it obtains in reality) was to master a method of projection, rather than to learn what entity the predicative fact is linked to. In his later works Wittgenstein extends this explicitly to the use of other expressions. We learn the meaning of common nouns and even names by learning, as it were, the method of projection. We can have a public way of telling whether there is something in reality corresponding to what is projected, but the method of projection itself does not come from a *prior* vertical link between a word and a bit of the world. Indeed if this were the case one could not give an account of how words which are used to refer to our sensations or experiences could ever have public use and meaning. This is one of the central themes of the *Philosophical Investigations* as Hintikka and Provence have said. I suspect that Wittgenstein already had views of this kind and objects and names in the *Tractatus* also. As he wrote in *Philosophical Remarks* Section 36 a bit mysteriously, what he meant by objects in the *Tractatus* was what one can indicate without having to fear that they may not exist. We may understand this much better in our talk of mathematical objects. We have objective ways of establishing the truth of propositions which refer to them and that is all that matters. Precisely because the name-object relation did not depend on establishing vertical links between independently identified entities, he did not provide any examples of objects. The fact that he used verbs like 'standing for' however falsely suggested the existence of a vertical link (i.e., of a link that came about not as a result of the possession of a method of projection but of something that explains the method of projection).

The picture view in the *Tractatus* made it however quite clear already that in the case of a predicative fact we can see that language can be used with a public criterion of application to describe reality, without a correlation between this fact about language and an element of reality. That is why he said later that this relation should not really be called correspondence (*Übereinstimmung*) but pictorial character (*Bildhaftigkeit*).

University College, London

NOTES

[1] *Vide* e.g. 'Quantification and the Picture Theory of Language', *The Monist* 53 (1969) and reprinted in *Logic, Language-Games and Information*; 'Language Games' in *Essays on Wittgenstein in Honour of G. H. von Wright*, Acta Philosophica Fennica, vol. 28, nos. 1–3, North-Holland, Amsterdam, 1976.

[2] There are less central uses of both 'stand for' and 'represent'. In the *Tractatus* notation a horizontal bar over a bracketed propositional variable stands for the conjunction of all the propositions which are the values of the variable in the bracket (5.501). The truth table represent formal concepts.

[3] 'Language Games', p. 106.

[4] See 'Quantification and the Picture Theory of Language', p. 33. "What represents this predicate is not the predicate symbol 'P' itself but rather the relation which obtains between n argument places in the expression '$P(x_1, x_2, \ldots, x_n)$'".

[5] *Philosophical Remarks*, Section 93.

[6] The fullest discussion on this point is given by Peter Long in 'Are Predicates and Relational Expressions Incomplete?', *Philosophical Review* 78 (1969). Also discussed on pp. 37–40 of G. E. M. Anscombe, *Introduction to Wittgenstein's Tractatus*, Hutchinson & Co., London, 1959, and in Hidé Ishiguro, 'Use and Reference of Names' in *Studies in the Philosophy of Wittgenstein*, P. Winch (ed.), Routledge & Kegan Paul, London, 1969. Each gives a different account of elementary propositions with only one name.

[7] Terence Parsons has criticised in *Philosophical Review* 79 (1970) what I believe is a correct view expounded by Peter Long in 'Are Predicates and Relational Expressions Incomplete?', *Philosophical Review* 78 (1969), and has concluded wrongly that there is a symmetry between the role of predicate and the name in the subject place.

[8] *Philosophical Remarks* Sections 93–94.

[9] 'Complex and Fact'. Appendix 1 of *Philosophical Grammar* and *Philosophical Remarks*.

[10] 'Concept and Object', in *Translations from the Philosophical Writings of Gottlob Frege*, P. T. Geach and M. Black (eds.), Basil Blackwell, Oxford, 1960, p. 45.

[11] *Notebooks 1914–1916*, p. 61.

[12] 4.123.

[13] E.g. Hilary Putnam, 'The Meaning of Meaning', *Philosophical Papers*, vol. 2, Cambridge University Press, Cambridge, 1975, p. 218.

[14] Michael Dummett, *Frege: Philosophy of Language*, Duckworth, London, 1973, p. 219.

[15] It is true that Hintikka, Lambert and others have given semantical theories in which even singular terms carry no existential presupposition. But this is made possible by assigning a singular term to an entity in some possible world. Whereas, here, no entity is involved. Corresponding to every different way in which we notice how things can be, either in themselves, or in relation to other things, we agree to use, or we find ourselves using, different predicative facts to represent them.

[16] Jaakko Hintikka and Merrill Provence, 'Wittgenstein on Privacy and Publicity' (p. 5 of typescript).

DAGFINN FØLLESDAL

HUSSERL AND HEIDEGGER ON THE ROLE OF ACTIONS IN THE CONSTITUTION OF THE WORLD*

In this paper, I shall be discussing Husserl's and Heidegger's views on the role that human activity plays in the constitution of the world. While the basic idea in Husserl's phenomenology is that we constitute the world through our consciousness, Heidegger's main contribution to philosophy, it seems to me, is to focus attention on the idea that all human activity, all our ways of relating to the world, to one another and to ourselves, contribute to constituting the world.

This is, I think, a major advance upon Husserl. Yet, there are passages in Husserl's writings too, mainly in the unpublished manuscripts, that indicate that he thought that our actions and practical activity did play a role in our constitution of the world. We shall consider these passages later. Let us, however, first discuss generally what Husserl's and Heidegger's views on constitution were, according to their published works, and how their views differ. Let us begin with Husserl.

1. HUSSERL'S VIEW ON CONSTITUTION: INTENTIONALITY

For Husserl, 'constitution' is just another label for the intentionality of consciousness. As you will all know, Husserl held, like this teacher Brentano, that consciousness is characterized by a certain kind of directed-ness, there always seems to be some object towards which consciousness is directed, *of* which we have consciousness. When we think there always seems to be something of which we think, when we perceive, something that we perceive, etc., and similarly for all other kinds of mental activity, or acts. Brentano tried to clarify this notion of directedness by focusing upon the object, the "intentional object," as he called it, but he came into serious difficulties in connection with cases where there is no such object, as in the case of hallucinations, or someone thinking of Pegasus. One might try to overcome these difficulties by holding that the objects of our consciousness are not real, but in some way are contained in our own

E. Saarinen, R. Hilpinen, I. Niiniluoto, and M. Provence Hintikka (eds.), Essays in Honour of Jaakko Hintikka, 365–378. All Rights Reserved.

consciousness, whatever that may mean. However, watering down the objects in this way leads to difficulties in the case of many other acts, for example, acts of normal perception. It seems that, on that view, what we see when we see a tree is not the real tree in front of us, but something else, which we would also have seen if we were hallucinating. So, the view that every act is directed toward an object leads to a dilemma.

Husserl resolved this dilemma by proposing an analysis of consciousness where it is not crucial that there be an object towards which the act is directed, but where attention is focused on what the directedness consists in, what features of consciousness it is that make consciousness always be as if it is consciousness of something. Thus, in the case of perception, Husserl is interested in those features of consciousness that make an act of perception be as if it is of an object of such-and-such a kind, located in such-and-such a manner with respect to other objects and with respect to the perceiver. Husserl is also interested in those features of the act that make it an act of perceiving and not, for example, one of remembering or imagining.

2. NOEMA

All these features of the act, both those that determine its objects, if it has any, and those that determine its kind, Husserl calls the noema of the act, from Greek νόημα, that which is thought, that which is grasped.

Husserl conceives of the noema as an intensional entity, a "generalization of the notion of meaning to the realm of all acts" (1952, p. 89, 1.2–4). Just as the meaning of a linguistic expression determines which object the expression refers to, so the noema determines what the object of an act is – if the act has an object; some acts have a noema for which there is no corresponding object.

The object of the act is a function of the noema, that is, given the noema, the object, if any, is uniquely determined. The converse, however, does not hold, to one and the same object there may correspond several different noemata, depending upon the various ways in which the object can be experienced, whether it be perceived, imagined, remembered, etc., and depending upon its orientation, our point of view, etc.

To take an example from perception, let us consider the act of seeing a tree. When we see a tree we do not see a collection of colored spots, for example brown and green distributed in a certain way; we see a tree, a material object with back, sides, and so forth. Part of it, for example the back, we cannot presently see, but we see a thing which has a back. That seeing is intentional, object-directed, means that the near side of the thing we have in front of us is regarded only as a side of a thing, and that the thing we are seeing has other sides and features which are co-intended to the extent that the full thing is regarded as something more than the one side. The noema is the complex system of determinations which unifies this multitude of features into aspects of one object. Note that the noema itself is an abstract entity. Its different components correspond to different features of the object, to its color, its flammability, its changeability etc., but the noema itself does not have these features. For example, the noema cannot burn, Husserl observed (*Ideen* I, 222.10 [= page 222, line 10 of the Husserliana edition]).

3. TWO MAIN DIFFERENCES BETWEEN HUSSERL'S NOEMA AND BRENTANO'S INTENTIONAL OBJECT

There are two features of Husserl's analysis of intentionality which are of crucial importance when we compare him with Brentano. First, and most obviously, the noema, that thanks to which the act is directed, is not that *towards* which the act is directed. If the noema were the object of the act, or contained the object as a part, then we would be back in Brentano's dilemma again, we would not have a satisfactory way of treating acts like thinking of Pegasus, that lack an object.

Secondly, by focusing on the structures of our consciousness, Husserl gives more weight than Brentano to what the intentionality of consciousness consist in. To say, as Brentano did, that intentionality consists in consciousness always having an object, is as we have seen, not quite true. Nor is it very illuminating. What is true, is that consciousness is always *as if* it has an object. This formulation preserves Brentano's valuable insight. The 'as if' also leads us to ask the questions that Husserl asked and attempted to answer by his phenomenology: What is it for consciousness to have an object?

4. CONSTITUTION

Let this be enough about Brentano and Husserl's analyses of intentionality. Given the Husserlian analysis, in terms of the noema, it is now easy to see what is meant by *constitution* in Husserl. That objects are constituted by us simply means that they are intended in the way that we have described, as having a great number of aspects and features, normally, as in the case of all material objects, many more than can ever be exhausted in our experience of them. That objects are constituted through our acts does not mean that they are caused by our acts or brought about by our acts, but just that in the act the various components of consciousness are interconnected in such a way that we have an experience as of one full-fledged object. All there is to the existence of an object hence corresponds to components in the act. In the case of physical objects, the inexhaustible character of what is experienced is a characteristic feature of the act and of what it is to *be* for a physical object. However, in the extended sense of the term "constitution," Husserl says, an object – whether it is physical or not – " 'constitutes' itself within certain connections of consciousness which bear in themselves a transparent unity so far as they carry with them essentially the consciousness of an identical *x*" (*Ideen* I, 332.25–28).

Incidentally, Husserl's use, here and many other places, of the reflexive form 'an object constitutes itself' is an indication that he did not regard the object as being brought about by the act. Husserl considered phenomenology as the first strictly scientific version of transcendental idealism, but he also held that phenomenology transcended the traditional idealism-realism distinction, and in 1934 he wrote in a letter to Abbé Baudin: "No ordinary 'realist' has ever been as realistic and as concrete as I, the phenomenological 'idealist' (a word which by the way I no longer use)." (Letter quoted in Iso Kern, *Husserl und Kant*, Nijhoff, The Hague, 1964, p. 276n.) Husserl did not attempt to 'reduce' reality to consciousness. According to Husserl, there is a certain givenness in our experience of the world, an ego-foreign element enters, the hyle. However, we shall not go into this aspect of his philosophy in this paper.

5. HEIDEGGER'S PHILOSOPHY AS A TRANSLATION OF HUSSERL'S

After this brief sketch of Husserl's views on constitution, let us now turn to Heidegger, before at the end of the paper we return to Husserl, to consider some of his unpublished writings.

Heidegger's philosophy seems at first sight very different from that of Husserl. The themes discussed, the vocabulary and the style of writing, all are different. However, Heidegger acknowledges that he has been strongly influenced by Husserl. He characterizes *Being and Time*, which is the work that we will concentrate on, as a phenomenological work, and he states explicitly that he is using the phenomenological method.

The key to this puzzle, and also, I think, the key to understanding what goes on in Heidegger's philosophy, is that Heidegger's philosophy is basically isomorphic to that of Husserl. Where Husserl speaks of the ego, Heidegger speaks of Dasein, where Husserl speaks of the noema, Heidegger speaks of the structure of Dasein's Being-in-the-world and so on. Husserl also observed this. Several places in his copy of *Being and Time* Husserl noted in the margin that Heidegger was just translating Husserl's phenomenology into another terminology. Thus, for example, on page 13 Husserl wrote: "Heidegger transposes or transforms the consttutive phenomenological clarification of all realms of entities and universals, the total region World into the anthropological. The whole problematic is translation, to the ego corresponds Dasein etc. Thereby everything becomes deep-soundingly unclear, and philosophically it loses its value." Similarly, on page 62, Husserl remarks: "What is said here is my own theory, but without a deeper justification."

In saying that Heidegger's philosophy is basically isomorphic to that of Husserl, I do not claim that there is isomorphism in every point. As we shall see, there are at least two major points where they differ. But the isomorphism comprises more than just some main points which Husserl and Heidegger have in common with e.g. Kant. The basic notions in their philosophy are interrelated down to small details that are not even found in Kant. There will not be place in this paper to support this claim by detailed analysis of the text. In the following we shall only make use of the main traits of the isomorphism, which will serve two basic purposes: to facilitate the presentation of some of Heidegger's ideas on the background of what has now been said about Husserl, and to make it easier to see and assess the major differences between them.

6. AN OUTLINE OF HEIDEGGER'S PHILOSOPHY

When interpreted in this way, as a translation of Husserl, the basic ideas in Heidegger's philosophy can be outlined as follows: Heidegger's aim is to clarify what it is to be. For a thing in the world to be is to be constituted. To understand what it is to be constituted, we have to focus on that which constitutes, viz. the ego, or Dasein. Heidegger uses the word 'Dasein' in order to emphasize the dual role of the ego, it is there (da), thing among things in the world, but it is also the source of Being (Sein) by being that which constitutes the things in the world. While for things to be is to be constituted, Dasein is both constituted and constituting, and this latter is its distinguishing mark. The most appropriate answer to the question "What is Dasein?" is "It constitutes."

There are hence two varieties of Being, to be constituted and to constitute. For the latter of these, Heidegger introduces the term 'existence'. Dasein's "essence", that is the appropriate answer to the question "What is Dasein?" is hence existence. These two aspects of Dasein in Heidegger, that it is in the world and that it constitutes the world, correspond to Husserl's empirical and transcendental ego, respectively. Heidegger in a letter to Husserl criticized Husserl for operating with two egos and thereby having a schizophrenic theory of the ego. Husserl replied that the empirical and the transcendental egos were not two egos but two aspects of the same ego. There is just one ego, Husserl said, it is in the world and it constitutes the world. On this point there hence seems to be a more complete parallelism between Heidegger and Husserl than Heidegger seemed to think.

7. HEIDEGGER'S VIEW ON CONSTITUTION

However, what is it now to constitute? We have already seen what Husserl meant by 'constitution.' Heidegger's theory is more complex. He means by 'constitution' Being-in-the-world in the special way in which Dasein is in the world. That is, not being in the world the way water is in a glass and thing are among things, but relating to the world in all the numerous ways that Dasein may relate to the world. *Being and Time* is largely a working out of all the different ways in which Dasein may relate to the world.

One of these ways of relating to the world, the one that Husserl considers, is to relate to it in a theoretical way, as in the sciences, when we study the various objects and their properties, perceive them and theorize about them. When dealt with in this way, objects are experienced as "present-at-hand" Heidegger says. Ever since Aristotle, this has been the favorite way for philosophers to conceive of our relationship to the world, so favorite, in fact, that most philosophers have not thought of other ways, and if they have considered them, they have tended to regard them as secondary, based on the theoretical attitude. It has commonly been held that practical activity presupposes theoretical understanding of the world, action presupposes that I have some idea of what is likely to happen if I do this or that. Heidegger rejects this. He regards our practical ways of dealing with the world as more basic than the theoretical. Things in the world are primarily experienced as ready-to-hand, he says, as tools and equipment that are used by us. They are what they are in virtue of the position they have within the full context of human activity, with its pattern of means and ends. To be a hammer, for example, is to be a tool for driving in nails, for crushing things etc. It is only when we reflect theoretically that a hammer is conceived of as an object with a certain shape, length, weight, color etc. The theoretical attitude is "parasitic" on the practical: it is through our practical dealing with the world that objects become separated out from their surroundings and become individuated, as objects that we may later in our theoretical moments subject to theoretical scrutiny.

I will not here go into any of the details of Heidegger's analyses of the many ways in which we may relate to the world. They anticipate in many respects some of Wittgenstein's later analyses of forms of life. In fact, much of what Wittgenstein says about the meanings of words being a product of their use, parallels Heidegger's observations concerning the "meanings" of objects, that is, what it is to be for various objects. This is what we should expect if, as Husserl claimed, questions of constitution and questions of meaning are intimately connected, the noema being as we noted "a generalization of the notion of meaning to the realm of all acts."

8. ANTICIPATIONS OF HEIDEGGER'S IDEAS IN HUSSERL

This idea of Heidegger's that all our human activity plays a role in our constitution of the world, and his analyses of how this happens, I regard, as I mentioned in my introduction, as Heidegger's main contribution to philosophy. This is also the main point where he goes beyond Husserl. (Another main point where he differs from Husserl, is that he does not accept Husserl's "transcendental reduction," i.e. the special reflection whereby, according to Husserl, we study the structure of our own consciousness, i.e. the noema, or constitution. Heidegger held that no such reduction is possible, and that we can only study this structure "from

inside," i.e. by becoming aware that we are in the midst of it. This awareness is normally brought about by some familiar tool's breaking down,
or by our facing death, etc. However, we shall not discuss Heidegger's
views on the transcendental reduction here, nor his alternative way of
doing phenomenological analysis.) Interpreting Heidegger in the Husserlian
way that I have done makes it rather easy to pinpoint just what Heidegger's
contribution consisted in. Husserl, as we have seen, in the works that he
published focused almost all his attention on our theoretical way of
relating to the world. Most of his examples deal with perception. He says
that this is because perception is a relatively simple and basic kind of act
that occurs as a component in other more complex acts, e.g. of evaluation
and volition. However, after he came to Freiburg in 1916, in the late
teens and especially in the early twenties, Husserl clearly became more
and more aware that our practical activity is an important part of our
relating to the world. Thus, in the notes that he made for a revised edition
of the *Ideas*, he remarks that the practical attitude should also be included
in several places where in the first edition he speaks only of our theoretical
attitude. In three lectures on Fichte that he gave in November 1917 and
repeated twice in 1918, he points out that for Fichte to be a subject is to
be one who acts, and correspondingly, "to be object for a subject is to be
a product of action. When we go back to the beginning, there is nothing
that lies before action. When we consider what we may call the *history of
the subject*, the beginning is not a matter of fact (Tatsache) but an action
(Tathandlung), and we have to think of a history here. To be a subject is
eo ipso to have a history, a development. To be a subject is not only to
act, but also necessarily to proceed from action to action, from the product
of one action through a new action to new products" (F I 22, 9a–20).
There is, according to Husserl, "an infinite chain of goals, aims and tasks"
that our actions and their products relate to. This idea and terminology
of Husserl's from 1917 strikingly resembles the idea of a chain of assignment (Verweisungen) or 'in order to' relations that our tools and activities
fit into, that Heidegger presented in *Being and Time* ten years later (especially in Sections 17 and 18 in *Being and Time*). Incidentally, there is
another striking terminological parallel between Heidegger and Husserl,
in that Heidegger's key term 'Sorge' (care) which Heidegger uses as a
common term for all of Dasein's ways of relating to the world, occurs in
a manuscript of Husserl's from 1925, where Husserl separates human
activity into two main kinds: Spiel (game) which is aimless, and "serious
praxis in the original and serious Sorge (care)" (B I 21, IV, 52).

We should also note that Fichte, whom Husserl discusses in the lectures from 1917, seems to have been the first to introduce the notion of action in connection with constitution and may have been a common source of inspiration for both Husserl and Heidegger on this topic.

9. HUSSERL ON HOW ACTIONS CONSTITUTE THE WORLD

The manuscript where Husserl most explicitly explains how he thinks that our practical actions constitute the world, dates from about the same time as the Fichte lectures, 1917–18, and is called 'Science and Life'. This is a very important manuscript, both because it is the most informative that Husserl wrote on the topic of how our practical actions constitute the world, and because it is so early, from 1917–18, i.e. well before Heidegger started writing *Being and Time*. *Being and Time* was published in 1927, ten years after Husserl's Fichte lectures. In this manuscript, in a section on 'Similarity,' Husserl first explains briefly how our theoretical conception of an object is based on analogies from earlier experiences. These analogies, or similarities, serve as guides (Leitfäden) for our expectations. An "analogical predelineation" (analogische Vorzeichnung) takes place. After this explanation of how constitution takes place in the theoretical realm, Husserl continues by introducing the notion of practical apperception: "This then carries over to the practical realm and to practical apperception: One relates to that which is apperceived by analogy in a way similar to that in which one relates to that to which it is analogous. One 'acts with' it ('behandelt' es) in a similar way, one tries to set up similar ends in connection with it, or to use it in a similar way as a means, to direct it, etc. The analogical extrapolation is confirmed through practical activity and leads to firmer theoretical and practical apperceptions than in the previous analogical apperceptions" (B I-21.16). We may conclude from this passage that, according to Husserl, constitution in the theoretical and in the practical fields work in parallel ways and are in both cases based on similarities and analogies. These analogies give rise to the extrapolations that make up the noema and thereby constitute the object. The difference between the theoretical and the practical realm is that in the theoretical realm the extrapolations concern factual properties of the object, while in the practical realm the extrapolations have to do with the object's servicability within our web of means and ends. This explanation of how our actions constitute the world by extrapolation based on similarity in our practical ways of dealing with things is in my opinion some of the most informative that either Husserl *or* Heidegger ever said on this issue.

Note that there is nothing to indicate that Husserl thought of practical apperception as being based on a *perception* of factual similarities between things, from which one infers or anticipates that they can also be treated in similar ways for practical purposes. Nor does he claim that we *perceive* praxis and then anticipate similar praxis by a *cognitive* process. He does not impart the order of seeing and knowing into a domain where doing and know-how are basic. Practical apperception for Husserl consists in our extrapolating practical features in a way similar to that in which we extrapolate theoretical features: when two things have turned out to be similar in some practical applications, we come to deal with them as similar also in many other practical applications. Perception and seeing do not have primacy over action for Husserl. Remember how a couple of pages ago we quoted him as saying: "the beginning is not a matter of fact, but an action."

10. PARALLELS BETWEEN THEORETICAL ACTIVITY AND PRACTICAL ACTIVITY

1. *Both Governed by Interests*

This parallelism between theoretical and practical constitution fits well in with Husserl's observation, in several other manuscripts, that theoretical activity is just one of the many kinds of human activity. Like all other activities it is, for example, intended to serve our ends and purposes. All activity, including theoretical activity, is governed by our interests, and we have a special purely theoretical or purely doxastic, interest (B I 21. IV.14, from 1925) in knowing and in forming concepts and judgments concerning them (A VI 26.103, from 1927). "The theoretical interest inhibits all other interests of a personal kind . . . one is the 'disinterested spectator,' who puts out of play interests in the common other sense." (B I 21.IV.14 again, from 1925). Husserl also says that "the 'theoretical', the purely observing attitude is . . . a special practical attitude . . ." (F I 44 [1. Teil] from 1926–27). Husserl points out that "the main interest in knowledge may serve some practical interest, but it may also be a pure theoretical interest, an interest in the thing itself 'as it really is'." (A VI 26, 18a from 1928).

2. *Anticipations and Fulfilment*

However, now back again to the parallelism between theoretical and practical anticipation. Husserl talks about "practical anticipations" also in a manuscript from 1922–23, 'Introduction to Philosophy' (F I 29.186. B1.94Rs.) and discusses in several manuscripts the practical horizon of possibilities connected with our actions (e.g. in A V 10 [1. Teil], 2 from 1931). Even in one of his published works, the *Crisis of the European Sciences*, Husserl talks about theoretical activity as a kind of Praxis and points out that the "Life-world" is the horizon within which all human activity theoretical and non-theoretical, takes place (*Krisis*, 145.24-31).

In another manuscript from 1922–23, 'Premeditations concerning the *Idea of Philosophy*' (Vormeditationen über die *Idee der Philosophie*), Husserl points out that one may speak of a fulfilment of the intention not only in the theoretical realm, but also in the practical realm. In the theoretical realm our intention becomes filled e.g. through perception, filling here is what we call evidence. In other realms, like those of actions and feelings, the filling is of other kinds, depending in each case on the kind of intention involved and what will satisfy it. Following Fichte, Husserl uses the word 'bliss' (Seligkeit) as a common term for every kind of fulfilment of the intention of an act. (B I 37.31a). There is here and some other places in Husserl a tendency to fuse together two notions of intention that Brentano kept well separated, intention in the sense of the directedness of an act upon an object, and the practical sense of the purpose, or aim, of an action. This fusion is especially apparent in a manuscript from 1928, where Husserl says: "Just as there is a 'practical' aim in the case of a volitional decision and an action by which something is done . . . so there is generally [an aim] in the case of every intentionality which is directed, that is, also in the case of representation and thought." (F I 44 [2. Teil], 123-b). This fusion of two notions that Brentano regarded as very different, might easily be considered as a *con*fusion. However, Husserl is careful to point out that there are important similarities between the two notions: "as long as the action has not yet led to the goal towards which it is directed we are still conscious of the goal, as that towards which the action is aimed, but the goal is transcendent – a transcendent reality that has not yet been made real." (*Ibid.*)

11. CONCLUSIONS

I will end my paper with one historical observation and one systematic one.

1. *Husserl and Heidegger*

The historical observation concerns the relation between Husserl and Heidegger. There seems to me to be every reason to hold, as Husserl did, that the main framework of Heidegger's philosophy in *Being and Time* is a translation of Husserl's. Regarding it as such a translation, it becomes easier to see where the two philosophers were in agreement and where they differed. What seems to me to be the main point of difference, is that Heidegger held that the world is constituted through all kinds of human activity, while Husserl, in his published works, thought of the world as constituted basically through our theoretical activity. However, it should be clear from all the quotes I have given from Husserl's unpublished manuscripts that Husserl had ideas similar to those of Heidegger long before *Being and Time* was published. These ideas started appearing in Husserl shortly after he arrived in Freiburg and met Heidegger in 1916. It is possible that Husserl influenced Heidegger in this "practical" direction. In any case, these ideas start appearing in Husserl's manuscript long before Heidegger began writing *Being and Time*. However, it is also possible that it was Husserl who was influenced in this direction through his discussions with the younger Heidegger. Neither of them tells, and there seems to be no other available information concerning what way the influence went. Most likely, it went both ways.

However, what is important, is that Heidegger made this idea of how we constitute the world through practical activity a main theme in his philosophy, around which everything else is organized, while for Husserl this idea seems to have been an afterthought that is only fragmentarily presented and not incorporated in the body of his writings.

2. *Is the Body Necessary for Explaining What Constitution Is?*

Secondly, my systematic point, that fits in closely with the historical one: Husserl insists throughout his writings, including all the manuscripts, that it is through our consciousness that we constitute the world. All of our actions and activities are regarded by him as "activities of consciousness" (Bewusstseinstätigkeiten). I quote one of his manuscripts from 1922–23

(Vormeditationen über die *Idee der Philosophie*) "Now we are considering man as not merely an inductive external unity of body and mind, as a two-levelled real object in a space-time-causal context; now we take him in the way in which we take *ourselves* when we say *I* or when in an I-you relationship I take the other as you, ask him for something etc. That is, we are now considering man as an I-subject, which *as such* 'relates to its surrounding world' as an I – that is, perceives, experiences, remembers, thinks, feels, wills, acts as an I and in all these 'activities of consciousness' has consciousness of a surrounding world, which he is conscious that affects him. To this world, of which he is conscious and which affects him, he consciously takes a position, theoretically as well as practically." (B. I 37.15). However, is consciousness all we need to consider when we seek to clarify how our actions contribute to constituting the world? Do not our actions also involve our bodies in a way that makes it necessary to bring the body into the picture when we discuss constitution? Several places, both in his published and unpublished writings, Husserl explains how the body does play a role in constitution. It is a "center of all orientation in the world" (Zentrum aller Orientierungen) (B I 37, 126), it is a "center of my bodily activities" (D 13 VII, 1) and "the direct object of all experiencing and acting-effectuating praxis" (handelnden-Wirkenden Praxis) (*ibid.*). Also, Husserl emphasizes repeatedly the role that kinesthesis plays in perception. Thus, for example, in the manuscript 'Introduction to Philosophy' from 1922-23, he says: "... my having a body (Leiblichkeit) plays an essential role in all perception, partly because of the kinestheses that are constantly carried out and partly because of what I call perfection of my senses, my sense of sight, of touch etc . . ." (F 1 29, 221). There are also many similar passages in Husserl's *Krisis*.

And what about the routine movements of my body, of which I am not normally conscious, the movements of my legs when I walk, etc.? Do not also these movements contribute to the way in which I constitute the world? And do they have any counterpart in consciousness? Perhaps should such routine movements not be called actions, but if so, does this not merely show that in our constitution of the world, not only actions, but also such movements play a role?

Husserl insists that all constitution is by way of consciousness. He is constantly occupied with the genesis of constitution, that is how we come to have the noemata that we have, and he may have held that while the body is important in the *process* that leads us to constitute the world the way we do, the *product* of the constitution, the world as I conceive of it

DAGFINN FØLLESDAL

at any given time, has to be a counterpart to the structure of my *conscious-ness* at that time. The body hence has to be brought in in order to explain the genesis of the structure of my consciousness, and Husserl brings it in. However, this does not disprove, he thinks, that the world is constituted by our consciousness.

Heidegger, and also Merleau-Ponty, seem to hold that the body has to be brought in not only to explain the genesis of constitution but also to account for what constitution is. The world for them is not constituted by consciousness alone, but by an ego that is both bodily and conscious, i.e. Dasein in Heidegger's terminology. Of course, for Husserl, too, the ego is both bodily and conscious, but for him only consciousness has to be appealed to in order to explain what constitution is.

University of Oslo and Stanford University

NOTE

* I am grateful to mag. art. Elling Schwabe-Hansen for his help in finding passages in Husserl's manuscripts that illuminate Husserl's views and to the Norwegian Research Council for Science and the Humanities for its generous support of this research. I am also indebted to Hubert Dreyfus, Harrison Hall and Samuel Todes for their comments on an earlier version of this paper that was read at a symposium at the American Philosophical Association Western Division Meeting in Cincinnati, April 27–29, where they served as commentators. I have also been helped by comments from Robert Nozick and John Perry.

INDEX OF NAMES

TABULA GRATULATORIA

Joseph Agassi, *Boston University and Tel-Aviv University*
Evandro Agazzi, *University of Genova*
Erik Allardt, *University of Helsinki*
Jens Allwood, *University of Göteborg*
Lennart Åqvist, *Universität Stuttgart*

Kent Bach, *California State University, San Francisco*
John Bacon, *New York University and School of Visual Arts*
E. M. Barth, *University of Groningen*
Jon Barwise, *University of Wisconsin*
Jonathan Bennett, *The University of British Columbia*
David Berlinski, *University of Puget Sound, Tacoma*
Jaakko Blomberg
Karel L. de Bouvère, *The University of Santa Clara*
Myles Brand, *University of Illinois at Chicago Circle*
Arthur W. Burks, *University of Michigan*
Robert E. Butts, *The University of Western Ontario*

Vincenzo Cappelletti, *Istituto della Enciclopedia Italiana, Roma*
Nancy Cartwright, *Stanford University*
Robert L. Causey, *University of Texas*
Brian F. Chellas, *University of Calgary*
Noam Chomsky, *Linguistics and Philosophy, MIT*
Niels Egmont Christensen, *Aarhus University*
Maurice Clavelin, *Université de Paris – Sorbonne*
Nino B. Cocchiarella, *Indiana University*
L. Jonathan Cohen, *The Queen's College, Oxford*

Donald Davidson, *University of Chicago*

Sveinn Eldon, *Helsinki*

383

W. K. Essler, *Universität Frankfurt*

Anne M. Fagot, *University of Paris*
Matts Furberg, *Stockholm University and Göteborg University*

Richard E. Grandy, *University of North Carolina – Chapel Hill*
C. David Gruender, *Florida State University*
Ingemund Gullvåg, *University of Trondheim*

Ian Hacking
Sören Halldén, *Lund*
Rudolf Haller, *Universität Graz*
Roland Hausser, *University of Pittsburgh*
Lars Hertzberg, *Academy of Finland*
Erwin Hiebert, *Harvard University*
Seppo Hiltunen, *Espoo*

R. C. Jeffrey, *Princeton University*

Stig Kanger, *Uppsala*
David Kaplan, *University of California, Los Angeles*
Asa Kasher, *Tel-Aviv University*
Raili Kauppi, *Tampere*
Edward L. Keenan, *University of California at Los Angeles and Tel-Aviv University*
Lorenz Krüger, *Universität Bielefeld*
Thomas S. Kuhn, *Princeton University*
Theo A. F. Kuipers, *University of Groningen*

Henri Lauener, *Universität Bern*
Harry A. Lewis, *University of Leeds*
Ingrid and Sten Lindström, *Stanford University*
Olli V. Lounasmaa, *Helsinki University of Technology*

Ronald McIntyre, *California State University, Northridge*
Jane L. McIntyre, *Cleveland State University*
Juha Manninen, *University of Oulu*

Jo Martens, *University of Utrecht*
Rex Martin, *University of Kansas*
Gerald J. Massey, *University of Pittsburgh*
Eino Mikkola, *Helsinki*
David Miller, *University of Warwick*
C. Ulises Moulines, *Instituto de Investigaciones Filosóficas, UNAM, México*
John Murdoch, *Harvard University*

J. D. North, *University of Groningen*

Charles D. Parsons, *Columbia University*
Terence Parsons, *University of Massachusetts, Amhurst*
Barbara Hall Partee, *University of Massachusetts, Amhurst*
Mark Pastin, *Indiana University*
Ilkka Patoluoto, *Helsinki*
Juhani Pietarinen, *Helsinki*
Jan Pinborg, *University of Copenhagen*
Dag Prawitz, *Stockholm University*
D. L. Provence, *Petaluma*

Irma Rantavaara, *University of Helsinki*
Richard Rorty, *Princeton University*
Nathan Rotenstreich, *The Hebrew University of Jerusalem*

Henrik Sahlqvist, *Stanford University*
Wesley C. and Merrilee H. Salmon, *University of Arizona*
David H. Sanford, *Duke University*
Eileen and Joseph Serene, *Yale University and State University of New York
    at Stony Brook*
Arto Siitonen, *University of Turku*
Matti Sintonen, *University of Helsinki and Queen's College, Oxford*
David Woodruff Smith, *University of California at Irvine*
Howard E. Smokler, *University of Colorado at Boulder*
Ernest Sosa, *Brown University*

Leila Taiminen, *University of Helsinki*
Liisa and Jussi Tenkku, *Helsinki*

Holger Thesleff, *Helsinki*
Richmond H. Thomason, *University of Pittsburgh*
Hakan Törnebohm, *University of Göteborg*
Eva and Knut Erik Tranöy, *University of Oslo*
Vitali Tselishchev, *Siberian Division of USSR Academy of Science, Novosibirsk*
Raimo Tuomela, *University of Helsinki*
Amos Tversky, *Stanford University*

Joseph S. Ullian, *Washington University, St. Louis*

Bas C. van Fraassen, *University of Toronto and University of Southern California*
Veli Verronen, *University of Tampere*
Jan von Plato, *University of Helsinki*

Thomas Wasow, *Stanford University*
Paul Weingartner, *Universität Salzburg*
Achilles Westling, *Hesperia Hospital, Helsinki*
Onni Wiberheimo, *Helsinki*
Osmo A. Wiio, *University of Helsinki*
Kathleen G. J. Wu, *The University of Alabama*

SYNTHESE LIBRARY

Studies in Epistemology, Logic, Methodology,
and Philosophy of Science

1. J. M. Bocheński, *A Precis of Mathematical Logic.* 1959, X + 100 pp.
2. P. L. Guiraud, *Problèmes et méthodes de la statistique linguistique.* 1960, VI + 146 pp.
3. Hans Freudenthal (ed.), *The Concept and the Role of the Model in Mathematics and Natural and Social Sciences. Proceedings of a Colloquium held at Utrecht, The Netherlands, January 1960.* 1961, VI + 194 pp.
4. Evert W. Beth, *Formal Methods. An Introduction to Symbolic Logic and the Study of Effective Operations in Arithmetic and Logic.* 1962, XIV + 170 pp.
5. B. H. Kazemier and D. Vuysje (eds.), *Logic and Language. Studies Dedicated to Professor Rudolf Carnap on the Occasion of His Seventieth Birthday.* 1962, VI + 256 pp.
6. Marx W. Wartofsky (ed.), *Proceedings of the Boston Colloquium for the Philosophy of Science 1961-1962,* Boston Studies in the Philosophy of Science (ed. by Robert S. Cohen and Marx W. Wartofsky), Volume I. 1963, VIII + 212 pp.
7. A. A. Zinov'ev, *Philosophical Problems of Many-Valued Logic.* 1963, XIV + 155 pp.
8. Georges Gurvitch, *The Spectrum of Social Time.* 1964, XXVI + 152 pp.
9. Paul Lorenzen, *Formal Logic.* 1965, VIII + 123 pp.
10. Robert S. Cohen and Marx W. Wartofsky (eds.), *In Honor of Philipp Frank,* Boston Studies in the Philosophy of Science (ed. by Robert S. Cohen and Marx W. Wartofsky), Volume II. 1965, XXXIV + 475 pp.
11. Evert W. Beth, *Mathematical Thought. An Introduction to the Philosophy of Mathematics.* 1965, XII + 208 pp.
12. Evert W. Beth and Jean Piaget, *Mathematical Epistemology and Psychology.* 1966, XII + 326 pp.
13. Guido Küng, *Ontology and the Logistic Analysis of Language. An Enquiry into the Contemporary Views on Universals.* 1967, XI + 210 pp.
14. Robert S. Cohen and Marx W. Wartofsky (eds.), *Proceedings of the Boston Colloquium for the Philosophy of Science 1964-1966, in Memory of Norwood Russell Hanson,* Boston Studies in the Philosophy of Science (ed. by Robert S. Cohen and Marx W. Wartofsky), Volume III. 1967, XLIX + 489 pp.

15. C. D. Broad, *Induction, Probability, and Causation. Selected Papers*. 1968, XI + 296 pp.
16. Günther Patzig, *Aristotle's Theory of the Syllogism. A Logical-Philosophical Study of Book A of the Prior Analytics*. 1968, XVII + 215 pp.
17. Nicholas Rescher, *Topics in Philosophical Logic*. 1968, XIV + 347 pp.
18. Robert S. Cohen and Marx W. Wartofsky (eds.), *Proceedings of the Boston Colloquium for the Philosophy of Science 1966-1968*, Boston Studies in the Philosophy of Science (ed. by Robert S. Cohen and Marx W. Wartofsky), Volume IV. 1969, VIII + 537 pp.
19. Robert S. Cohen and Marx W. Wartofsky (eds.), *Proceedings of the Boston Colloquium for the Philosophy of Science 1966-1968*, Boston Studies in the Philosophy of Science (ed. by Robert S. Cohen and Marx W. Wartofsky), Volume V. 1969, VIII + 482 pp.
20. J.W. Davis, D. J. Hockney, and W. K. Wilson (eds.), *Philosophical Logic*. 1969, VIII + 277 pp.
21. D. Davidson and J. Hintikka (eds.), *Words and Objections: Essays on the Work of W. V. Quine*. 1969, VIII + 366 pp.
22. Patrick Suppes, *Studies in the Methodology and Foundations of Science. Selected Papers from 1911 to 1969*. 1969, XII + 473 pp.
23. Jaakko Hintikka, *Models for Modalities. Selected Essays*. 1969, IX + 220 pp.
24. Nicholas Rescher *et al.* (eds.), *Essays in Honor of Carl G. Hempel. A Tribute on the Occasion of His Sixty-Fifth Birthday*. 1969, VII + 272 pp.
25. P. V. Tavanec (ed.), *Problems of the Logic of Scientific Knowledge*. 1969, XII + 429 pp.
26. Marshall Swain (ed.), *Induction, Acceptance, and Rational Belief*. 1970, VII + 232 pp.
27. Robert S. Cohen and Raymond J. Seeger (eds.), *Ernst Mach: Physicist and Philosopher*, Boston Studies in the Philosophy of Science (ed. by Robert S. Cohen and Marx W. Wartofsky), Volume VI. 1970, VIII + 295 pp.
28. Jaakko Hintikka and Patrick Suppes, *Information and Inference*. 1970, X + 336 pp.
29. Karel Lambert, *Philosophical Problems in Logic. Some Recent Developments*. 1970, VII + 176 pp.
30. Rolf A. Eberle, *Nominalistic Systems*. 1970, IX + 217 pp.
31. Paul Weingartner and Gerhard Zecha (eds.), *Induction, Physics, and Ethics: Proceedings and Discussions of the 1968 Salzburg Colloquium in the Philosophy of Science*. 1970, X + 382 pp.
32. Evert W. Beth, *Aspects of Modern Logic*. 1970, XI + 176 pp.
33. Risto Hilpinen (ed.), *Deontic Logic: Introductory and Systematic Readings*. 1971, VII + 182 pp.
34. Jean-Louis Krivine, *Introduction to Axiomatic Set Theory*. 1971, VII + 98 pp.
35. Joseph D. Sneed, *The Logical Structure of Mathematical Physics*. 1971, XV + 311 pp.
36. Carl R. Kordig, *The Justification of Scientific Change*. 1971, XIV + 119 pp.
37. Milič Čapek, *Bergson and Modern Physics*, Boston Studies in the Philosophy of Science (ed. by Robert S. Cohen and Marx W. Wartofsky), Volume VII. 1971, XV + 414 pp.

38. Norwood Russell Hanson, *What I Do Not Believe, and Other Essays* (ed. by Stephen Toulmin and Harry Woolf). 1971, XII + 390 pp.
39. Roger C. Buck and Robert S. Cohen (eds.), *PSA 1970. In Memory of Rudolf Carnap*, Boston Studies in the Philosophy of Science (ed. by Robert S. Cohen and Marx W. Wartofsky), Volume VIII. 1971, LXVI + 615 pp. Also available as paperback.
40. Donald Davidson and Gilbert Harman (eds.), *Semantics of Natural Language*. 1972, X + 769 pp. Also available as paperback.
41. Yehoshua Bar-Hillel (ed.), *Pragmatics of Natural Languages*. 1971, VII + 231 pp.
42. Sören Stenlund, *Combinators, λ-Terms and Proof Theory*. 1972, 184 pp.
43. Martin Strauss, *Modern Physics and Its Philosophy. Selected Papers in the Logic, History, and Philosophy of Science*. 1972, X + 297 pp.
44. Mario Bunge, *Method, Model and Matter*. 1973, VII + 196 pp.
45. Mario Bunge, *Philosophy of Physics*. 1973, IX + 248 pp.
46. A. A. Zinov'ev, *Foundations of the Logical Theory of Scientific Knowledge (Complex Logic)*, Boston Studies in the Philosophy of Science (ed. by Robert S. Cohen and Marx W. Wartofsky), Volume IX. Revised and enlarged English edition with an appendix, by G. A. Smirnov, E. A. Sidorenka, A. M. Fedina, and L. A. Bobrova. 1973, XXII + 301 pp. Also available as paperback.
47. Ladislav Tondl, *Scientific Procedures*, Boston Studies in the Philosophy of Science (ed. by Robert S. Cohen and Marx W. Wartofsky), Volume X. 1973, XII + 268 pp. Also available as paperback.
48. Norwood Russell Hanson, *Constellations and Conjectures* (ed. by Willard C. Humphreys, Jr.). 1973, X + 282 pp.
49. K. J. J. Hintikka, J. M. E. Moravcsik, and P. Suppes (eds.), *Approaches to Natural Language. Proceedings of the 1970 Stanford Workshop on Grammar and Semantics*. 1973, VIII + 526 pp. Also available as paperback.
50. Mario Bunge (ed.), *Exact Philosophy – Problems, Tools, and Goals*. 1973, X + 214 pp.
51. Radu J. Bogdan and Ilkka Niiniluoto (eds.), *Logic, Language, and Probability. A Selection of Papers Contributed to Sections IV, VI, and XI of the Fourth International Congress for Logic, Methodology, and Philosophy of Science, Bucharest, September 1971*. 1973, X + 323 pp.
52. Glenn Pearce and Patrick Maynard (eds.), *Conceptual Change*. 1973, XII + 282 pp.
53. Ilkka Niiniluoto and Raimo Tuomela, *Theoretical Concepts and Hypothetico-Inductive Inference*. 1973, VII + 264 pp.
54. Roland Fraïssé, *Course of Mathematical Logic* – Volume 1: *Relation and Logical Formula*. 1973, XVI + 186 pp. Also available as paperback.
55. Adolf Grünbaum, *Philosophical Problems of Space and Time*. Second, enlarged edition, Boston Studies in the Philosophy of Science (ed. by Robert S. Cohen and Marx W. Wartofsky), Volume XII. 1973, XXIII + 884 pp. Also available as paperback.
56. Patrick Suppes (ed.), *Space, Time, and Geometry*. 1973, XI + 424 pp.
57. Hans Kelsen, *Essays in Legal and Moral Philosophy*, selected and introduced by Ota Weinberger. 1973, XXVIII + 300 pp.
58. R. J. Seeger and Robert S. Cohen (eds.), *Philosophical Foundations of Science. Proceedings of an AAAS Program, 1969*, Boston Studies in the Philosophy of

Science (ed. by Robert S. Cohen and Marx W. Wartofsky), Volume XI. 1974, X + 545 pp. Also available as paperback.

59. Robert S. Cohen and Marx W. Wartofsky (eds.), *Logical and Epistemological Studies in Contemporary Physics*, Boston Studies in the Philosophy of Science (ed. by Robert S. Cohen and Marx W. Wartofsky), Volume XIII. 1973, VIII + 462 pp. Also available as paperback.

60. Robert S. Cohen and Marx W. Wartofsky (eds.), *Methodological and Historical Essays in the Natural and Social Sciences. Proceedings of the Boston Colloquium for the Philosophy of Science 1969-1972*, Boston Studies in the Philosophy of Science (ed. by Robert S. Cohen and Marx W. Wartofsky), Volume XIV. 1974, VIII + 405 pp. Also available as paperback.

61. Robert S. Cohen, J. J. Stachel and Marx W. Wartofsky (eds.), *For Dirk Struik. Scientific, Historical and Political Essays in Honor of Dirk J. Struik*, Boston Studies in the Philosophy of Science (ed. by Robert S. Cohen and Marx W. Wartofsky), Volume XV. 1974, XXVII + 652 pp. Also available as paperback.

62. Kazimierz Ajdukiewicz, *Pragmatic Logic*, transl. from the Polish by Olgierd Wojtasiewicz. 1974, XV + 460 pp.

63. Sören Stenlund (ed.), *Logical Theory and Semantic Analysis. Essays Dedicated to Stig Kanger on His Fiftieth Birthday*. 1974, V + 217 pp.

64. Kenneth F. Schaffner and Robert S. Cohen (eds.), *Proceedings of the 1972 Biennial Meeting, Philosophy of Science Association*, Boston Studies in the Philosophy of Science (ed. by Robert S. Cohen and Marx W. Wartofsky), Volume XX. 1974, IX + 444 pp. Also available as paperback.

65. Henry E. Kyburg, Jr., *The Logical Foundations of Statistical Inference*. 1974, IX + 421 pp.

66. Marjorie Grene, *The Understanding of Nature: Essays in the Philosophy of Biology*, Boston Studies in the Philosophy of Science (ed. by Robert S. Cohen and Marx W. Wartofsky), Volume XXIII. 1974, XII + 360 pp. Also available as paperback.

67. Jan M. Broekman, *Structuralism: Moscow, Prague, Paris*. 1974, IX + 117 pp.

68. Norman Geschwind, *Selected Papers on Language and the Brain*, Boston Studies in the Philosophy of Science (ed. by Robert S. Cohen and Marx W. Wartofsky), Volume XVI. 1974, XII + 549 pp. Also available as paperback.

69. Roland Fraïssé, *Course of Mathematical Logic – Volume 2: Model Theory*. 1974, XIX + 192 pp.

70. Andrzej Grzegorczyk, *An Outline of Mathematical Logic. Fundamental Results and Notions Explained with All Details*. 1974, X + 596 pp.

71. Franz von Kutschera, *Philosophy of Language*. 1975, VII + 305 pp.

72. Juha Manninen and Raimo Tuomela (eds.), *Essays on Explanation and Understanding. Studies in the Foundations of Humanities and Social Sciences*. 1976, VII + 440 pp.

73. Jaakko Hintikka (ed.), *Rudolf Carnap, Logical Empiricist. Materials and Perspectives*. 1975, LXVIII + 400 pp.

74. Milič Čapek (ed.), *The Concepts of Space and Time. Their Structure and Their Development*, Boston Studies in the Philosophy of Science (ed. by Robert S. Cohen and Marx W. Wartofsky), Volume XXII. 1976, LVI + 570 pp. Also available as paperback.

75. Jaakko Hintikka and Unto Remes, *The Method of Analysis. Its Geometrical Origin and Its General Significance*, Boston Studies in the Philosophy of Science (ed. by Robert S. Cohen and Marx W. Wartofsky), Volume XXV. 1974, XVIII + 144 pp. Also available as paperback.

76. John Emery Murdoch and Edith Dudley Sylla, *The Cultural Context of Medieval Learning. Proceedings of the First International Colloquium on Philosophy, Science, and Theology in the Middle Ages – September 1973*, Boston Studies in the Philosophy of Science (ed. by Robert S. Cohen and Marx W. Wartofsky), Volume XXVI. 1975, X + 566 pp. Also available as paperback.

77. Stefan Amsterdamski, *Between Experience and Metaphysics. Philosophical Problems of the Evolution of Science*, Boston Studies in the Philosophy of Science (ed. by Robert S. Cohen and Marx W. Wartofsky), Volume XXXV. 1975, XVIII + 193 pp. Also available as paperback.

78. Patrick Suppes (ed.), *Logic and Probability in Quantum Mechanics*. 1976, XV + 541 pp.

79. Hermann von Helmholtz: *Epistemological Writings. The Paul Hertz/Moritz Schlick Centenary Edition of 1921 with Notes and Commentary by the Editors.* (Newly translated by Malcolm F. Lowe. Edited with an Introduction and Bibliography, by Robert S. Cohen and Yehuda Elkana), Boston Studies in the Philosophy of Science (ed. by Robert S. Cohen and Marx W. Wartofsky), Volume XXXVII. 1977, XXXVIII+204 pp. Also available as paperback.

80. Joseph Agassi, *Science in Flux*, Boston Studies in the Philosophy of Science (ed. by Robert S. Cohen and Marx W. Wartofsky), Volume XXVIII. 1975, XXVI + 553 pp. Also available as paperback.

81. Sandra G. Harding (ed.), *Can Theories Be Refuted? Essays on the Duhem-Quine Thesis.* 1976, XXI + 318 pp. Also available as paperback.

82. Stefan Nowak, *Methodology of Sociological Research: General Problems.* 1977, XVIII + 504 pp.

83. Jean Piaget, Jean-Blaise Grize, Alina Szeminska, and Vinh Bang, *Epistemology and Psychology of Functions*, Studies in Genetic Epistemology, Volume XXIII. 1977, XIV+205 pp.

84. Marjorie Grene and Everett Mendelsohn (eds.), *Topics in the Philosophy of Biology*, Boston Studies in the Philosophy of Science (ed. by Robert S. Cohen and Marx W. Wartofsky), Volume XXVII. 1976, XIII + 454 pp. Also available as paperback.

85. E. Fischbein, *The Intuitive Sources of Probabilistic Thinking in Children.* 1975, XIII + 204 pp.

86. Ernest W. Adams, *The Logic of Conditionals. An Application of Probability to Deductive Logic.* 1975, XIII + 156 pp.

87. Marian Przełęcki and Ryszard Wójcicki (eds.), *Twenty-Five Years of Logical Methodology in Poland.* 1977, VIII + 803 pp.

88. J. Topolski, *The Methodology of History.* 1976, X + 673 pp.

89. A. Kasher (ed.), *Language in Focus: Foundations, Methods and Systems. Essays Dedicated to Yehoshua Bar-Hillel*, Boston Studies in the Philosophy of Science (ed. by Robert S. Cohen and Marx W. Wartofsky), Volume XLIII. 1976, XXVIII + 679 pp. Also available as paperback.

90. Jaakko Hintikka, *The Intentions of Intentionality and Other New Models for Modalities.* 1975, XVIII + 262 pp. Also available as paperback.

91. Wolfgang Stegmüller, *Collected Papers on Epistemology, Philosophy of Science and History of Philosophy*, 2 Volumes, 1977, XXVII + 525 pp.
92. Dov M. Gabbay, *Investigations in Modal and Tense Logics with Applications to Problems in Philosophy and Linguistics.* 1976, XI + 306 pp.
93. Radu J. Bogdan, *Local Induction.* 1976, XIV + 340 pp.
94. Stefan Nowak, *Understanding and Prediction: Essays in the Methodology of Social and Behavioral Theories.* 1976, XIX + 482 pp.
95. Peter Mittelstaedt, *Philosophical Problems of Modern Physics*, Boston Studies in the Philosophy of Science (ed. by Robert S. Cohen and Marx W. Wartofsky), Volume XVIII. 1976, X + 211 pp. Also available as paperback.
96. Gerald Holton and William Blanpied (eds.), *Science and Its Public: The Changing Relationship*, Boston Studies in the Philosophy of Science (ed. by Robert S. Cohen and Marx W. Wartofsky), Volume XXXIII. 1976, XXV + 289 pp. Also available as paperback.
97. Myles Brand and Douglas Walton (eds.), *Action Theory. Proceedings of the Winnipeg Conference on Human Action, Held at Winnipeg, Manitoba, Canada, 9-11 May 1975.* 1976, VI + 345 pp.
98. Risto Hilpinen, *Knowledge and Rational Belief.* 1979 (forthcoming).
99. R. S. Cohen, P. K. Feyerabend, and M. W. Wartofsky (eds.), *Essays in Memory of Imre Lakatos*, Boston Studies in the Philosophy of Science (ed. by Robert S. Cohen and Marx W. Wartofsky), Volume XXXIX. 1976, XI + 762 pp. Also available as paperback.
100. R. S. Cohen and J. J. Stachel (eds.), *Selected Papers of Léon Rosenfeld*, Boston Studies in the Philosophy of Science (ed. by Robert S. Cohen and Marx W. Wartofsky), Volume XXI. 1978, XXX + 927 pp.
101. R. S. Cohen, C. A. Hooker, A. C. Michalos, and J. W. van Evra (eds.), *PSA 1974: Proceedings of the 1974 Biennial Meeting of the Philosophy of Science Association*, Boston Studies in the Philosophy of Science (ed. by Robert S. Cohen and Marx W. Wartofsky), Volume XXXII. 1976, XIII + 734 pp. Also available as paperback.
102. Yehuda Fried and Joseph Agassi, *Paranoia: A Study in Diagnosis*, Boston Studies in the Philosophy of Science (ed. by Robert S. Cohen and Marx W. Wartofsky), Volume L. 1976, XV + 212 pp. Also available as paperback.
103. Marian Przełęcki, Klemens Szaniawski, and Ryszard Wójcicki (eds.), *Formal Methods in the Methodology of Empirical Sciences.* 1976, 455 pp.
104. John M. Vickers, *Belief and Probability.* 1976, VIII + 202 pp.
105. Kurt H. Wolff, *Surrender and Catch: Experience and Inquiry Today*, Boston Studies in the Philosophy of Science (ed. by Robert S. Cohen and Marx W. Wartofsky), Volume LI. 1976, XII + 410 pp. Also available as paperback.
106. Karel Kosík, *Dialectics of the Concrete*, Boston Studies in the Philosophy of Science (ed. by Robert S. Cohen and Marx W. Wartofsky), Volume LII. 1976, VIII + 158 pp. Also available as paperback.
107. Nelson Goodman, *The Structure of Appearance*, Boston Studies in the Philosophy of Science (ed. by Robert S. Cohen and Marx W. Wartofsky), Volume LIII. 1977, L + 285 pp.
108. Jerzy Giedymin (ed.), *Kazimierz Ajdukiewicz: The Scientific World-Perspective and Other Essays, 1931 - 1963.* 1978, LIII + 378 pp.

109. Robert L. Causey, *Unity of Science*. 1977, VIII+185 pp.
110. Richard E. Grandy, *Advanced Logic for Applications*. 1977, XIV + 168 pp.
111. Robert P. McArthur, *Tense Logic*. 1976, VII + 84 pp.
112. Lars Lindahl, *Position and Change: A Study in Law and Logic*. 1977, IX + 299 pp.
113. Raimo Tuomela, *Dispositions*. 1978, X + 450 pp.
114. Herbert A. Simon, *Models of Discovery and Other Topics in the Methods of Science*, Boston Studies in the Philosophy of Science (ed. by Robert S. Cohen and Marx W. Wartofsky), Volume LIV. 1977, XX + 456 pp. Also available as paperback.
115. Roger D. Rosenkrantz, *Inference, Method and Decision*. 1977, XVI + 262 pp. Also available as paperback.
116. Raimo Tuomela, *Human Action and Its Explanation. A Study on the Philosophical Foundations of Psychology*. 1977, XII + 426 pp.
117. Morris Lazerowitz, *The Language of Philosophy. Freud and Wittgenstein*, Boston Studies in the Philosophy of Science (ed. by Robert S. Cohen and Marx W. Wartofsky), Volume LV. 1977, XVI + 209 pp.
118. Tran Duc Thao, *Origins of Language and Consciousness*, Boston Studies in the Philosophy of Science (ed. by Robert S. Cohen and Marx. W. Wartofsky), Volume LVI. 1979 (forthcoming).
119. Jerzy Pelč, *Semiotics in Poland, 1894 - 1969*. 1977, XXVI + 504 pp.
120. Ingmar Pörn, *Action Theory and Social Science. Some Formal Models*. 1977, X + 129 pp.
121. Joseph Margolis, *Persons and Minds, The Prospects of Nonreductive Materialism*, Boston Studies in the Philosophy of Science (ed. by Robert S. Cohen and Marx W. Wartofsky), Volume LVII. 1977, XIV + 282 pp. Also available as paperback.
122. Jaakko Hintikka, Ilkka Niiniluoto, and Esa Saarinen (eds.), *Essays on Mathematical and Philosophical Logic. Proceedings of the Fourth Scandinavian Logic Symposium and of the First Soviet-Finnish Logic Conference, Jyväskylä, Finland, 1976*. 1978, VIII + 458 pp. + index.
123. Theo A. F. Kuipers, *Studies in Inductive Probability and Rational Expectation*. 1978, XII + 145 pp.
124. Esa Saarinen, Risto Hilpinen, Ilkka Niiniluoto, and Merrill Provence Hintikka (eds.), *Essays in Honour of Jaakko Hintikka on the Occasion of His Fiftieth Birthday*. 1978, IX + 378 pp. + index.
125. Gerard Radnitzky and Gunnar Andersson (eds.), *Progress and Rationality in Science*, Boston Studies in the Philosophy of Science (ed. by Robert S. Cohen and Marx W. Wartofsky), Volume LVIII. 1978, X + 400 pp. + index. Also available as paperback.
126. Peter Mittelstaedt, *Quantum Logic*. 1978, IX + 149 pp.
127. Kenneth A. Bowen, *Model Theory for Modal Logic. Kripke Models for Modal Predicate Calculi*. 1978, X + 128 pp.
128. Howard Alexander Bursen, *Dismantling the Memory Machine. A Philosophical Investigation of Machine Theories of Memory*. 1978, XIII + 157 pp.
129. Marx W. Wartofsky, *Models: Representation and Scientific Understanding*, Boston Studies in the Philosophy of Science (ed. by Robert S. Cohen and Marx W. Wartofsky), Volume XLVIII. 1979 (forthcoming). Also available as a paperback.
130. Don Ihde, *Technics and Praxis. A Philosophy of Technology*, Boston Studies in

the Philosophy of Science (ed. by Robert S. Cohen and Marx W. Wartofsky), Volume XXIV. 1979 (forthcoming). Also available as a paperback.

131. Jerzy J. Wiatr (ed.), *Polish Essays in the Methodology of the Social Sciences*, Boston Studies in the Philosophy of Science (ed. by Robert S. Cohen and Marx W. Wartofsky), Volume XXIX. 1979 (forthcoming). Also available as a paperback.

132. Wesley C. Salmon (ed.), *Hans Reichenbach: Logical Empiricist*. 1979 (forthcoming).

133. R.-P. Horstmann and L. Krüger (eds.), *Transcendental Arguments and Science*. 1979 (forthcoming). Also available as a paperback.

SYNTHESE HISTORICAL LIBRARY

Texts and Studies
in the History of Logic and Philosophy

Editors:

N. KRETZMANN (Cornell University)
G. NUCHELMANS (University of Leyden)
L. M. DE RIJK (University of Leyden)

17. Arpád Szabó, *The Beginnings of Greek Mathematics*. 1979 (forthcoming).

18. Rita Guerlac, *Juan Luis Vives Against the Pseudodialecticians. A Humanist Attack on Medieval Logic*. Texts, with translation, introduction and notes. 1978, xiv + 227 pp. + index.

SYNTHESE LANGUAGE LIBRARY

Texts and Studies
in Linguistics and Philosophy

Managing Editors:

JAAKKO HINTIKKA
Academy of Finland, Stanford University, and Florida State University (Tallahassee)

STANLEY PETERS
The University of Texas at Austin

Editors:

EMMON BACH (University of Massachusetts at Amherst)
JOAN BRESNAN (Massachusetts Institute of Technology)
JOHN LYONS (University of Sussex)
JULIUS M. E. MORAVCSIK (Stanford University)
PATRICK SUPPES (Stanford University)
DANA SCOTT (Oxford University)

1. Henry Hiż (ed.), *Questions.* 1977, xvii + 366 pp.
2. William S. Cooper, *Foundations of Logico-Linguistics. A Unified Theory of Information, Language, and Logic.* 1978, xvi + 249 pp.
3. Avishai Margalit (ed.), *Meaning and Use. Papers Presented at the Second Jerusalem Philosophical Encounter, April 1976.* 1978 (forthcoming).
4. F. Guenthner and S. J. Schmidt (eds.), *Formal Semantics and Pragmatics for Natural Languages.* 1978, viii + 374 pp. + index.
5. Esa Saarinen (ed.), *Game-Theoretical Semantics.* 1978, xiv + 379 pp. + index.
6. F. J. Pelletier (ed.), *Mass Terms: Some Philosophical Problems.* 1978, xiv + 300 pp. + index.